PLANETS X AND PLUTO

PLANETS X AND PLUTO

William Graves Hoyt

THE UNIVERSITY OF ARIZONA PRESS

TUCSON, ARIZONA

About the Author...

WILLIAM GRAVES HOYT, reporter, writer, and editor, has for the past several years conducted research on the history of astronomy as a research associate in the Lowell Observatory at Flagstaff. Director of the Public Information Office at Northern Arizona University, he holds an M.A. degree in history and has published numerous articles on the history of astronomy as well as the book *Lowell and Mars,* 1976.

Second printing 1981

THE UNIVERSITY OF ARIZONA PRESS

This book was set in 10/12 V-I-P Times Roman

Copyright © 1980
The Arizona Board of Regents
All Rights Reserved
Manufactured in the U.S.A.

Library of Congress Cataloging in Publication Data

Hoyt, William Graves.
   Planets X and Pluto.

   Bibliography: p.
   Includes index.
   1. Pluto (Planet) 2. Planets. I. Title.
QB701.H69   523.4'82   79-15665
ISBN 0-8165-0684-1
ISBN 0-8165-0664-7 pbk.

For Ellen, Deb
and Bill Jr.

CONTENTS

List of Illustrations   ix
Foreword by Bart J. Bok   xi
Acknowledgments   xiii

Introduction: Of Time and Number   1
1. Uranus and the Asteroids   9
2. Neptune   32
3. Beyond Neptune   60
4. The First Search for Planet X   83
5. The Second Search for Planet X   105
6. A Trans-Neptunian Planet   126
7. Lowell's Legacies   142
8. The Third Search for Planet X   161
9. A Young Man From Kansas   178
10. Pluto: The Early Controversies   198
11. The Longer Controversy   220
12. More Planets X?   248

Chapter Notes   263
Bibliography   291
Index   293

ILLUSTRATIONS

Sir William Herschel   13
Herschel's Newtonian Reflector   14
Carl Friedrich Gauss   19
Sir George Biddell Airy   36
John Couch Adams   41
Urbain Jean Joseph Leverrier   44
Benjamin Peirce   64
Benjamin Apthorp Gould   67
David Peck Todd   75
See, Cogshall and Pickering   79
Lowell Observatory Staff in 1905   88
Earl Carl Slipher and Kenneth P. Williams   89
The 5-Inch Brashear Camera   90
Percival Lowell   107
Camille Flammarion With Lowell   109
Lowell Observatory 40-Inch Reflector and Dome   114
C. A. Robert Lundin   116, 117
Thomas Gill   131
Constance S. Lowell   143

Guy Lowell   144
Carl Otto Lampland   145
Vesto Melvin Slipher   148, 149
Stanley Sykes   165
Dome of the 13-Inch Lawrence Lowell Refractor   166
Clyde William Tombaugh   180
The 13-Inch Lowell Photographic Refractor   186
Clyde Tombaugh, 1931   187
Discovery Plates of Pluto   191
Vesto Melvin Slipher   199
John A. Miller   202
Armin O. Leuschner   203
Wilbur A. Cogshall   211
Ernest William Brown   227
Dirk Brouwer   242
Gerard P. Kuiper   243
Pluto's Satellite, Charon   245
James W. Christy and Robert Harrington   246
Clyde Tombaugh, 1938   251
Clyde Tombaugh, 1978   252
John Hall, Roger Lowell Putnam,
    Vesto Melvin Slipher and Michael Putnam   259

FOREWORD

In my younger years as an astronomer, I read most of the books and articles that related to the discovery of Uranus and Neptune. My respect for Sir William Herschel, who discovered Uranus, increased as I read in this book about his role in that discovery. My sympathies grew for John Couch Adams, who as a young man was pushed about by the villian in the piece, Sir George Airy, the Astronomer Royal at Greenwich during the 1830s and 1840s. I was impressed with the manner in which the great French Académie des Sciences and the French astronomical establishment supported their man, Leverrier, in his studies of predicted positions for the yet undiscovered Neptune. I was the more amazed to learn how little support Adams received from the British establishment. And then, it was exciting to read how Johann Gottfried Galle, the Berlin astronomer, found Neptune.

The stories of the discoveries of Uranus and Neptune are vividly told in this book. In the early pages we meet for the first time the question that haunts us through the rest of the book. Did Leverrier and Adams actually predict the existence of the planet Neptune, or was it a "happy accident" as asserted by distinguished Harvard mathematician Benjamin Peirce (who was incidentally Percival Lowell's much admired teacher)? The argument which follows is enlightening indeed.

At the turn of the century—from 1902 up to the time of World War I—astronomers showed an increasing interest in one or more possible trans-Neptunian planets. The stage is set in chapter 3 for the story of the discovery of Pluto, and here we obtain the background material for the work by Percival Lowell, William H. Pickering and others on perturbations in the orbits of

Uranus and of Neptune that suggested the presence of Pluto, and possibly one other far-out planet. The story is one of troubles and tribulations. Mr. Hoyt provides the clearest coverage yet available on the subject.

Of the utmost importance is the author's most lucid explanation of the complex mathematics involved in the calculations in which unexplained deviations from the standard orbits of Uranus and Neptune were interpreted in terms of perturbations caused by a remote Planet X. It was Lowell's aim to obtain a position in the sky for Planet X. It is well to note that during these busy years the equipment at Flagstaff was not really up to the standards needed for an adequate photographic search.

With the death of Percival Lowell on November 12, 1916, the haphazard and stumbling searches for Planet X were ended. It is to the everlasting credit of Lowell's distinguished astronomer "assistants," V. M. Slipher and C. O. Lampland, however, that in 1927 the search was resumed. Strong support for a renewal of the search came from my good friend and Lowell's nephew Roger Lowell Putnam who, in 1927, became Lowell Observatory's Sole Trustee, a post in which he served for 40 years. President Abbott Lawrence Lowell of Harvard, Percival's brother, provided the funds to complete and mount a very fine 13-inch photographic refractor. What was lacking was a competent and dedicated observer who would search for the new planet on the photographs made with the 13-inch refractor. The remarkable story of the selection of the right observer, Clyde Tombaugh, and the subsequent discovery of Pluto is a climax of the book and certainly makes for enjoyable reading. I remember well the great excitement the discovery produced among the astronomers at Harvard College Observatory. It surely must have been the same all over the world.

Still another story well worth telling is handled superbly by Mr. Hoyt. The period of confusion which followed the discovery centered around the business of getting a reliable first orbit for Pluto. Once again, astronomers young and old, all over the world, argued the question. Some claimed the new planet was really a comet. Then, once its planetary status was established, a mild and dignified debate ensued on the naming of the planet.

Off and on since the discovery of Pluto in 1930, still another controversy among astronomers, which the author discusses in the final chapters of the book, has centered on the question of credit due Lowell for the discovery. Did Lowell actually predict the presence of Pluto for which he had erroneously assumed a mass of seven times that of the earth? No new planet was found during Lowell's lifetime, but personally I feel full credit should go to Lowell and his collaborators for the discovery. Whatever measure of credit is finally granted, it is certain that Lowell did the best he could to calculate an orbit for Planet X. He marked areas of the sky that he felt should be especially searched. He initiated searches for Planet X, and the final successful search leading to the discovery of Pluto was certainly carried out in his spirit.

BART J. BOK

ACKNOWLEDGMENTS

A study such as this owes much to many people.

I am in debt to all the people of the Lowell Observatory in Flagstaff, and particularly to the late Roger Lowell Putnam, for forty years the observatory's sole trustee, and to his wife, Caroline; to their son and the present trustee, Michael C. J. Putnam; to former Lowell director John S. Hall, and to his successor, Arthur A. Hoag. Members of the Lowell staff who have assisted me in many ways, large and small, include Henry L. Giclas, William A. Baum, Otto G. Franz, Stuart E. Jones, Robert L. Millis, Jean Scheele, Mary Lou Kantz, Helen Horstmann, and Susan Kephart. All have been good friends and have taken an active interest in my work.

I am deeply grateful, too, to A. M. J. (Tom) Gehrels of the University of Arizona's Lunar and Planetary Laboratory, who lent his very considerable astronomical and editorial talents to this endeavor. Others who have contributed in various ways to this study include Sara Lee Lippincott of Swarthmore College's Sproul Observatory, Frank K. Edmondson of Indiana University, Donald E. Osterbrock of Lick Observatory, George Greenstein of Amherst College Observatory, Kaj Aa. Strand, former scientific director of the U.S. Naval Observatory, and Harold Ables, the director of the Naval Observatory's Flagstaff station.

My special appreciation goes to Clyde W. Tombaugh, the discoverer of Pluto and now professor emeritus of Astronomy at New Mexico State University, who has shared his recollections of his discovery and of his work at

Lowell Observatory with me from time to time since 1974. Dr. Tombaugh read and commented upon the completed manuscript of this study in November 1977, and subsequently provided additional details that have been incorporated in the final text and several of the photographs that illustrate it.

I am indebted to Bart Jan Bok, the "astronomers' astronomer," for his enthusiasm for this study, and for providing it with a suitable Foreword.

Friends and colleagues at Northern Arizona University have also been helpful in the course of this work. In particular, I thank Arthur Adel, professor emeritus of science and mathematics and a former Lowell astronomer, and Lawrence Perko, associate professor of mathematics, for their critical reading of early versions of the manuscript; and President J. Lawrence Walkup for his quiet encouragement of my avocation as a historian of science.

Various authors and publishers have been wonderfully cooperative in granting me permission to use material for which they hold, or control, publication rights, and I thank them.

They include the Cambridge University Press, publishers of Constance A. Lubbock's *The Herschel Chronicle* (1933), and Dr. Morton Grosser, author of *The Discovery of Neptune* (1962), originally published by Harvard University Press and recently reissued by Dover Publications. I have also used references from A. F. O'D. Alexander's *The Planet Uranus* (1965), published by American Elsevier Publishing Co., with permission of Faber & Faber Ltd., London.

Sky and Telescope has permitted me to quote from "Current Problems of Pluto" (1950) by Dirk Brouwer, from "The Motions of the Five Outer Planets" (1951) by Brouwer and G. M. Clemence, from "Reminiscences of the Discovery of Pluto" (1960) by C. W. Tombaugh, and from "The Titius-Bode Law: A Strange Bicentennial" (1972) by S. J. Jaki. *Scientific American* granted permission to quote from two 1930 articles by Henry Norris Russell, "Planet X," and "More About Pluto." Material from "Pluto: Evidence for Methane Frost," by D. Cruikshank, C. Pilcher and D. Morrison, which appeared in *Science* November 19, 1976, copyright 1976 by the American Association for the Advancement of Science, is used with the permission of *Science* and Dr. Cruikshank. Faber and Faber, Ltd., of London, England, gave me permission to quote from Gunter D. Roth's *The System of Minor Planets* (1962), and David Higham Associates of London, authors' agents, allowed the use of material from *William Herschel— Explorer of the Heavens* by J. D. Sidgwick (Faber and Faber, Ltd.).

I am indebted to the staff of the University of Arizona Press—to Director Marshall Townsend, Editorial Coordinator Elizabeth Shaw and particularly Assistant Editor Patricia Shelton—for the professional manner in which they prepared this study for publication.

I also thank Mrs. Sara Hall Coughanour, and Mrs. Jadene Wardrip for their secretarial services, and my wife, Beverly, for her effective editorial assistance and sustained encouragement of my work.

W. G. H.

INTRODUCTION: OF TIME AND NUMBER

The sun, the moon and the other planets, Plato declared more than two thousand years ago in the *Timaeus*, were placed in the firmament by the Creator to teach the lesser gods and men time and number. And so they have. From the sun and moon, man has counted his hours, his days, his months and his years, and from his study of the planets, he has learned some fundamental truths about the universe in which he lives—truths discovered in part through those sciences, known collectively as mathematics, which deal with quantities, magnitudes and forms and their relationships and attributes in terms of number and symbol.

The "X" in the title of this study is such a symbol, traditionally standing for the unknown quantity in mathematical equations. But here, it also represents something more. Generally, it symbolizes all the unknown planets that have been postulated by astronomers since William Herschel (1738–1822) stumbled on Uranus in 1781 in the course of one of his systematic telescopic sweeps of the heavens. Specifically, it refers to the trans-Neptunian planet postulated by Percival Lowell (1855–1916) and which he called "Planet X." Lowell spent the last eleven years of his life searching observationally and analytically for this elusive body, and died without knowing that his long, exhaustive effort had been successful. But his enthusiasm, and the intensity of his belief in its existence, inspired his assistants and successors to continue his work and led directly to the discovery in 1930 of Pluto, by Clyde William Tombaugh (1906–), at the observatory Lowell had founded at Flagstaff, Arizona, in 1894.

What follows is the first comprehensive history of the discovery of Pluto, from Lowell's early explorations of trans-Neptunian space in the first decades of the twentieth century through the various controversies that the discovery precipitated both in and out of astronomy. This, of necessity, is set in the matrix of the earlier history of planetary discovery, for the broader history is, or at least ought to be, all of a piece and the dramatic discoveries of Uranus, the first asteroids and Neptune must be summarized here in some detail as essential prolegomena to the more recent, but no less dramatic event.

Of necessity, too, is a brief review of Lowell's life and other astronomical work included as background for his work on Planet X. Much of the Lowell history has been presented previously and in far greater detail in the author's *Lowell and Mars,** to which the present study is, perhaps, a logical sequel.

The major portion of this study, and particularly the part devoted to Lowell and the work of his observatory, is based primarily on original documentary material contained in the Lowell Observatory Archives. This consists of approximately 35,000 pages of correspondence, as well as scores of notebooks, manuscript texts, observation logs, data sheets and other miscellaneous records dating from 1894, when Lowell established his observatory, through 1935, five years after Tombaugh's discovery of Pluto. Most of the earlier material dated prior to Lowell's death in 1916 can be found in the microfilm edition of *The Early Correspondence of the Lowell Observatory 1894–1916.*† But the later documents from 1917 on have not yet been formally organized, much less indexed and microfilmed, and thus are not readily accessible to scholars. Although part of this material has been arranged in rough chronological order during the course of the research for this study, retrieval of some specific items may still require considerable time and effort. Perhaps this situation will be remedied in the not-too-distant future.

To some degree, of course, this study also relies on secondary sources, and this is especially true of the early chapters on the discoveries of Uranus, the first asteroids and Neptune. The literature here is quite extensive and in most cases is both excellent and definitive. For Uranus, the primary sources have been supplemented by *The Herschel Chronicle* (1933), edited by Sir William's granddaughter, Constance A. Lubbock; by A. F. O'D. Alexander's *The Planet Uranus* (1965); and by J. B. Sidgwick's *William Herschel—Explorer of the Heavens* (1953). For the asteroids, I have relied on F. G. Watson's *Between the Planets* (1941) and Gunter D. Roth's *The System of Minor Planets* (1962). And for Neptune, secondary sources include Morton Grosser's *The Discovery of Neptune* (1962), Benjamin A. Gould's *Report on the History of the Discovery of Neptune* (1850) and W. M. Smart's fine centennial essay on "John Couch Adams and the Discovery of Neptune" in the *Occasional Notes of the Royal Astronomical Society* (1947). Together, these works tell all there is to be told, and the reader interested in greater

---

*Tucson: The University of Arizona Press, 1976.
†W. G. Hoyt and A. Babbitt, eds; Flagstaff: The Lowell Observatory, 1973.

detail than is contained herein is urged to consult these well-written, well-documented studies.

In the case of the discovery of Pluto, however, the available literature is sparse indeed, somewhat repetitious, and almost entirely confined to the circumstances immediately surrounding the event itself, without prologue or epilogue. This study, then, is an attempt to correct this deficiency in the historical record, as well as to provide scholars, students and laymen alike with a convenient, accurate summary of the whole of the history of planetary discovery.

When and by whom the brighter planets—Mercury, Venus, Mars, Jupiter and Saturn—were discovered is not, and never will be, known. They are naked-eye objects in the night sky, and most certainly they were discovered independently by many prehistoric peoples, at many times, and in many parts of the world. Some of the oldest known records contain references to these planets. Cuneiform tablets found in the library of the Assyrian king, Assurbanipal, at ruined Nineveh, for example, record observations of Venus —the goddess Ishtar to the ancient Babylonians—made perhaps as early as 2000 B.C.

Philosophically, the history of planetary discovery begins with Copernicus (1473-1543), the effective discoverer of the planetary status of the earth. Others in antiquity—the Pythagorean Philolaus (*fl.* 450 B.C.) for one, and the early Greek astronomers Heraclides (*fl.* 373 B.C.) and Aristarchus (*fl.* 250 B.C.)—had suggested that such might be the state of affairs, but men were not persuaded by their ideas.

In the physical sense with which this study is concerned, however, the history of planetary discovery properly begins with Galileo Galilei (1564-1642), Copernicus' great advocate, who in 1610 discovered the four largest satellites of Jupiter with his crude, homemade version of the newly invented telescope and thus became the first person in history to add substance, so to speak, to the solar system, indeed, to the universe itself. In the years since Galileo's careful observations, astronomers, many of them gifted amateurs, have added three new major planets, more than thirty planetary satellites and three ring systems, and numberless thousands of asteroids to the sun's family, and have learned that comets and meteors, long believed to be phenomena of the earth's upper atmosphere, are part and parcel of this same great system as well.

Importantly, the history of planetary discovery is marked by a strong mathematical strain that can also be traced back to Galileo's time—to his contemporary Johannes Kepler (1571-1630) and Kepler's enunciation of the laws of planetary motion in the first two decades of the seventeenth century, and to Isaac Newton (1642-1727) and his discovery, later that same century, of the law of universal gravitation. To a large degree, in fact, this history is concerned with attempts to compute, through the application of Keplerian-Newtonian mechanics, accurate orbits for newly discovered planets on one hand, and to predict, by what is in effect an inversion of the problem, possible orbits for hypothetical, undiscovered planets on the other. The calculations involved in these attempts are more often than not highly complex and particularly before the advent of the electronic computer required extraordinary amounts of time and tedious labor. Fortunately, however, the concepts on which these calculations rest are not difficult to grasp, and thus the mathematical aspects of planetary discovery can be discussed herein largely in conceptual terms and in such a way that they can be understood by the nonprofessional reader.

Celestial mechanicians seeking to determine the motions of planetary bodies deal essentially in the geometry of conic sections, first worked out by Appollonius of Perga (*fl.* 220 B.C.). Under Kepler's first law, such bodies move not in circles, as the ancients and even Copernicus believed, but in ellipses, one of whose foci is the sun. The determination of a planetary orbit, then, is first a matter of finding the right ellipse and its orientation in space in reference to, say, the plane of the ecliptic, i.e. the apparent path of the sun in the sky, or the so-called "invariable plane," i.e. the mean of the orbital planes of the planets. Since the work of mathematicians Pierre Simon Laplace (1749–1827) and Carl Friedrich Gauss (1777–1855) in the late eighteenth and early nineteenth centuries, there have been very elegant methods for accomplishing this.

Five elements are necessary to define a planetary orbit, and it will be useful to cite them here along with their mathematical symbols because they appear with progressive frequency in this study. The semi-major axis, or mean distance from the sun (a), and the eccentricity (e) give the size and shape of the orbital ellipse respectively. The inclination (i) of the orbit to the reference plane, the longitude of the ascending node (Ω) where the orbit intersects the reference plane,* and the longitude of the planet's perihelion ($\bar{w}$) together give the orientation of the orbit in space and in relation to the sun. To find a planet's position in such an orbit, two additional elements are needed: its longitude at epoch (ϵ), that is, on some specific date; and its mean daily or annual motion (μ). In the case of a hypothetical planet, these symbols are usually designated as primes, that is a' and e'.

There are, nevertheless, a number of factors which seriously complicate the basic problem. A major difficulty in defining planetary motion precisely is the fact that each planet as it moves around the sun is irregularly perturbed from its course by the variable gravitational influences of all its neighbors in space, so that its elliptical path must always fall somewhat short of perfection. Furthermore, the observations from which an elliptical orbit is constructed, from which the so-called "theory" of a planet is computed, are inevitably and unavoidably subject to error, so that perfection is further marred. The problem, which is a form of the classic three-body problem, in consequence,

---

*There are, of course, two nodes, and occasionally the descending node is used in orbital calculations.

permits only a closely proximate solution, although this is usually sufficient for most practical purposes.

As it is in the nature of a theory of a planet to be somewhat imprecise, so is its nature also to be adjustable within limits, and thus celestial mechanicians tinker with its elements from time to time with an eye toward achieving a better fit between the theoretical orbit and the actual observations. Such tinkering usually involves slight refinements in the values used for one or more of the orbital elements, or adopting slightly different values for the mass of one or more of the perturbing planets. Estimated value for Pluto's mass, for example, which for many years was not known accurately from direct observation, has been revised downward several times since its discovery to bring theories of Neptune's motion into closer conformity with observed fact.

When such tinkering will not do, and when what are called the "residuals" between theory and observation remain larger than error alone can explain, it may be assumed that the unexplained perturbations are caused by some unknown gravitational factor which has not been taken into account. Twice in the history of planetary discovery, when men of talent and tenacity have made this assumption, the discovery of a planet has been the result.

The history of planetary discovery provides insights into the nature and processes of scientific discovery in general.

The discoveries of Uranus, Neptune and Pluto, for instance, have all for various reasons been called "accidental" at one time or another. Yet the word "accidental" here does not do justice to the events and is unfortunate in that it fosters the idea in many minds, at least, that these discoveries were made wholly without cause or intent, that they somehow just happened, and that with a little luck they might have been made by any astronomer.

Herschel, however, was deliberately and systematically scouring the sky for something new, *anything* new, when he found Uranus, and though an "amateur" in the eyes of his professional contemporaries, he was experienced enough as an observer to recognize its unusual character when he saw it. And when Johann Gottfried Galle (1812–1910) and Tombaugh, another "amateur," discovered Neptune and Pluto respectively, they were looking specifically for planets whose existence and positions in the sky had been predicted—with surprising accuracy as it turned out—by Urbain J. J. Leverrier (1811–1877) and Percival Lowell. It can be, and has been argued, that any similarity between these predictions and subsequent fact was simply coincidental. But that these predictions, however scientifically suspect after the event, led directly to the discoveries of Neptune and Pluto is beyond all dispute.

This is semantics, perhaps, for this is not to say that an element of chance was not present in these discoveries, as it is in all discovery. Rather, it is to say that they were not "accidental" in the accepted sense of the word, that they were not casual, unexpected or, in the cases of Neptune and Pluto, unintended as required by Websterian definitions. In each instance, the odds

Introduction 5

in favor of discovery were substantially enhanced by the implementation of a purposeful, systematic plan of observation which was, if you will, specifically designed for discovery. And in each case, the discoverers were intellectually prepared for discovery, and thus were predisposed to recognize the unusual, the improbable and even the incredible.

Moreover, when chance alone was operative, these planets were *not* discovered many times, a fact that prompted Lowell to remark on "how vital scientific imagination is to science."[1] Indeed, what is truly "accidental" in the history of planetary discovery is that Uranus, Neptune and Pluto were not found long before they were.

Uranus was recorded by experienced astronomers as a star more than twenty times before Herschel recognized its unstarlike aspect. Three of these prediscovery observations came in a single month in 1715 by England's first Astronomer Royal, John Flamsteed (1646–1720). Another six were made over the span of only nine nights in January 1769 by the French astronomer Pierre Charles Lemonnier (1715–1799). "Lemonnier," Lowell once noted, "discovered that he himself had not discovered it several times.... Flamsteed was spared a like mortification by being dead. For both these observers had recorded it two or more nights running from which it would seem almost incredible not to have suspected its character from its change of place."[2]

Neptune, too, was recorded twice over three nights in May 1795, more than fifty years before its discovery, by Joseph J. F. Lalande (1732–1807) who, apparently never thinking that it might be a planet, attributed its slightly different positions to observational error. The lapse of "scientific imagination" here is all the more surprising, perhaps, in view of the fact that Uranus had been discovered only fourteen years before, and Lalande had since been particularly active in observing the new planet and in subsequent efforts to compute its orbit and devise accurate tables of its motion. And in 1846, less than two months before Neptune's discovery, James Challis (1803–1882), Plumian professor of astronomy and director of the Cambridge Observatory, recorded the planet twice—on August 4 and 12—without realizing it, and he was looking for just such a planet!*

Finally, faint, remote Pluto was photographed at least sixteen times between 1914 and 1930 when Tombaugh recognized its image blinking back at him from plates he had exposed during the Lowell Observatory's third and final search for Percival Lowell's postulated Planet X. These more recent missed opportunities, perhaps, are at least more understandable. Pluto's

---

*English astronomer John Russell Hind reported in 1850 that a "Dr. Lamont" of Munich had observed "Neptune as a star" on October 25, 1845, and on September 7, 1846, and the following year he reported a third observation by Lamont on September 11, 1846. Of the latter, he wrote: "A notice of this additional observation may not prove uninteresting, as it shows that an *immediate reduction* of the zones of Sept. 7th and 11th could not have failed to point out the planet; and the discovery might have been effected prior to the 23rd of September, when it was recognized by Dr. Galle." See J. R. Hind, "Unnoticed Observations of Neptune," *Monthly Notices of the Royal Astronomical Society*, 10:42, 1850; and "On Lamont's Observations of Neptune as a Fixed Star," *M.N.R.A.S.* 11:11, 1851.

pale, starlike image is, in comparison to those of Uranus and Neptune, far more difficult to find and recognize for what it is in the star-crowded heavens. Ironically photography, which makes such a discovery possible in the first place, compounds the difficulty by multiplying the number of possibilities to be explored.

The history of planetary discovery is, in fact, very much a history of missed opportunities. The most sensational instance of this, and surely the best known, is the uproar that followed the discovery of Neptune in 1846 when it was revealed that England's John Couch Adams (1819–1892) had made an earlier but unpublished prediction for a trans-Uranian planet that was comparable, at least as far as the planet's position was concerned, with Leverrier's published prediction. One can make a case here that chance deprived Adams of the full measure of credit for the prediction and discovery of Neptune and argue that he was simply the victim of a set of unfortunate circumstances, as indeed he was. But this begs the question. Certainly, a singular lack of scientific imagination on the part of the Astronomer Royal, George Biddell Airy (1801–1892), and Challis, to both of whom Adams had communicated his results, had something to do with it. Nor was Adams, in sharp contrast to the confident Leverrier, particularly aggressive in seeking publication of his results or in seeing to it that they were properly followed up observationally. The opportunities, made both evident and urgent by Leverrier's published analyses of the problem, were there and were missed, and not by chance alone.

Planet-hunting, indeed, has proven to be a frustrating business. Only a handful of the score or more astronomers who have seriously and deliberately set out to increase the sun's family have been at all successful, and then their success has usually been tainted by controversies of one kind or another—over the nature of the discovery, over priorities or over the scientific validity of the work which produced the discovery.

There have been lesser frustrations, too. Take the matter of naming new discoveries, traditionally the prerogative of the discoverers who have, nonetheless, been notably unsuccessful in this regard. Galileo sought in vain to name his four Jovian moons "the Mediccean Stars" after his mentor, Cosimo II de Medici, fourth grand duke of Tuscany; Herschel, happily, failed in his attempt to call Uranus "Georgium Sidus" or "the Georgian Star" after his benefactor, King George III of England; Giuseppe Piazzi (1746–1826) of Palermo was only partly successful in naming the first asteroid "Ceres Ferdinandea" after *his* sovereign, King Ferdinand IV of Sicily and Naples; and Leverrier connived futilely to have Neptune named after himself. Even the name Pluto was formally proposed by Lowell Observatory astronomers only after extensive public discussion and wide professional usage left them but little choice.

The history of planetary discovery is rich in drama, in ironies large and small, and in coincidences that even the most conservative of astronomers have referred to as "remarkable" or "incredible." Inevitably, so it seems, the discovery of each new planet has brought immediate speculation and subsequent predictions of still other planets farther from the sun, and indeed it *is* incredible that twice over a span of less than one hundred years, such predictions have borne fruit.

The discovery of Pluto is no exception here. Almost at once there was speculation about possible new discoveries, and within months the conscientious Tombaugh had resumed his systematic search for new planets at the Lowell Observatory. Over the intervening years, a number of astronomers have attempted, with widely varying degrees of analytical rigor, to predict the existence of a trans-Plutonian planet and to suggest its probable nature and position in the sky.

They have so far failed and it may be they are doomed to failure. For one thing, the finding of new planets becomes vastly and progressively more difficult, both theoretically and observationally, with increasing distance from the sun. For another, there is obviously some limit to the number of "major" planets in the solar system, and it may well be that this limit has been reached. Yet there are some tantalizing hints that there may be something more in trans-Neptunian space than meets the eye. The recent discovery of the Plutonian satellite, Charon, as will be seen, is one of them.

That there will be more objects found orbiting the sun is beyond doubt, but whether they will be "planets" is a question that hinges ultimately on the matter of definition. Astronomers have not yet grappled with the taxonomic problem of defining precisely just what is, or is not, a "planet" or an "asteroid" or what have you. Everyone's idea of a "planet," incidentally, seems to be roughly the definition given in Webster's unabridged New Twentieth Century Dictionary, which divides the category into nine "major" planets, from Mercury out to Pluto, and the "minor" planets, or asteroids. This definition has, at least, the benefit of wide acceptance and common usage both by astronomers and laymen, and it is generally relied on in this study.

Whether there is still another "major" planet to be discovered in the outer reaches of the solar system may be determined eventually, as in the past, by time and number. When in time the perturbations of slow-footed Neptune and Pluto become better known and better understood, it may be possible for astronomers to count with confidence the number of "major" planets under the sun.

CHAPTER 1

URANUS AND THE ASTEROIDS

"In the quartile near ζ [zeta] Tauri the lowest of two is a curious either nebulous star or perhaps a comet. A small star follows the comet at ⅔ of the field's distance."[1]

Thus did William Herschel, sometime between 10 and 11 P.M. on March 13, 1781, while systematically observing stars with a homemade reflecting telescope from the yard of his home in Bath, England, record the discovery of Uranus, the seventh planet from the sun and the first to be added to the solar system since unremembered antiquity.

Over the next three nights Herschel, either from lack of opportunity or from preoccupation with other concerns, did not observe the "curious" object again. But on the night of March 17, he was again at the eyepiece of his telescope, a small Newtonian reflector with a speculum metal* mirror. "I looked for the comet or nebulous star," he wrote in his journal, "and found it is a comet, for it has changed its place."[2]

Sporadically, through March and into April, he continued to observe the "comet" as it moved slowly away from the constellation of Taurus and into Gemini, making measurements of its angular motion and apparent diameter and concluding, erroneously as it turned out, that it was accelerating and its apparent size increasing, indicating that it was approaching the sun.

---

*Speculum metal is a highly reflective alloy of copper and tin, at an approximate ratio of 2:1, with traces of other metals, such as arsenic and antimony.

Herschel formally announced his discovery in a brief paper, dated March 22, 1781, and entitled "Account of a Comet," which was presented later that month to the Bath Literary and Philosophical Society of which he was a member, and then on April 26, to the Royal Society in London through the good offices of his friend, Dr. (later Sir) William Watson, F.R.S.

In this, Herschel reported that "while examining the small stars in the neighborhood of H Geminorum [1 Geminorum], I perceived one that appeared visibly larger than the rest; being struck with its uncommon magnitude, I compared it to H Geminorum and the small star in the quartile between Auriga and Gemini, and finding it so much larger than either of them, suspected it to be a comet...."[3]

He then described how he had tested his suspicion observationally. Noting that he had been engaged in a series of observations on the parallax* of fixed stars that required high magnifying powers and had been using an eyepiece with a power of 227, he wrote: "From experience I knew that the diameters of fixed stars are not proportionately magnified with higher powers, as planets are; therefore I now put on the powers of 460 and 932, and found the diameter of the comet increased in proportion to the power.... The sequel has shown that my surmises were well founded, this proving to be the comet we have lately observed."[4]

Once he had concluded that the object was a comet, Herschel had dutifully reported his discovery through Watson to England's two most eminent professional astronomers—the Rev. Dr. Nevil Maskelyne, fifth Astronomer Royal, at Greenwich Observatory, and the Rev. Thomas Hornsby, director of Oxford Observatory. As he later noted in his "Account," he felt that there should be "regular, constant and long-continued observations with fixed instruments, the excellence of them is too well known to say anything upon that subject; for which reason I failed not to give immediate notice of this moving star, and I was happy to surrender it to the care of the Astronomer Royal and others, as soon as I found that they had begun their observations upon it."[5]

Herschel may have had another reason, too, for he was not primarily interested in comets, particularly when they interfered with his grand plans for surveying the heavens, and he may well have been "happy to surrender" his comet to others to free himself for what he considered to be more important projects. His granddaughter, Constance A. Lubbock, in her biography of Herschel and his sister Caroline, has suggested as much. "In comparison with the wonderful vision which was ever before his mind, that of probing into the hitherto invisible depths of space and unravelling the problem of the structure of the heavens, the discovery of a strange comet, added to the solar system, seemed a minor matter," she has written.[6]

In any case, others had quickly taken up observations of the new comet. Hornsby initially had difficulty locating it from Herschel's generalized

---

*Parallax here refers to a nearby body's apparent movement relative to background stars due to the change in an observer's position in space as the earth moves around the sun.

descriptions of its position, but Maskelyne found it readily enough and seems to have been the first to suggest that it might be a primary planet, an idea that did not occur to Herschel. In a letter to Watson on April 4, Maskelyne declared that his first observations had convinced him "that it is a comet or a new planet, but very different from any comet I ever read a description of or saw."[7]

On April 23, the Astronomer Royal wrote Herschel acknowledging "my obligation to you for the communication of your discovery of the present comet, or planet. I do not know which to call it. It is as likely to be a regular planet moving in an orbit very nearly circular around the sun as a comet moving in a very eccentric ellipsis. I have yet to see any coma or tail to it." He also reported that he had observed it on April 6 with a reflector of six-foot focal length and his "greatest" magnifying power of 270, and "saw it of a very sensible size but not well defined. This however showed it to be a planet and not a fixt star, or of the same kind as the fixt stars as to possessing native light with an insensible diameter."[8]

Hornsby, on the other hand, once he found the object, quickly became convinced that it was, indeed, a comet, and in a letter to Herschel on April 14 he speculated that it might be the Comet of 1770, known as Lexell's "lost" comet.[9] With Herschel, Hornsby was among the last to concede the object's planetary status. Nearly a year after its discovery, and even after a number of circular orbits for it had been computed which showed it to be more than seventeen times more distant from the sun than the earth, he was still reluctant to consider it a planet. "It is the fashion," he wrote Herschel in February 1782, "to call it a new star or a planet, but I cannot help thinking that it will prove to be a comet; but it does not necessarily follow that it should be either a Planet or a Comet."[10]

Herschel himself, even at this late date, seems to have been of a similar mind. For as other astronomers began to doubt the cometary nature of his discovery soon after the promulgation of his "Account," Herschel became increasingly circumspect in characterizing the object, preferring to call it "the new star." Still another six months would pass before he would concede unequivocally that it was a planet, and by then his fame as the discoverer of Uranus had already won him high honors and the patronage of his sovereign, King George III.[11]

Herschel's belief that his "curious" object was a comet was at first reinforced by a series of measurements he had made of its position in March and April 1781 which showed, as he reported to Hornsby in May, "a very considerable parallax which was sufficient to prove it to be on our side of the sun," that is, within the earth's orbit. A passage to this effect, in fact, had been included in his "Account" as read before the Royal Society, but was deleted from the text when it was later printed in the *Philosophical Transactions*. In the interim, others had failed to find any such parallactic effect, and Herschel had found a flaw in the homemade micrometer he had used that had resulted in "a small but pretty regular error" in his measurements.[12]

Word of Herschel's discovery spread rapidly and soon stirred wide interest among astronomers on the continent as well. French astronomer Charles Messier, a leading discoverer and observer of comets who would soon complete his famous catalog of nebulae, wrote Herschel on April 29 that he had learned that "it is to you, Sir, that we owe this discovery. It does you the more honor, as nothing could be more difficult than to recognize it, and I cannot conceive how you were able to return several times to this star—or comet—as it was absolutely necessary to observe it several days in succession to perceive that it had motion, since it had none of the characteristics of a comet."[13]

But along with admiration for his discovery, there was also skepticism, and for a number of reasons. For one thing, some astronomers were dubious of Herschel's claims for the high magnifying powers of his telescopic eyepieces, and he subsequently felt obliged to explain how he derived these powers to Sir Joseph Banks, president of the Royal Society, in a letter later printed in its *Transactions*.[14] Nor did the vagueness of his reports of the object's location in the sky, or his mistaken measures of its movement and position reassure the astronomical fraternity as to the efficacy of his methods or the rigor with which he applied them. And finally, there was undoubtedly something of the skeptical attitude which the credentialed professional not infrequently holds for the amateur, the dilettante, no matter how talented. For Herschel was an amateur, with no formal training in astronomy, mathematics or optics. His was a self-disciplined mind, and what knowledge he had of these disciplines, and it was considerable, he had painstakingly acquired over time through voracious reading and practical experience. He was, from the very beginning of his interest in astronomy, an indefatigable observer, and throughout his career spent long hours on every clear night at the telescope, usually with his sister Caroline at his side to record his observations. Because he could not afford to buy telescopes, he made his own—scores of them, in fact, that were as fine or finer than the "fixed instruments" whose excellence was too well known to warrant comment in his "Account." He was later able to record, no doubt with considerable satisfaction, that when the homemade reflector with which he had discovered Uranus was taken to the Greenwich Observatory and examined by Maskelyne and others, they "declared it to exceed in distinctness and magnifying power all they had seen before."[15]

This early skepticism over his work, and particularly the doubts about the high magnifying powers he used, were brought directly to Herschel's attention in December 1781 by Watson in a letter in which he noted that Maskelyne and Alexander Aubert, an amateur observer with a small observatory near Deptford, had questioned his claims to be able to see stars as "round and well defined."[16] This, in turn, led to a series of letters from Herschel in January 1782, culminating in his communication to Banks, in which he explained something of his observing techniques, and discussed the persistent problem of astronomical "seeing."

To Watson, for example, he wrote on January 9, 1782, that "Seeing is in some respects an art which must be learnt. To make a person see with such a

Sir William Herschel, holding a diagram of the planet Uranus and its satellites. From an engraving of a pastel by John Russell.

Ann Ronan Picture Library

Sir William Herschel's Newtonian reflector of seven-foot focal length with which he discovered Uranus. From Charles F. Partington, *The British Cyclopaedia*, London, 1835.

power is nearly the same as if I were asked to make him play one of Handel's fugues upon the organ. Many a night have I been practicing to see, and it would be strange if one did not acquire a certain dexterity by such constant practice.''[17]

To Aubert, he commented along similar lines: ''I shall beg leave to give a hint or two that experience has taught me in the use of high powers. One of no small consequence is that when you have looked for a considerable time with a small power and change it for a higher, not to be too ready to lay it aside if you should find it not so distinct as you could wish. I remember the time when I could not see with a power above 200, with the same instrument which now gives me 460 so distinct that in fine weather I can wish for nothing more so. When you want to practice seeing (for believe me, Sir—to use a musical phrase—you must not expect to see at sight or *a livre ouvert*) apply a power higher than what you can see well with, and go on increasing it after you have used it some time.... The consequence of this was that every time I tried again, my eyes acquired more practice, and I can now see with powers I used to reject for a long time."[18]

And again to Aubert, in an oft-quoted passage, he wrote:

"It would be hard to be condemned because I have tried to improve telescopes and practiced continually to see with them. These instruments have played me so many tricks that I have at last found them out in many of their humours and have made them confess to me what they would have concealed, if I had not with such perseverence and patience courted them. I have tortured them with powers, flattered them with attendance to find out the critical moments when they would act, tried them with specula of short or long focus, a large aperture or a narrow one; it would be hard if they had not been kind to me at last."[19]

The musical analogies in these letters, incidentally, including the clear message that practice makes perfect in astronomy as well as in music, came easily enough to Herschel. For, at 44 years of age, he had spent almost the whole of his life as a professional musician.

Herschel, of course, became a towering figure in the history of astronomy, and not for his discovery of Uranus alone. In a field notable for the contributions made by dedicated, imaginative amateurs, he has been called the "greatest of all amateur astronomers."[20] During his long and productive career, his pioneering studies of planets, stars and nebulae, and of "the structure of the heavens," laid the basic foundations for much of modern astronomical research. Even such a relatively recent and rapidly developing area of investigation as infrared astronomy can trace its beginnings back to Herschel's work and discoveries.[21]

Friedrich Wilhelm Herschel was born November 15, 1738, in Hanover, Germany, one of the six surviving children of Isaac and Anna Ilse Maritzen Herschel. He did not become an English subject until 1793, when he was naturalized as William Herschel, and he was not knighted as Sir William by the Prince Regent for the then insane George III until 1816, thirty-five years after his discovery of Uranus had made him world-famous. Other honors acquired during his astronomical career included honorary doctoral degrees (LL.D.) from the University of Edinburgh in 1786 and the University of Glasgow in 1792.

He is said to have excelled in arithmetic in school, and later to have been an avid linguist, becoming fluent in French and English in addition to his native German, and eventually acquiring Italian and Latin and attempting Greek. Under the tutelage of his father, an army musician, he began to learn to play the violin. He did so, he says in his own words, "as soon as I was able to hold a small one made on purpose for me." Thereafter, he became an accomplished musician.[22]

At 14, he was engaged as a bandsman in the Hanoverian Guards, serving for more than four years. During this period, he studied French and other subjects under an apparently excellent teacher named Hofschlager who,

Herschel later recalled, "encouraged the taste he found in his pupil for the study of philosophy, especially logic, ethics and metaphysics...." At 19, after the French occupation of Hanover during which he came under fire, Herschel obtained his discharge from the Guards and went to England, where he had visited briefly the previous year. There he pursued a career as a professional musician, working first as a copyist, and then as a teacher, a composer of "symphonies," and a performer at recitals and concerts. In 1766, he received an appointment as organist for a fashionable church at Bath, and this post provided a basic living for him for the next sixteen years.[23]

It is in 1766, too, that his interest in astronomy first appears in the record in the form of brief notes of observations of Venus made on February 19 and of an eclipse of the moon on February 24. Apparently this interest was casual at first and only one of several that occupied his versatile mind and driving curiosity. Not until seven years later is there an indication that he had determined to take up astronomy as a serious avocation. "In the spring of 1773," he later recorded, "I began to provide myself with the materials for Astronomical purposes. The 19th of April I bought a Hadley's quadrant and soon after Ferguson's *Astronomy*. In May I procured some short object glasses and had tubes made for them, beginning with a 4-feet one of Huyghenian construction. With this [refractor] I began to look at the planets and stars. It magnified 40 times. In the next place I attempted a 12-feet one and contrived a stand for it. I saw Jupiter and its satellites with it. After this, I made a 15-feet and also a 30-feet refractor and observed with them." But Herschel soon found that the great length of these instruments made them unwieldy and impractical. "The great trouble occasioned by such long tubes, which I found almost impossible to manage, induced me to turn my thoughts to reflectors, and about the 8th of September I hired a Gregorian one. This was so much more convenient than my long glasses that I soon resolved to try whether I could not make myself such another...."[24]

In the fall of 1773, thus, he began experimenting with making and polishing speculum metal mirrors, using a formula of "32 copper, 13 tin and one Regulus of Antimony," and by mid-December had "got so far as to give a tolerable gloss to some metals...."[25] This work, it may be noted, he carried on in addition to an arduous daily schedule in music which involved not only his organist chores but teaching up to forty-six "scholars" a week, or "nearly 8 per day."[26]

Some years later, in an autobiographical letter, Herschel described how his interest in astronomy came about. "After a fatiguing day of 14 or 16 hours spent in my vocation," he recalled, "I retired at night with the greatest avidity to unbend the mind (if it may be so called) with a few propositions in Maclaurin's *Fluxions* or other books of that sort.... Among other mathematical subjects, Optics and astronomy came in turn, and when I read the many charming discoveries that had been made by means of the telescope, I was so delighted with the subject that I wished to see the heavens and the planets with my own eyes thro' one of those instruments."[27]

In 1774, Herschel began keeping an astronomical journal to record his observational and instrument-making activities. His whole life now was music by day and astronomy by night. In 1777, he noted in a memorandum: "Musical business carried on as usual. All my leisure time was given to preparing telescopes and contriving proper stands for them." But by 1779, his avocation had begun to infringe on his vocation, for he "gave up so much time to astronomical preparations that I reduced the number of my scholars so as seldom to attend more than three or four per day." After his discovery of Uranus, incidentally, he continued to "attend many scholars," but he wrote: "Some of them made me give them astronomical instead of musical lessons."[28]

In 1778, Herschel completed what he considered to be a "most capital" reflector of the Newtonian type, with a speculum mirror 6.2 inches in diameter and a focal length of 7 feet. With this, he launched his first systematic survey, or "sweep," of the heavens, observing stars out to the fourth magnitude. The following year, he began a second survey, this time out to eighth magnitude stars, which required two years to complete and yielded a catalog of 269 double stars and the planet Uranus.[29]

During these years, a seemingly trivial incident occurred that proved to have great significance both for Herschel's personal life and his astronomical career. On a clear December evening in 1779, after setting up his reflector on the street in front of his home, he courteously allowed an inquisitive passerby to view the moon through the instrument. Thus began his long friendship with Dr. William Watson, a Fellow of the Royal Society. Within a month, Watson had introduced him into the newly formed but short-lived Bath Literary and Philosophical Society, and thus into formalized science as it was carried on in that day.[30]

From January 1780 until March 1781, when he discovered Uranus, Herschel submitted thirty-one papers to the Bath Society on subjects ranging from astronomy and physics to natural history and philosophy, and giving his considered opinions on such topics as "On the Utility of Speculative Enquiries," and "On Liberty and Necessity." Herschel's work as an observer and telescope-maker was already known to at least a few astronomers—Maskelyne had visited him at Bath in 1777, for example, and he had met Hornsby that same year—but Watson now made it more widely known by communicating some of his scientific papers to the membership of the Royal Society.[31]

Herschel's discovery of Uranus came at a time when the dominant thrust in astronomy was mathematical, rather than observational, and when most astronomers were preoccupied with computing planetary orbits and working out the mechanism of the solar system under Keplerian-Newtonian laws. Observation of the stars was minimal, and the relatively poor quality of the telescopes of the day probably had something to do with this circumstance. It is not the least of Herschel's accomplishments that his practical talent for optics, as well as his discoveries in stellar astronomy, would soon help restore a balance. Lubbock points out that very little was then known of the "fixed" stars, that less than 3,000 were then listed in the British catalog, and that Ferguson's *Astronomy*, Herschel's first self-study textbook, devoted only one of its twenty-two chapters to subjects unrelated to the solar system. "Observation had given way to calculation," she has observed. "Astronomy had become almost entirely a deductive science."[32]

But precisely because of this, Herschel's new "comet" presented astronomers and mathematicians with a particularly fascinating problem, and no sooner had the news of the discovery gone out than attempts were made to calculate its motion. In these initial attempts, the computers accepted the object's cometary status and, to simplify their problem, assumed that its path was a parabola (eccentricity = 1). The French mathematician Pierre François André Méchain was the first to derive such an orbit, finding the perihelion* distance from the sun to be 0.46 astronomical units† or well within the orbit of the earth, and the date of this closest approach to be May 23, 1781, only a few weeks hence. Messier forwarded Méchain's results to Herschel.[33]

But others soon found that a parabolic orbit would not do at all. On May 8, J. B. G. Bouchart de Saron, a well-known computer of cometary orbits and friend of the French mathematical astronomer Pierre Simon Laplace, reported that his efforts to devise a parabolic orbit indicated that a perihelion distance of at least 14 a.u. was necessary to account for the observed positions of the "comet," putting it far beyond the orbit of Saturn, then the outermost of the known planets, and making it a very unusual comet indeed.[34] Soon after this, mathematician Anders J. Lexell of St. Petersburg, who was visiting in England, calculated a perihelion distance of 16 a.u. and perihelion passage on April 10, 1789.[35] Still another computer, the Italian Jesuit Abbé Ruggiero Giuseppe Boscovich, also took up the problem at the urging of French astronomer Joseph J. F. Lalande who informed him that Méchain's elements were unsatisfactory. By early June, Boscovich had computed four parabolic orbits, each from a different set of observations, and found that no two of them agreed on the object's path.[36]

Boscovich, parenthetically, may well have employed an innovative method in these and subsequent computations of Uranus' orbit that he had devised more than ten years before. This was, in effect, a "method of least sums" for assessing the most probable value for unknowns, a concept which was independently developed in the first years of the nineteenth century by Carl Friedrich Gauss and Adrien-Marie Legendre into the fruitful "method of least squares."[37] This latter method, as will be seen, was to be used to some extent by both Urbain J. J. Leverrier and John Couch Adams in their mathematical predictions for a trans-Uranian planet, and was applied massively by Percival Lowell in his analytical search for a hypothetical trans-

---

*Perihelion is the point in the orbit of a planet, comet or other circumsolar object that is nearest the sun.
†An astronomical unit is the mean distance from the earth to the sun, or about 150,000,000 kilometers (93,000,000 miles).

Neptunian Planet X in the early years of the twentieth century. Boscovich seems to be a sadly neglected figure in the history of science and astronomy. He was one of the first European scientists to accept Newton's law of gravitation, and he was an excellent observer who published extensively on a wide variety of astronomical subjects, including the aurora borealis, the observation of the fixed stars, the inequities of terrestrial gravitation, the application of mathematics to the theory of the telescope, the theory of comets, and the limits of certainty in astronomical observations. In 1771, he published an atomic theory.[38]

Carl Friedrich Gauss, from the obverse of a commemorative medal struck after his death in 1855.

Ann Ronan Picture Library

Both Boscovich and Lexell had concluded from their preliminary calculations that a circular orbit ($e = 0$) was a better assumption for the problem of Herschel's "comet," but Lexell seems to have been the first to put this assumption to a test. Using Herschel's March 17 observation and one by Maskelyne on May 11, he computed such an orbit with a radius, or distance from the sun (a), of 18.93 a.u. and a mean annual motion (μ) of 4° 20', or a period of something over 82 years. These estimates, as A. F. O'D. Alexander has pointed out in his definitive study, *The Planet Uranus* (1965), were very good, falling within 1.3 percent of modern values, and they were generally confirmed by subsequent circular orbits worked out by Boscovich, Méchain, Lalande and others.[39]

After Lexell's result, and by the fall of 1781, astronomical opinion clearly favored planetary status for Herschel's discovery although, as noted above, it would be many more months before some astronomers, Herschel included, would accept Uranus without qualification as a full-fledged planet of the solar system.

Uranus and the Asteroids   19

Circular orbits, of course, were not the complete answer, because planets travel in ellipses ($e = <1, >0$), and continuing observations of Herschel's new planet soon showed that further refinements of its orbit were in order. Thus, beginning in 1783 with Barnabo Oriani of the Observatory of Brera, a number of computers published ellipitical elements with eccentricities around 0.04 and mean distances from the sun of slightly more than 19 a.u. In 1789, Oriani was also the first to publish an elliptical orbit which took into account the gravitational influences of the other outer planets, and again he was followed by others.[40]

For more than a decade after Herschel's discovery, the problem occupied some of the finest mathematical minds of the day. The great Leonhard Euler, for example, outlined his computations of a Uranian orbit over dinner with Lexell and his family on September 18, 1783—only hours before he died.[41]

But these efforts still did not settle the matter of Uranus' motion. For in this work, the computers could use the first few of the more than twenty prediscovery sightings of Uranus as a star that are now known, including one in 1690 by England's first Astronomer Royal, John Flamsteed. Such "ancient" observations, as they are called, greatly extended the time span of the data available for the calculations and should, therefore, have increased the accuracy of the orbits and the tables of Uranus' motion derived from them. They did not, however. And, as will be seen, the problem of Uranus's motion became, over time, progressively worse.

Some of the delay in determining the true nature of Herschel's discovery undoubtedly stemmed from a forced suspension of observations in the late spring and early summer of 1781. For Gemini is a winter constellation and by the end of May, the "comet" had vanished with it in the evening twilight. Its reappearance in the predawn sky was apparently first observed by A. Darquier at Toulouse, France, July 20, and then again on August 1 by German observer Johann Elert Bode, director of the Berlin Observatory and editor of the *Berliner Astronomisches Jahrbuch*. Bode, a major if somewhat mischievous figure in the history of planetary discovery whose role will be discussed at greater length later, seems to have concluded almost immediately from his observations that the object was a planet, or at least planetlike, and he promptly suggested that it be called Uranus.[42]

But whatever its actual nature, Herschel's was an unusual and interesting discovery from the first. "Your success, you perceive," Watson wrote him at the end of August, "has set both the English and French astronomers on exerting themselves to the utmost; a comet which is likely to be visible for at least 12 years longer is a very extraordinary phenomenon."[43] And with Lexell's circular orbit, which was communicated to him by Watson in mid-October, Herschel began to receive formal recognition for his discovery.[44]

On November 15, Sir Joseph Banks informed Herschel that the Royal Society intended to present him with its annual Copley Medal for his achievement, and requested some details concerning his discovery. "Some of our astronomers here incline to the opinion that it is a planet and not a comet," he added. "If you are of that opinion it should forthwith be provided with a name, or our nimble neighbors, the French, will certainly save us the trouble of baptizing it."[45]

Herschel was now so inclined, although still tentatively, and he indicated as much in a long letter to Banks four days later. He took special pains to point out that his discovery had not been accidental:

"This new star could not have been found even with the best telescopes had I not undertaken to examine every star in the heav'ns including such as are telescopic, to the amount of at least 8 or 10 thousand. I found it at the end of my second review after a number of observations not less than 15 thousand; so that the discovery can be said to be owing to chance only it being almost impossible that such a star should escape my notice.... The first moment I directed my telescope to the New Star, I saw with the power of 227 that it differed sufficiently from other celestial bodies; and when I put on the higher powers of 460 and 932 was quite convinced that it was not a fixed star. By the help of my new invented micrometer for taking the angle of position I could in a single day (viz. from the 18 to the 19 of March, 1781) determine that the new star moved then at a rate of 2¼ seconds per hour; that its apparent course was according to ye signs; and that its orbit was but little declined from the ecliptic...."[46]

Herschel, incidentally, would have occasion to deny that his discovery was accidental again, and in much the same words. "In the regular manner," he later wrote, "I examined every star in the heavens, not only of that magnitude but many far inferior, it was that night *its turn* to be discovered. Had business prevented me that evening, I must have found it the next, and the goodness of my telescope was such that I perceived its visible planetary disc as soon as I looked at it."[47]

In his letter to Banks, however, Herschel was still cautious about the status of his discovery. "With regard to the New Star I may still observe that tho' we are not sufficiently acquainted with it to ascertain its nature, yet enough has been seen already to shew that it differs in many essential particulars from Comets and rather resembles the condition of Planets." He then cited four such particulars: The slowness and direction of its motion, the fact that the edges of its disc appeared "so well defined," its color which was "not fiery or cloudy like that of comets but of an equable bright uniform lustre, something between the light of Jupiter and Venus;" and its path "which is exactly that of the planets." He did not respond to Banks' suggestion about a name.[48]

Herschel was presented with the Copley Medal at a meeting of the Royal Society late in November at which Banks expounded on the importance and possible implications of his discovery. "Who can say but what your new star, which exceeds Saturn in its distance from the sun, may exceed him as much in magnificence of attendance?" he wondered. "Who can say what new rings, new satellites, or what other nameless and numberless phenomena remain behind, waiting to reward future industry?"[49]

Sir Joseph was being rhetorical here, of course, but subsequent events, one of them quite recent, have given his questions pertinence. For in January and February 1787, Herschel himself discovered the first two of the five known Uranian satellites, later named Titania and Oberon, determining that their orbits "Make a considerable angle with the ecliptic," and making excellent estimates of their periods of revolution of, respectively, 8 days, 17 hours, and 13 days, 11 hours.[50]

In 1797, in a paper summarizing his last ten years of observations of Uranus, Herschel announced two other discoveries relating to the planet, one of them spurious. First, he found that the direction of the orbital motion of Titania and Oberon around the planet was retrograde,* thus providing the first example of such anomalistic motion in the solar system. And second, he reported sighting four additional Uranian satellites, none of which have been seen since, by Herschel or anyone else. All four of these he placed outside the orbit of Oberon, the outermost of Uranus' moons, while the three satellites that have been subsequently discovered—Ariel and Umbriel in 1851 by amateur observer William Lassell, and Miranda photographically in 1948 by Gerard P. Kuiper—all circle Uranus well inside Oberon's path.[51]

Herschel also thought Uranus might have two rings, and importantly, nearly two hundred years after his discovery of the planet, it has been found that there are indeed rings. In March 1977, astronomers James L. Elliot of Cornell University and Robert L. Millis of the Lowell Observatory and their colleagues, during photometric observations of an occultation† by Uranus of a 9th magnitude star, discovered that the planet is girdled by a system of at least five, and very probably more, thin, narrow rings. By 1979, nine rings were known. These, however, are apparently unlike Saturn's broad, flat rings, whose width is measured in thousands of kilometers. Rather, they are better described as ribbonlike, only a few tens of kilometers wide, and are visually and photographically invisible to the earthbound observer. Planetary rings, in fact, may be the rule in the outer solar system. Voyager spacecrafts 1 and 2 in March and July 1979 revealed a single, faint ring around Jupiter.[52]

On December 6, 1781, Herschel was elected a Fellow of the Royal Society which, as Watson later informed him, waived all fees on the ground that "the money which was withheld from the Society would still be expended on such objects which would best answer the view of the Society."[53] Herschel remained active for thirty-seven years and between 1780 and 1818 published one or more papers in the Society's *Philosophical Transactions* nearly every year. These concerned not only his observations of planets, stars and nebulae, but ranged from his discovery of infrared radiation to an exposition of his belief in the habitation of the sun, both by-products of his studies of

---

*That is, in a direction opposite to that of the earth, the other planets, and the then known satellites, namely counter-clockwise as seen from the North ecliptic pole.

†An occultation occurs when an astronomical body passes in front of a more distant body in the line of sight from the earth. An eclipse of the sun is an occultation of the sun by the moon.

sunspots. He also discovered Enceladus and Mimas, the sixth and seventh satellites of Saturn, in 1787 and 1789 respectively. It is a curious sidelight to his long career as an observer, however, that he never did discover an actual comet.[54]

After surrendering his "comet" to the care of the Astronomer Royal and others, Herschel seems to have observed it only sporadically, and in fact by the time of his Royal Society honors, he had already turned his attention to his third systematic review of the stars. In October and November 1781, however, and again in the fall of 1782, he did make some new measures of his "new star" from which he concluded that it was somewhat smaller than he had originally thought. From his new measurements, and with Lalande's latest calculation of a distance of 18.913 a.u., he now deduced a real diameter of 4.454 times that of the earth, or more than 34,000 miles (54,800 kilometers), a figure that compares with the modern value of 32,400 miles (52,200 kilometers). These results he incorporated into a paper presented to the Royal Society in November 1782, his first relating to his observations of Uranus since his "Account."[55]

Herschel's fame continued to grow, but it was not until May 1782, fourteen months after his discovery, that he learned that he was to receive royal recognition from George III, who was something of an astronomy enthusiast himself and who maintained a small observatory for his own use in the Deer Park at Richmond, near Windsor. The delay may perhaps be explained in part by the king's preoccupation with such matters as the end of the war with his rebellious American colonists, fighting at sea and in India, and political unrest in Ireland and in his own cabinet.[56]

Herschel's friends had apparently been working to secure royal patronage for him for some months, although there is little documentation of their efforts. It seems clear, however, that these behind-the-scenes negotiations concerned a name for the new planet. Banks, as noted above, was anxious for Herschel to give it a name, and in February 1782, as Bode's suggestion of "Uranus" began to gain wide acceptance among European astronomers, Banks confided ruefully to Watson that "I wished the new star, so remarkable a phenomenon, to have been sacrificed somehow to the king."[57]

Finally, on May 10, 1782, court official John Walsh wrote Herschel that "in a conversation I had the honor to hold with His Majesty the 30th inst. concerning you and your memorable discovery of a new planet, I took occasion to mention that you have a two-fold claim as a native of Hanover and a Resident of Great Britain where the discovery was made, to be permitted to name the Planet for His Majesty. His Majesty has since been pleased to ask me when you would be in Town...."[58]

A few days after May 20, Herschel had his first audience with the king, subsequently recording this and later events in a series of memoranda. "The king said that my telescope in three weeks was to go to Richmond and meanwhile was to be put up at Greenwich," he wrote in an undated entry. There, as related earlier, it was tested and its excellence acclaimed. "My telescope having undergone this kind of examination, the King desired it to be brought to the Queen's Lodge at Windsor,'' he added.[59]

On July 2, he recorded that "I had the honor of showing the King and Queen and the Royal Family the planets Jupiter and Saturn and other objects...." And in his very next entry, also unfortunately without a date, he reported that "about the last week of the month it was settled by His Majesty that I should give up my musical profession and, settling somewhere in the neighborhood of Windsor, devote my time to astronomy."[60]

The king had provided Herschel with an income of £200 a year, with no specific duties attached. Later he gave £4,000 for the construction of Herschel's 40-foot telescope with a 48-inch speculum mirror, and provided an allowance for its upkeep, as well as £50 a year for Herschel's sister and assistant, Caroline. Almost immediately, Herschel moved to Datchet, where he lived three years, then to Clay Hall at Old Windsor, remaining only a few months before moving in April 1786 to Slough, where he lived for the rest of his life.[61]

Apparently, Watson knew as early as July 14 that Herschel had been made "independent of music," and in a letter a week later, he discussed at some length the name of "Georgium Sidus" (the Georgian Star) and the proper form for proposing it according to the usage of the famous Latin poets, Horace and Virgil.[62] It is not known whether the name was Herschel's idea. But nevertheless, shortly after being granted the king's largess, he wrote Banks formally proposing the name "Georgium Sidus," and declaring that "the new star which I had the honor of pointing out . . . is a Primary Planet of our Solar System."

The name, he wrote, "presents itself to me, as an appellation which will conveniently convey the information of the time and country where and when it was brought into view. But as a subject of the best of Kings, who is the liberal protector of every art and science;—as a native of the country from which this illustrious Family was called to the British throne;—as a member of that Society which flourishes by the distinguished liberality of its Royal Patron;—and last of all, as a person now more immediately under the protection of this excellent monarch, and owing everything to his unlimited bounty;—I cannot but wish to take this opportunity of expressing my sense of gratitude, by giving the name Georgium Sidus . . . to a star, which (with respect to us) first began to shine under His Auspicious Reign."[63]

As word of the new name and the king's beneficence went out, letters from eminent astronomers came in, first from England and then from the continent.[64] In January 1783, for example, Johann Heironymous Schroeter, who has been called the "Herschel of Germany," wrote from Lilienthal: "Is it true that you have given the name of George, or Sidus Georgium as the newspapers have it, to the new planet, which in Germany from the first has been called Uranus? It is to be hoped that His Majesty has presented you entirely to Astronomy and provided you with a handsome pension. With all lovers of astronomy I wish this with my whole heart."[65] And in July, Bode wrote "to congratulate you heartily on your fortunate discovery of the new planet.... You know perhaps that I was the first person in Germany to see the new star; this was on the 1st of August, 1781.... I have proposed the name of Uranus for the new planet as I thought that we had better stick to mythology and for several other reasons. However, I assure you that had I been in your situation I should have felt it proper to do as you have done. I have further proposed the sign ♁ for our new planet."[66]

Herschel's brief reply to this was explanatory: "When I named it Geo-Sidus I hardly expected that this name would become generally accepted, because we already know by experience that the first names of the satellites of Jupiter and Saturn were changed."[67]

There were, in fact, a number of other suggestions for a name, including "Herschel's Planet," a favorite with Lalande and in France; "Hypercronius," "Neptune," with such variations as "Neptune de George III"; "Cybele," wife of Saturn; "Astraea," and "Minerva."[68] But Bode's "Uranus," of course, has prevailed. It is interesting to note here that after the discovery of Neptune in 1846, Leverrier temporarily reverted to "Herschel's Planet" in a not-too-subtle ploy to gain acceptance for his proposal to name Neptune "Leverrier."[69] Bode's symbol also prevailed, although in 1784, another symbol, ♅ , was devised in France and this has since been used on occasion, especially by British astronomers.[70]

The discovery of Uranus was, as it turned out, a considerable triumph for Bode. For the planet, at a distance of slightly more than 19 a.u. from the sun, seemed to confirm a curious arithmetic progression relating the distances of the various planets usually known as Bode's law, not because Bode discovered it, but because he had vigorously promoted it for many years. With this apparent confirmation, Bode's law would influence the history of planetary discovery for the next sixty-five years. During this period, it would be apparently confirmed once again by the discovery of the asteroids before Neptune relegated it to the status of an astronomical curiosity.

It has sometimes been said that Herschel did not think that his "comet" might be a planet because there was simply no reason at the time for him or anyone else to think such a thing, the number of planets having remained the same for thousands of years and since before the dawn of history.[71]

But this is not true, strictly speaking. As early as 1596, Johannes Kepler declared in his *Misterium Cosmographicum* that an undiscovered planet must exist between Mars and Jupiter. In 1741, the German observer Christian Freiherr von Wolff suggested that something was not quite right with the spacing of the planets and that perhaps one was missing. A few years later philosopher Immanuel Kant speculated along similar lines.[72]

In 1772, as a footnote in an edition of Charles Bonnet's *Betrachtung über die Natur* (Reflections on Nature), editor Johann Daniel Titius, professor of mathematics at Wittenburg, formulated an arithmetic rationale for such speculations. His footnote read: "Divide the distance from the Sun to Saturn into 100 parts; then Mercury is separated by 4 such parts from the Sun; Venus by

4 + 3 = 7 such parts; the earth by 4 + 6 = 10; Mars by 4 + 12 = 16. But notice that from Mars to Jupiter there comes a deviation from the exact progression. After Mars, there follows a distance of 4 + 24 = 28 parts, but so far no planet or satellite has been sighted there. Let us assume that this space belongs to the still undiscovered satellites of Mars.... Next to this still unexplored space lies Jupiter's sphere of influence at 4 + 48 = 52 parts; and that of Saturn at 4 + 96 = 100."[73]

That same year, Bode picked this up and quoted it almost verbatim in a new edition of his *Anleitung zur Kenntniss des gestirnen Himmels* (Introduction to the Study of the Starry Sky), discarding, however, Titius' assumption of undiscovered satellites in favor of postulating another planet for the gap at 28 parts.[74] Titius' values correspond roughly to the distances of the planets in astronomical units for which, of course, the earth-sun distance is taken as 1, rather than 10. The progression then comes out:

Planet	Bode's Law	Actual
Mercury	0.4	0.39
Venus	0.7	0.72
Earth	1.0	1.0
Mars	1.6	1.52
?	2.8	?
Jupiter	5.2	5.2
Saturn	10.0	9.5

If this progression is carried one step further, by again doubling the second term in the equation for Saturn, it yields a distance for a possible planet at 4 + 192 = 196 or, in astronomical units, 19.6. And Uranus, at 19+ a.u. from the sun, was close enough to this figure for Bode and others to hail Herschel's discovery as confirmation of the "law." Bode himself, in 1784, expressed satisfaction over the circumstance in a paper pointing out the close agreement between the prediction and the fact.[75]

For ten years before Herschel's discovery of Uranus, Bode had been arguing for the existence of an undiscovered planet in the gap at 2.8 a.u., and the fact that Uranus fit so well into the progression now gave new impetus to the search for such a body. In 1785, Baron Franz Xavier von Zach, director of the Ernestine Observatory at Seeburg, near Gotha, attempted to compute an orbit for the suspected planet, but found that the one element that might reveal its location—the longitude—eluded him. In September, he confided to Bode: "I am having much the same success as the alchemists in their search for gold; they had everything except the vital factor. Thus I too believe that I am in possession of all the elements of the orbit of this still unknown planet, except for one, namely the Epocham Longitudinus; this alone now keeps me amused, and although one may not find gold in one's wanderings, one does occasionally come across a chemical process."[76]

For more than ten years after Uranus' discovery, von Zach kept an eye out for the postulated planet, but the gap in Bode's law remained unfilled. In 1796, von Zach called a conference of astronomers at Gotha on the problem at which Lalande suggested that a systematic search be mounted by a number of observers, each to be responsible for a particular segment of the sky. Four more years passed before von Zach, during a visit to Schroeter at Lilienthal, was able to organize a group of astronomers, known as the "Lilienthal detectives," to carry out such a plan. But before the new search could be implemented, a conscientious Sicilian astronomer apparently found what they were seeking.[77]

Giuseppe Piazzi of Palermo had not yet been informed of von Zach's newly-organized search when, on January 1, 1801, in the course of observations for a new star catalog, he noticed what appeared to be an 8th magnitude star in the constellation of Taurus. Over the next two nights, as was his custom, he observed it again and determined that it had changed position. Thinking it might be a comet, he continued to follow it and on January 14 noted that it changed from a retrograde to a direct motion. A few days later, he notified Oriani, Lalande and Bode of his discovery, confiding only to his friend Oriani, however, his thought that it might possibly be a new planet.[78]

But such was the state of postal services in that day that it was some two months before Piazzi's letters reached their destinations. Bode, for example, did not receive the news until March 20. This delay, incidentally, provided the German philosopher Georg Friedrich Wilhelm Hegel with the opportunity to publish his *Dissertatio philosophia de orbitus planetarum* in which he argued on logical grounds that there could be no more than seven planets orbiting the sun, and the folly of predicting the existence of another planet merely to fill out an arithmetical series.[79]

When Bode finally received Piazzi's letter, he immediately concluded that the Sicilian had found his long-postulated trans-martian planet, and he quickly spread the word. Von Zach was elated, and reported the news in his *Monatliche Correspondenz*, the first astronomical journal, under the heading: "On a long supposed, now probably discovered, new major planet of our solar system between Mars and Jupiter."[80]

During the time when Piazzi's announcements of his discovery were en route, however, the object had become lost, for Piazzi himself had become ill and was unable to observe it after February 11. And by the time others learned of his discovery, Taurus had disappeared into the light of the sun and attempts to recover the object had to be postponed for several months. Herschel was among the first to undertake an observational search for it in July, and he continued his work after September 1 after receiving a written plea from Piazzi himself for help "in verifying this discovery which I do not doubt interests you as much as me." But Herschel's efforts, as well as those of other observers, were to no avail. They were, of course, looking for something quite different—a planet with an appreciable disk—from what Piazzi had discovered, although they could not know this at the time. Von Zach and

Johann Karl Burckhardt also sought to compute circular and parabolic orbits respectively, which did not, however, recover the object.[81]

It required the mathematical genius of Carl Friedrich Gauss to bring Piazzi's lost object back into view. Gauss, from Newton's laws of motion, had devised a new method for deriving elliptical orbits from only a few observations covering only a very small arc of a total orbit.[82]

Gauss set to work on the problem of Piazzi's lost discovery in October, using the Sicilian's positions of January and early February as his data base. He quickly produced a set of elements and an ephemeris which allowed von Zach, on December 7, to glimpse the object through the clouds and, on December 31, to confirm the rediscovery. On January 1, 1802, both von Zach and Heinrich Wilhelm Matthias Olbers, an enthusiastic amateur observer in Bremen, observed it. They found Piazzi's object only a lunar diameter away from the position predicted by Gauss' ephemeris, an extraordinary feat of computation and a landmark event in the history of planetary discovery.[83]

Astronomers, and particularly Herschel, now eagerly turned their telescopes to the new object, first named "Ceres Ferdinandea" by Piazzi after the tutelary goddess of Sicily and the then Sicilian monarch, King Ferdinand IV, but soon shortened to "Ceres." Ceres' mean distance from the sun, as calculated by Gauss, was 2.767 a.u. and thus it fell neatly into the planetless gap at 2.8 a.u. in the Titius-Bode progression. Most astronomers quickly concluded that this was indeed the postulated missing planet. But observations soon showed that Ceres was a most unusual planet, being too small to show a discernible disk, and within three months of its rediscovery Olbers, on March 28, 1802, found a second object with a mean distance, again as calculated by Gauss, of 2.67 a.u., almost the same as that of Ceres. Olbers, a medical doctor by profession, named it "Pallas."[84]

Astronomers now were not at all sure what they were dealing with. On April 9, Banks wrote Herschel that Olbers had discovered "another Ceres," adding that "it is to appearance larger than the Georgium. Messrs. Schroeter and [his assistant Karl] Harding have seen it since April 2nd Mr. Harding saw a star near it which he thinks will prove a satellite."[85]

Herschel, from micrometer measurements, soon disposed of Banks's estimate of the object's size, deducing diameters for Ceres of only 161.6 miles and for Pallas of only 110.3 miles, well below modern determinations of 1020 and 540 kilometers, respectively, and reported his results to the Royal Society. Schroeter's diameter measurements were higher, 2620 and 2790 kilometers, but were still far short of the diameter of "the Georgium." Gauss' computations also gave the objects other characteristics that made them quite unlike the known major planets. The inclination of Ceres' orbit, for example, exceeded 10°, greater than any of the known planets, and the orbital inclination of Pallas was a startling 34° 39'. The eccentricity of Pallas's orbit was 0.2476 and this, too, was far greater than any of the then known planets.[86]

Bode was particularly upset with this sudden duplicity of objects in orbit between Mars and Jupiter, and with their unusual nature. "Olbers has found a moving star which I take to be a very distant comet," he wrote Herschel in May. "Olbers suggests most strange things; that it is a planet travelling with Ceres in the same orbit, at the same distance, round the sun. Such a thing is unheard of !"[87] Three months later, however, he was able to accept the fact and to claim still another confirmation of Bode's law. To Herschel again, he declared that "I hold myself convinced that Ceres is the eighth primary planet of our solar system and that Pallas is a special exceptional planet—or comet—in her neighborhood, circulating around the sun. So there would be two planets between Mars and Jupiter, where ever since 1772, I have expected only one; and the well-known progressive order of the distances of the planets from the sun, is by this fully proved."[88]

Part of the difficulty for Bode was the apparent size of the two bodies. In this same letter, he noted that he had read Herschel's diameter computations "with a certain astonishment; as it remains to me inexplicable how you can suggest such small diameters for both these bodies when you agree, as I think you do, with Dr. Gauss about their real distance."[89]

Toward the end of April 1802, Herschel had observed the new objects long enough to become convinced that they represented a new and different class, and thus ought to have a name. On April 25, he wrote to Watson that it "appears to me much more poor in language to call them planets than if we were to call a *rasor* a *knife*, a *cleaver* a *hatchet*, etc.... I surmise (again) that possibly numbers of such small bodies that have not matter enough in them to hurt one another by attraction, or to disturb the planets, may possibly be running through the great vacancies, left perhaps for them, between the other planets, especially Mars and Jupiter.... Now as we already have Planets, Comets, Satellites, pray help me to another dignified name as soon as possible."[90]

What name, if any, Watson came up with is not known, but Herschel shortly proposed the name "asteroids" for the class to the Royal Society,[91] a name which then and since has drawn protests from nomenclature purists. Laplace was among the first to question the choice, writing to Herschel on June 17 that he saw no reason not to call the new objects planets as Ceres differed from the other planets only in the inclination of its orbit, which he conceded was "a little large."[92] And Piazzi himself, on July 4, suggested "planetoids," pointing out that "asteroids" would be more appropriate for "little stars."[93] But the term "asteroids" has persisted and remains in common use, although these bodies are also referred to, more precisely, as the "minor planets."

Olbers thought the name "asteroids" a good one and said so in a long and insightful letter to Herschel in June 1802 in which he outlined several ideas that were to dominate thinking about the asteroids for many years. "I agree with you, honoured Sir, in your sagacious suggestion that Ceres and Pallas differ from the true planets in several respects, and the name *asteroid* seems to me to fit these bodies very well.... Yet taking all the particulars together there seems to me to be a real difference between the asteroids and the true planets."[94]

The Bremen amateur came up with an intriguing idea that was entertained seriously by some astronomers for a number of years. "The similarity in the period of their revolution, of their long axes [i.e. mean distances], and the remarkable position of both orbits in relation to each other, have suggested to me an idea which I hardly dare put forward as a hypothesis.... I mention it to you in confidence. How might it be, if Ceres and Pallas were just a pair of fragments, or portions of a once greater planet which at one time occupied its proper place between Mars and Jupiter, and was in size more analogous to the other planets, and perhaps millions of years ago had, either through the impact of a comet, or from internal explosion, burst into pieces? I repeat that I give this idea as nothing more than, hardly as much as a hypothesis."[95]

But Olbers nonetheless felt his hypothesis sufficient for prediction. "If it is true, we may expect to find other fragments of the broken down planet; their orbits must intersect that of Ceres in the same spot." These points of intersection, he thought, would be in the constellations of Cetus and Virgo.[96] And, as it turned out, Harding found the third asteroid, Juno, in Cetus, on September 2, 1804, while Olbers himself discovered the fourth, Vesta, in Virgo, on March 29, 1807.[97] No more were found until 1845 (Astraea), but thereafter their number steadily increased by visual observations to 322. On December 22, 1891, the German astronomer Max Wolf discovered No. 323 (Brucia) photographically, and since then thousands have been recorded, and some two thousand numbered and named. Estimates of the total number that can be photographed with, say, a 1.5-meter telescope have reached as high as 44,000, yet the total estimated mass of all the known and presumed asteroids is still only about one thousandth of that of the earth, and this is one of the major reasons why Olbers's provocative hypothesis has been generally discarded.[98]

Olbers's letter to Herschel, however, contained still another idea that has fared better in subsequent observations over the years. Noting that the brightness of Ceres and Pallas appeared to vary from one observation to another, he wrote that "Schroeter explains it by variations in the atmosphere of the asteroid; and it appears to me it would be better accounted for by supposing for both these asteroids an irregular rather than a round figure; by reason of which at one time a broad surface and at another a narrow profile would be turned to us."[99] An analogy that has often been used for this is that of a large brick tumbling through space.

The decades immediately following Herschel's discovery of Uranus were a particularly fruitful period in the history of planetary discovery. In a span of less than thirty years, a new major planet, four new satellites—two each for Uranus and Saturn—and a whole new class of objects—the asteroids—were added to the solar system. It may be noted also that the interplanetary nature of meteors first began to be recognized at this time. In 1798, two young German students observed the same meteor from different sites some distance apart and, comparing notes, realized that what they had seen could not be an atmospheric phenomenon, as had long been believed. They computed an altitude of nearly fifty miles for the meteor, and concluded that it thus must have originated somewhere in space beyond the orbit of the moon.[100]

The period is also notable for its major intellectual advances in the field of celestial mechanics, advances that were largely incorporated in two monumental publications. The first of these, in 1799, was Laplace's *Méchanique Céleste* in which he laid out in great detail basic perturbation theory. The second, in 1809, was Gauss' *Theoria Motus Corporum Coelestium in Sectionibus Conicus Solem Ambientum*, which contained his innovative methods for computing orbits from a limited number of observations, and the method of least squares. With these publications, mathematical astronomers henceforth would have sophisticated and elegant analytical tools at hand for the problem of computing the motions of planets known and unknown.

CHAPTER 2

NEPTUNE

In the course of their earliest investigations into the orbit of Uranus, Ruggiero Giuseppe Boscovich and Anders Johann Lexell independently came to the conclusion that there were probably other undiscovered planets orbiting the sun in the farther reaches of the solar system, and subsequently published their conclusions.[1] Undoubtedly, the thought also occurred to others as well. But it would be nearly fifty years before astronomers would seriously entertain the suggestion of a possible trans-Uranian planet.

In a very general way, the idea had been advanced even before Uranus's discovery, as the Jesuit astronomer Christian Mayer of Mannheim reminded William Herschel in a congratulatory letter in August 1782. Mayer recalled that in his *De novis in coelo sidero phenomenis* in 1779 he had made a "clear and precise declaration that I hold it for certain that among the fixed stars are a large number of planets, which we have hitherto taken for fixed stars."[2]

By August 1782, moreover, there was evidence for Mayer's conviction. For just a year earlier, Johann Elert Bode had discovered that another Mayer—Tobias of Göttingen—had observed Uranus on September 26, 1756, twenty-five years before Herschel's discovery, and had recorded it then as a star in the constellation of Aquarius.[3] "If Mayer had observed it twice at a short interval of days," Boscovich commented, "he would have made the interesting discovery."[4]

As a practical matter, the discovery of planets is not quite so simple, and in not too many more years it was found that several eminent and experienced observers had recorded Uranus "at a short interval of days" but had failed to recognize its planetary nature.

Mayer's was still the first of the twenty-three prediscovery or "ancient" observations of Uranus known as of 1981. Bode, with computational assistance from the Rev. Placidus Fixlmillner, soon found a second, made by England's first Astronomer Royal, John Flamsteed, on December 23, 1690, when he had recorded the planet as the star, 34 Tauri.[5] Six more such observations by Flamsteed have since been reported—for 1712, 1714, and 1715. The single 1714 sighting, incidentally, is the most recent of the ancient observations to be found—in 1968 by an American astronomer, Dennis Rawlins.[6]

But even more surprisingly, in 1788, Pierre Charles Lemonnier discovered that he had observed Uranus on three occasions in 1764 and 1769, and some twenty years later, after Lemonnier's death, Alexis Bouvard found nine more such observations by the French astronomer, including six made over only nine nights in January 1769. Thus Lemonnier accounts for twelve, or more than half, of the known prediscovery observations of Uranus. Morton Grosser, in his history of *The Discovery of Neptune* (1962), remarks that apparently Lemonnier "did not keep very tidy notes."[7]

The first prediscovery observations were eagerly received by astronomers attempting to compute orbits for Uranus. Flamsteed's 1690 sighting was particularly welcome for it extended the time span of observation of the planet well beyond its then estimated revolution period of 82+ years.

But these ancient observations soon proved to be part of the problem, rather than a key to its solution. As early as May 1782, the French astronomer Joseph J. F. Lalande noted that Uranus was not strictly adhering to the path and time schedule set for it in the earliest theories of its motion. And as more refined orbits and tables were calculated with both Mayer's 1756 and Flamsteed's 1690 positions, the difficulty persisted. Father Fixlmillner's 1784 tables, for example, by 1788 no longer adequately represented the planet's movement. The computations of other astronomers fared little better. In 1790, after Lemonnier's first three prediscovery observations became known, J. B. J. Delambre published new tables for Uranus, but these, too, proved deficient by the turn of the century.[8]

For almost three decades after Delambre's work, astronomers and mathematicians seem to have largely abandoned the problem of finding a better orbit for the planet, and to have turned their attention to other, perhaps more promising projects. The difficulty, as Fixlmillner was among the first to point out, was that it was apparently impossible to devise a theory of Uranus that would satisfy both the ancient and modern observations.[9]

Not until 1820 was another serious attempt made to settle the matter of Uranus's motion. Now the computer was Bouvard who, in 1808, had published some rather undistinguished tables for Jupiter and Saturn. He had also, as noted, kept up a search in old observation records and star catalogs for more prediscovery sightings of Uranus, along with Johann Karl Burckhardt, who turned up the additional Flamsteed observations, and Friedrich W. Bessel who, in 1810, found that Dr. James Bradley, the third Astronomer Royal, had recorded Uranus as a star in Aquarius on December 3, 1753. Two more of Bradley's observations of Uranus, in 1748 and 1750, were discovered in 1864.[10]

A total of seventeen ancient observations—one each by Mayer and Bradley, three by Flamsteed, and twelve by Lemonnier—were available when Bouvard began his revision of the troublesome theory of Uranus. At the same time, he undertook to correct his own earlier tables for Jupiter and Saturn, and in this he was more or less successful. But in the case of Uranus he ran into the same problem that others had encountered before him. As he explained in the introduction to his *Tables astronomiques*, published in Paris in 1821:

The construction of the tables of Uranus involves this alternative:—if we combine the ancient observations with the modern ones, the first will be adequately represented, but the second will not be described within their known precise tolerances; while if we reject the ancient positions and retain only modern observations, the resulting tables will accurately represent the latter, but will not satisfy the old figures. We must choose between the courses. I have adopted the second as combining the most probabilities in favor of truth, and I leave to the future the task of discovering whether the difficulty of reconciling the two systems results from the inaccuracy of the ancient observations, or whether it depends on some extraneous and unknown influence which may have acted on the planet.[11]

Bouvard's decision provoked considerable controversy, particularly over what Grosser has called his "shabby plea" for the rejection of the ancient observations on the basis of their supposed inaccuracy—"in effect accusing four first-class observers of coincidentally making extraordinary errors." Bouvard, Grosser adds, "attributed errors of enormous percentages higher than were ever found in the records of Flamsteed, Bradley or Lemonnier," while simultaneously allowing nearly twice the accepted standard of accuracy for the post-1781 positions. For the ancient observations, he had assigned errors of up to 65 seconds of arc in the case of Flamsteed, while allowing up to 9″ for the modern observations, compared with the then tolerable margin for error of no more than five seconds of arc.[12]

Others, it may be noted, have been somewhat more charitable to Bouvard. American astronomer Benjamin Apthorp Gould, in his 1850 history of the discovery of Neptune and the controversies that followed it, conceded that Bouvard had erred in "adducing arguments against the accuracy of the prediscovery observations." But, he added: "It is an easy thing to censure Bouvard for the readiness with which he abandoned the ancient observations, —now that we know that the discrepancies were caused by the action of an exterior planet, and that the maximum of error in the ancient observations amounted to only nine seconds."[13]

Parenthetically, some astronomers in more recent years have had occasion to challenge the accuracy of the ancient observations of Uranus, notably after the discovery of Pluto and in connection with investigations into the validity of Lowell's analytical prediction for a trans-Neptunian planet. It has thus been necessary for other astronomers to again show that the older observers were not as careless in their work as their critics have contended, and that the ancient observations of Uranus were reasonably sound.

But more immediately, the accuracy of Bouvard's own work soon came into question. By 1825, the German astronomer Boniface Schwartzenbrenner found that Uranus was deviating significantly from the positions given for it in Bouvard's 1821 tables, and specifically that the planet had gotten ahead of its predicted heliocentric longitude. The following year, he found similar discrepancies.[14]

This state of affairs stirred considerable interest among astronomers, but perversely, or so it might seem, the differences between Uranus's predicted and observed longitudes now began to decrease so that by 1828–29, Bouvard's tables and observations were again in acceptable agreement. Yet very soon it became evident that there were deviations developing in the other direction, that Uranus was now progressively falling behind its tabular positions. By 1832, England's soon-to-be Astronomer Royal, George Biddell Airy, reported to the British Association for the Advancement of Science that the discrepancies had reached the unacceptably large value of nearly half a minute of arc.[15]

Astronomical interest was again revived and there were various attempts to explain Uranus's laggard behavior. It was suggested that the planet was being slowed by the Cartesian "cosmic fluid" through which it moved, that a massive undiscovered satellite was perturbing it, that Uranus had been struck by a comet or some other large body, and even that Newton's law of universal gravitation might not be so universal after all, and worked in different ways at such great distances from the sun.[16]

But soon another possibility began to be discussed—that Uranus was being disturbed in its orbit by some unseen exterior planet, that "some extraneous and unknown influence," in Bouvard's phrase, was at work.

The origin of this idea, like that of so many ideas, is difficult to pin down. Bouvard's suggestion clearly was no more than just that. Grosser notes that it was not even new, and that it had been advanced by the French astronomer Alexis Claude Clairaut as early as 1758 in connection with a prediction relating to Halley's comet. But as this was twenty-three years before the discovery of Uranus, Clairaut's suggestion probably should not be considered in the same context as Bouvard's which, of course, emerged directly from his work on the observations of that planet.[17]

The first documented statement of the idea of an exterior planet in relation to Uranus apparently came in a letter, dated November 17, 1834, from an English amateur astronomer, the Rev. Thomas J. Hussey, to Airy, then director of the Cambridge Observatory. Hussey wrote:

Ann Ronan Picture Library

Sir George Biddell Airy. From the
Illustrated London News, 1868.

The apparently inexplicable discrepancies between the ancient and modern observations suggested to me the possibility of some disturbing body beyond Uranus, not taken into account, because unknown.... Subsequently, in a conversation with Bouvard, I inquired if the above might not be the case: his answer was, that, as might be expected, it had occurred to him, and some correspondence had taken place between [Peter Andreas] Hansen and himself respecting it. Hansen's opinion was, that one disturbing body would not satisfy the phenomena, but that he conjectured there were two planets beyond Uranus[*]. Upon my speaking of obtaining the places empirically, and then sweeping closely for the bodies, he fully acquiesced in the propriety of it, intimating that the previous calculations would be more laborious than difficult; that, if he had the leisure, he would undertake them and transmit the

---

*Hansen, the director of Seeberg Observatory at Gotha, later denied he had entertained a belief in the existence of two trans-Uranian planets. See B. A. Gould, *Report on the History of the Discovery of Neptune* (Washington: Smithsonian Institution, 1850), footnote, pp. 11–12.

result to me, as the basis for a very close and accurate sweep.... If the whole matter do not appear to you a chimera, which, until my conversation with Bouvard I was afraid it might, I shall be very glad of any sort of hint respecting it.[18]

Airy did find Hussey's proposal chimerical, and his reply was pessimistic, an attitude he steadfastly maintained on the question of a trans-Uranian planet until the problem of Uranus's deviations was dramatically resolved by the discovery of Neptune. "It is a puzzling subject," he informed Hussey, "but I give it as my opinion, without hesitation, that it is not yet in such a state as to give the smallest hope of making out the nature of any external action on the planet.... But if it were certain that there were any extraneous action, I doubt much the possibility of determining the place of the planet which produced it. I am sure it could not be done till the nature of the irregularity was well determined from several successive revolutions."[19]

Airy himself, at this time, had begun a series of observations of Uranus which led him to the discovery of another discrepancy between Bouvard's tables and the planet's actual behavior. From 1833–35 at Cambridge and in 1836 at Greenwich, he had observed the planet systematically and found that not only was Uranus seriously lagging behind its predicted longitude, but that its radius vector, i.e. mean distance, as given in Bouvard's tables, was "too small by a quantity considerably greater than the moon's distance from the earth."[20] As will be seen, Airy's subsequent preoccupation with this error in the tabular radius vector would blind him to the light eventually shed on the problem both by England's John Couch Adams and France's Urbain J. J. Leverrier.

Meanwhile, the idea of a possible trans-Uranian planet was advanced in another connection, that of Halley's comet, which made its perihelion passage around the sun in the fall of 1835. Astronomers J. E. B. Valz at Marseilles and F. B. G. Nicolai at Vilna both concluded from the comet's motion and its late arrival at perihelion that some unknown force was acting upon it. "I would prefer to have recourse to an invisible planet beyond Uranus," Valz wrote, "its orbit according to the progression of planetary distances [Bode's law] would be at least triple that of the comet.... Would it not be admirable thus to ascertain the existence of a body which we cannot even observe?" Nicolai's report was similar. "One immediately suspects that a trans-Uranian planet (at a radial distance of 38 astronomical units, according to the well-known rule) might be responsible," he wrote.[21] Halley's comet, as will be seen, would inspire planetary predictions in later years, too.

There were, indeed, two reports about this time that a new planet had actually been sighted. In May 1835 Niccolo Cacciatore at Palermo observed a star which, over three nights, moved an estimated 10 seconds of arc. Weather conditions for his observations were poor, and he lost the "star" in the evening twilight after the third night. His report, however, prompted another observer, Louis François Wartmann, to recall that he, too, had seen a moving star in September 1831 that he believed to have been a trans-Uranian planet.

He estimated its distance at 38.8 a.u., the distance predicted by Bode's law, and its period at about 243 years. It was later shown, incidentally, that neither of these two alleged "planets" could have been Neptune.[22]

As time went on, Uranus continued to ignore Bouvard's tables and to get farther behind its theoretical positions. In 1837, Bouvard's nephew, Eugene, informed Airy that he was undertaking the task of revising his uncle's 1821 tables for Uranus. Airy was not encouraging, advising him that he "would gain much ... by waiting a short time," and adding that if the discrepancies in Uranus' motion "be the effect of any unseen body, it will be nearly impossible to find its place."[23]

Bouvard, however, began the work and labored for seven years on the computations, reporting to Airy in May 1844 that whereas his uncle's 1821 tables for Uranus were then in error nearly two minutes of arc, his own new reductions for the planet gave a maximum error for the modern observations of only 15 seconds of arc—"still five times the amount we are warranted in assuming," Gould would later note. Nor did the younger Bouvard's revisions adequately account for the ancient observations.[24]

The much-respected German astronomer, Friedrich W. Bessel, was also keenly interested in the problem of Uranus's motion and was probably the first to take a strong public stand on the question of a trans-Uranian planet. In a lecture at Königsberg on February 28, 1840, he both defended the ancient observations, and insisted that only the discovery of a planet beyond Uranus would resolve the anomalies in its motion. The elder Bouvard, he declared, "made much too light" of the ancient observations. "I have subjected them to a more careful investigation, and have thereby attained the full conviction that the existing differences, which, in some cases, exceed a whole minute, are by no means to be attributed to the observations.... We have here to deal with discordances, whose explanation can only be found in a new physical discovery.... Further attempts to explain them must be based upon the endeavor to discover an orbit and a mass for some unknown planet, of such a nature, that the resulting perturbations of Uranus can actually be explained in this way." Such an approach to the problem, he added emphatically, would necessarily be both difficult and laborious.[25]

Bessel set his pupil, Friedrich Wilhelm Flemming, to the task of reducing the Uranian observations, and in 1842, the same year that Göttingen's Royal Academy of Sciences proposed the problem as its prize question, he determined to undertake an analytical search for the suspected planet himself, using Flemming's computations. But Flemming died suddenly, and Bessel's own illness, which culminated in his death in 1846, aborted the attempt before it could fairly begin.[26]

Bessel's 1840 public statement was followed a year later by the publication of Johann Heinrich von Maedler's *Populaire Astronomie* in which the German observer set down the case for a trans-Uranian planet and voiced the hope that it might be discovered eventually through the application of mathematical analysis.[27]

Thus progressively through the late 1830s and early 1840s, the idea that Uranus was being perturbed by some unknown exterior planet gained wide currency among astronomers. "The problem," Gould wrote a dozen years later, "became, from this time forth, one of the most important questions in Physical Astronomy. Astronomers in various countries busied themselves with it, and spoke of it without reserve."[28]

Not all astronomers, of course, for there was still George Biddell Airy, who had become England's Astronomer Royal in 1835, and whose continuing skepticism where a trans-Uranian planet was concerned would now play a dominant role in what Gould would later describe as "a strange series of wonderful occurrances ... utterly unparalleled in the whole history of science."[29]

Airy himself is a prime source for historians of these "wonderful occurrances" which he set down in considerable and confessional detail in an unusual report to the Royal Astronomical Society less than two months after the discovery of Neptune. Entitled "Account of some circumstances historically connected with the discovery of the planet exterior to Uranus," it was published both in the Society's *Monthly Notices* and in *Astronomische Nachrichten*.[30]

Gould, the first historian of the discovery of Neptune after Airy, has summarized these events thusly: "—the brilliant analysis which was the direct occasion of the search for a trans-Neptunian planet,—the actual detection of an exterior planet in almost precisely the direction indicated,—the immediate and most unexpected claim to an equal share of merit in the investigation, made in behalf of a mathematician till then unknown to the scientific world,—and finally the startling discovery, that in spite of all this, the orbit of the new planet was totally irreconcilable with those computations which led immediately to its detection, and that, although found in the direction predicted, it was by no means in the predicted place, nor yet moving in the predicted orbit."[31]

The principal actors in this drama were two brilliant mathematicians— Urbain Jean Joseph Leverrier of France and John Couch Adams of England. Airy was somewhat the villain of the piece, and the supporting characters included James Challis, Airy's successor in 1835 as Plumian professor of astronomy and director of the Cambridge Observatory; François Arago, director of the Paris Observatory and secretary of the French Academy of Sciences; Sir John F. W. Herschel, only child of the 1788 marriage of William Herschel and Mary Baldwin Pitt, and a distinguished astronomer in his own right; and Johann Gottfried Galle of the Berlin Observatory, who used Leverrier's predicted position to discover Neptune on September 23, 1846. Critics, as it were, were two American mathematical astronomers—Sears Cook Walker of the U.S. Coast Survey, and Benjamin Peirce of Harvard University.

Adams became interested in the problem of Uranus's motion in 1841 while he was still a student at Cambridge and after reading Airy's 1832 comments before the British Association for the Advancement of Science about the planet's unpredictable behavior. On July 3, 1841, he later recalled, he wrote himself this memorandum:

> Formed a design, in the beginning of this week, of investigating, as soon as possible after taking my degree, the irregularities in the motion of Uranus which are yet unaccounted for; in order to find whether they may be attributed to the action of an undiscovered planet beyond it; and if possible thence to determine the elements of its orbit, etc. approximately, which would probably lead to its discovery.[32]

Shortly thereafter, he told a fellow student: "You see, Uranus is a long way out of his course. I mean to find out why. I think I know."[33]

Adams took his degree with top honors in mathematics in January 1843,[34] and as a twenty-four-year-old post-graduate fellow at St. John's College at Cambridge immediately set to work on the Uranian problem. He subsequently wrote:

> Accordingly, in 1843, I attempted a first solution of the problem, assuming the orbit to be a circle, with a radius equal to twice the mean distance of Uranus from the Sun. Some assumption as to the mean distance was clearly necessary in the first instance, and Bode's Law appeared to render it probable that the above would not be too far from the truth. This investigation was founded exclusively on the modern observations.... The result showed that a good general agreement between theory and observation might be obtained; but the larger differences occurring in years where the observations used were deficient in number ... I applied to Mr. Airy, through the kind intervention of Professor Challis, for the observations of some years in which the agreement appeared least satisfactory. The Astronomer Royal, in the kindest possible manner, sent me, in February 1844, the results of all the Greenwich Observations of Uranus.[35]

These newly reduced observations spanned the period from 1754 to 1830 and, along with the fact that the Royal Academy of Sciences at Göttingen had proposed the Uranian problem as its prize question, inspired Adams to undertake a new solution to the problem, now assuming an elliptical orbit:

> I now took into account the most important terms depending on the first power of the eccentricity of the disturbing planet, retaining the same assumption as before with respect to the mean distance.... After obtaining several solutions differing little from each other, by gradually taking into account more and more terms of the series expressing the Perturbations, I communicated to Professor Challis, in September 1845, the final values which I had obtained for the mass, heliocentric longitude, and elements of the assumed planet. The same results, slightly corrected, I communicated the following month to the Astronomer Royal.[36]

The communication process, as it turned out, was considerably more complex, as Grosser has explained in detail.[37] Here, it is sufficient to say that on September 22, Challis wrote Adams a letter of introduction to Airy in

Ann Ronan Picture Library

John Couch Adams. From Robert Ball,
Great Astronomers, London, 1895.

which he noted that Adams wished to present his results in person or, if this were not possible, to write the Astronomer Royal on the subject. Adams then went to Greenwich and, finding that Airy was in Paris, left Challis's letter. A week later Airy returned and replied to Challis, declaring that he was "very much interested in the subject" and that he would be "delighted" to hear from Adams directly "by letter."[38]

On October 21, 1845, Adams made a second journey to Greenwich, calling twice at Airy's home. But he found that the Astronomer Royal was out at his first call, and on his second he was informed, apparently through some domestic misunderstanding, that Airy was at dinner and could not be disturbed. Thus frustrated, Adams returned to Cambridge after leaving his card and a written summary of his work for Airy's subsequent perusal.[39]

This summary contained the following elements for his assumed trans-Uranian planet:

a'	(mean distance)	38.4 a.u.
e'	(eccentricity)	0.1610
$\bar{\omega}'$	(longitude of perihelion)	315°55'
	(mean annual motion)	1°30'.9
m'	(mass; sun = 1)	0.0001656
	(heliocentric longitude on October 1, 1845)	323°34'

Adams also listed the errors left by his theory in both the ancient and modern observations of Uranus, by far the largest being 44.4 seconds of arc for Flamsteed's 1690 observation. "This being an isolated observation, very distant from the rest," he added, "I thought it best not to use it in forming the equations of condition. It is not improbable, however, that this error might be destroyed by a small change in the assumed mean motion of the planet."[40]

Airy replied to Adams' summary on November 5, 1845:

> I am very much obliged by the paper of results which you left here a few days since, shewing the perturbations on the place of *Uranus* produced by a planet with certain assumed elements. The latter numbers are extremely satisfactory: I am not enough acquainted with Flamsteed's observations about 1690 to say whether they bear such an error, but I think it extremely probable.
> But I should be very glad to know whether this assumed perturbation will explain the error of the radius vector of Uranus. This error is now very considerable.[41]

Airy would later explain in his "Account" that "I considered the establishment of this error of the radius vector to be a very important determination. I therefore considered that the trial, whether the error of radius vector would be explained by the same theory which explained the error in longitude, would be truly an *experimentum crucis*."[42]

Adams did not answer Airy's query on the radius vector for more than a year, and when he did, he clearly indicated that he had thought the question of little consequence, and that the correction of the error in longitude necessarily would result in the correction of the error in the radius vector. He also later apologized for his delay in replying to Airy's letter, but added that he had been "much pained at not having been able to see you when I called at the Royal Observatory the second time, as I felt the whole matter might be better explained by half an hour's conversation than by several letters."[43]

Nor did Airy follow up on his correspondence with Adams, or on Adams's elements for a hypothetical trans-Uranian planet, and thus the stubbornness of one man, and the sensitivity of another, may well have delayed the discovery of Neptune for nearly a year. Airy, as will be seen, would concede as much in his later "Account."

In November 1845, importantly, the second principal in the drama appears on stage—Leverrier. On November 10, five days after Airy's letter to Adams, Leverrier presented the first of three masterful memoirs on the problem of Uranus's motion to the French Academy of Sciences.[44]

Leverrier had been a brilliant student at the École Polytechnique and had graduated in 1833 with highest honors. His first position was as a government chemist in Paris under the great Joseph Louis Gay-Lussac. Over the next three years, he published two papers concerning the reactions of phosphorus with hydrogen and oxygen. In 1836, however, he resigned rather than accept a transfer to the provinces, and took a teaching post at the Collège Stanislaus in Paris. But Gay-Lussac had recognized his talent for higher mathematics and in 1837, he recommended the 26-year-old Leverrier for a vacancy as an assistant in astronomy at the École Polytechnique. Leverrier thus began his long and distinguished career in celestial mechanics.[45]

Leverrier's first project was to re-investigate the stability of the solar system, and after two years of reworking and refining the earlier work on this problem by Laplace and Lagrange, he presented his results to the Academy of Sciences in a well-received paper entitled "Sur les variations seculaire des orbites planetaires." The following year, he presented a second paper on the same general subject which contained two important methodological innovations. The first of these was that in computing planetary orbits it is necessary to take into account the higher power terms of the eccentricity and inclination, that contrary to what had long been assumed, these higher power terms were not always negligible and sensibly affected the final results. The second was a method for directly substituting corrections in the equations which greatly reduced computational time and labor.[46]

Leverrier's rigorous work attracted the attention of François Arago, then the most influential astronomer in France. Arago suggested that he consider the theory of Mercury, then and since a troublesome planet. Leverrier worked on the problem for three years, presenting his preliminary findings to the Academy in an 1843 paper entitled "Determination nouvelle de l'orbite de Mercure et de ses perturbations." His results were not as complete as this title might indicate, however, and it would be another sixteen years before he provided a provocative conclusion to his investigation by postulating a possible intra-Mercurian planet which he called "Vulcan."[47]

Leverrier next began to study short period comets, and particularly

Urbain Jean Joseph Leverrier. From the *Illustrated London News,* February 2, 1847.

Lexell's "lost" comet of 1770. The St. Petersburg mathematician had computed elliptical elements for this bright comet which gave it a period of 5.5 years, but astronomers had not observed it before 1770, nor did it appear in 1776 or 1782 as Lexell's orbit predicted. Lexell attempted to explain this by the perturbative action of Jupiter, and his calculations to this end were supported in 1806 by Burckhardt's review of his work. Leverrier in 1844, however, reworked the entire problem, and concluded that while the comet could not have been captured as a satellite of Jupiter, it could have been swung into a hyperbolic orbit ($e > 1$) by that planet and thus have left the solar system, or it could have been swung into any one of a series of elliptical orbits. Leverrier computed these possible orbits as an aid to the future identification of Lexell's comet, should it reappear again. Subsequently, he investigated the motions of comets discovered by Hervé Faye in November 1843 and by Francesco de Vico in August 1844, and concluded that neither was Lexell's, but that de Vico's was probably the same comet observed by Philippe de La Hire in 1678. Adams also worked on both Faye's and de Vico's comets.[48]

Leverrier had already given some attention to Uranus, but it was not until mid-1845 when, again at Arago's behest, he began to work on the problem of its motion in earnest. His results were not only included in his three memoirs on the subject that were published in *Comptes Rendus*, but in a lengthy summary entitled "Recherches sur le mouvement de la planète Herschel (dites Uranus)" which was published in the *Additions* to the *Connaissance des Temps* for 1849.[49] This summary Gould considered to be "among the most remarkable mathematical works of the age."[50]

Gould opens his discussion of Adams's and Leverrier's work by noting that "it is not easy for those who are not versed in the study of Physical Astronomy, to form any adequate idea of the difficulty of the problem which Messrs. Le Verrier and Adams proposed to themselves. The difficulties in the development of the proper methods were exceedingly great.... Not only the orbit and mass of the suspected planet, but the elements of Uranus also, were to be regarded as unknown quantities. The limits of error of the ancient observations were also undetermined, but must yet exercise an important influence on the result."[51]

The problem was not only difficult, but it was also completely new, as the noted English historian of astronomy Agnes M. Clerke has pointed out. "No problem in planetary disturbance had heretofore been attacked, so to speak, from the rear. The inverse method was untried, and might well be deemed impracticable. For the difficulty of determining the perturbations produced by a given planet is small compared with the difficulty of finding a planet by its resulting perturbations."[52]

Very generally, the problem involves first determining as accurately as possible the residual perturbations of Uranus, that is, the differences between its observed and theoretically predicted positions, after making allowances for observational error. Then equations of condition are formed containing elements of the perturbed and suspected perturbing planets as unknowns, which are set equal to the residuals for specific observations over time. There is one equation for each observation, and thus as many equations as there are observations used. These equations are then manipulated and solved by elimination, approximation and the method of least squares to derive the most probable numerical values for the unknown orbital elements. The effect of the disturbing planet so defined is then applied to determine if theory is thus reconciled to observation, or at least if the residuals of the new theory have been substantially reduced to values within allowable observational error.

Adams' and Leverrier's approaches to the problem were considerably different. Adams proceeded on a straightforward course and from the outset sought only to determine a probable orbit and mass for an exterior planet that would account for the unexplained perturbations of Uranus. As noted, he assumed a distance of 38.4 astronomical units from Bode's law and initially, to simplify the problem, a circular orbit. The results he obtained by October 1843 were encouraging and subsequently, after receiving the full set of newly reduced Greenwich observations from Airy, he pursued the problem on the basis of an elliptical orbit, using what time he could spare from his duties as a St. Johns Fellow.

Adams took his basic data from the Greenwich observations, supplemented after 1830 by observations at Cambridge. For the ancient observations, he relied on Bouvard's work in which, after recomputing the perturbations by Jupiter and Saturn, he "found no differences of any consequence" except an error in a single equation earlier pointed out by Bessel.[53] Leverrier, as will be seen, did not find Bouvard's work so faultless.

From these data, Adams derived a list of the residual differences in mean longitude for Uranus for both the ancient and modern observations. The latter he grouped into equidistant three-year periods to reduce the equations of conditions to a manageable number. In each of the resulting twenty-one groups, spanning the period from 1780 to 1840, he selected observations that were either made near Uranus's opposition, or could be combined with others within the group in such a way that the result would be nearly free from the effects of the tabular radius vector error that so concerned Airy. After forming mathematical expressions for the perturbations in mean longitude in terms of mass and eccentricity, he set up equations of condition for the modern observations alone. These, he found, separated themselves naturally into two groups, each with five unknowns. After computing coefficients of the various terms, he reduced these by a process of elimination to two equations which "would be sufficient for determining the mass of the disturbing planet and its longitude at epoch, if the eccentricity of the orbit were neglected."[54]

Then, using the ancient observations, he set up another set of equations to determine the eccentricity and longitude of perihelion, rejecting, however, Flamsteed's 1690 position as "unsafe ... since it is derived from the single observation ... twenty years anterior to any other observation." These equations he similarly reduced to two equations which, with the first two, suffice for the solution of our problem."[55]

After further manipulation, he derived three equations, each containing the same three unknowns. These he solved by approximation. As the coefficients of two of the unknowns in one equation were small, he found the approximate value for the third term which, when substituted in the second and third equations gave approximate values for the other unknowns. These, in turn, were substituted back into the first equation to give a more accurate value for the first unknown—"the process being repeated will enable us to satisfy all the equations as nearly as we please." From these lengthy and laborious calculations emerged the position and elements of the hypothetical trans-Uranian planet that Adams communicated to Airy on October 21, 1845.[56]

The following year, Adams undertook still another analysis of the problem, now basing his computations on the assumption that the disturbing planet's mean distance was one-thirtieth less than the former value, or 37.25 a.u. "The method employed," he later wrote, "was, in principle, exactly the same as that given before; but the numerical calculations were somewhat shortened by a few alterations in the process which had been suggested by the previous solution."[57] This work he completed during the summer of 1846, and on September 2, he sent the result to Airy by letter.[58] He had arrived at the following elements for an unknown exterior planet:

a' (mean distance) = 37.25 a.u.
e' (eccentricity) = 0.120615
ϖ' (longitude of perihelion) = 299° 11′
ϵ' (longitude at epoch 1810.328) = 264° 50′
m' (mass; sun = 1) = 0.00015003

From these elements he deduced that the planet's mean longitude on October 6, 1846—its position on that date and the "x" of his problem—would be 323° 2′.[59]

Adams had also attempted to determine the inclination of the orbit (i') and the longitude of the node (Ω), but found his results unsatisfactory. "The perturbations of the latitude [of Uranus] are in fact exceedingly small," he later explained, "and during the comparatively short period of three-fourths of a revolution, are nearly confounded with the effects of a constant alteration in the inclination and the position of the node ... so that very small errors in the observations may entirely vitiate the result."[50]

The residuals in Uranus's motion left by his two solutions, he found, "agree very closely with each other and with observation" except in the later years beginning in the mid-1820s—the first solution, based on his original Bode's law assumption of a distance of 38.4 a.u., less so than the second. Consequently, he applied both orbits to the most recent available observations of Uranus for 1843, 1844 and 1845 and concluded, as he wrote Airy, that an even further reduction in the mean distance to $a' = 33.42$ a.u. "would probably satisfy all the observations very nearly."[61]

In his September 2 letter to Airy, Adams also took note of the Astronomer Royal's concern over the error in the tabular radius vector for Uranus and suggested that the effect of reducing the mean distance would be to automatically correct the longitude errors. But he later conceded that this conclusion had been "hastily inferred" and was the result of "not making sufficient allowances for the increase in the mean motion" that would follow from lesser distances.[62]

Leverrier's investigation, begun in June 1845, was at once grander in scale and broader in scope, and involved the rigorous reinspection and restatement of the problem itself as well as its solution. The first phase, contained in his first memoir to the French Academy in November 1845, consisted of an exhaustive review and reconstruction of the theory of Uranus. His basic methods were primarily those of Laplace and Lagrange, but with some distinctly Leverrierian embellishments. Of these Gould commented, "It may with safety be said that almost all of the modifications are improvements." Leverrier had devised, for example, what he called the "simultaneous determination of the inequalities" to derive the perturbations of Uranus by Saturn to a precision of 0.01 second of arc. "The method is such," Gould explained, "that all the inequalities are determined at the same time, and are so mutually dependent, that an error in one part of the calculation vitiates the entire result. If, then, after the numerical computation is completed, any one part of the result is found to be exact, it may be fairly concluded that the remainder is exact also." Leverrier double-checked his results by calculating some of the same Saturnian perturbations with the method of Lagrange.[63]

In the course of reinvestigating the theory of Uranus, Leverrier showed that Bouvard's 1821 work was shamefully wanting in accuracy. He found that Bouvard's values for eccentricity were inconsistent, that his tables were discordant, that his equations of conditions were both analytically and numerically defective, and that his results, as published, contained an incredible number of typographical errors, no less than sixteen in a single table, and "the majority of them of grave importance."[64]

After laboriously correcting Bouvard's errors, he set out to determine the residuals for 262 of the "best" observations for the period from 1781 to 1845, and for the ancient observations, in both longitude and latitude. As the latter were small, he used only the longitude residuals in forming his equations of condition. He then proceeded to determine whether it was possible to represent Uranus's motion theoretically at all with the observational data alone. He found that he could not, without accepting impossible error. Even allowing for all maximum observational errors, he could only account for 92 of the total outstanding residuals of Uranus in longitude of 356 seconds of arc, leaving a total of 264" or 4.4 minutes of arc unexplained. Something else, something not taken into account, must be perturbing Uranus.[65]

Airy considered Leverrier's November 1845 memoir "a new and most important investigation" of the problem of which it could "be truly said that the theory of Uranus was now, for the first time, placed on a satisfactory foundation." But Leverrier's memoir, like Adams's letter two weeks before, inspired the Astronomer Royal to nothing more than comment.[66]

In the next phase of his work, Leverrier took up the question of what might be causing the unexplained perturbations that he had so ably and laboriously proved to exist. He first eliminated on logical grounds various possible causes that had been suggested, such as the impact of a comet, or an unknown satellite. As he was reluctant even to question Newton's law of universal gravitation, he found he was left with the hypothesis of an unknown exterior planet—exterior, he reasoned, because if inside Uranus's orbit its effect on Saturn must be perceptible.[67]

Initially, he assumed a distance for this planet of twice that of Uranus, reasoning not specifically from Bode's law, but "from the fact that at any appreciably greater distance, say three times that of Uranus, its mass necessarily would be such as to sensibly affect the motion of Saturn more than theory or observation allowed." Later, he set inner and outer limits of 35 and 37.9 a.u. respectively which, it may be noted, provided a basis for one of the several controversies that followed the discovery of Neptune. He also reasoned that as Uranus's orbit was only slightly inclined to the plane of the ecliptic, he could safely assume that the orbital inclination of the unknown planet was negligible for his purposes.[68]

His first attempt at a solution involved equations founded on observations representing eight equidistant periods from 1845 back to 1747. But after reducing these, he concluded that the ninety-eight-year interval was not sufficient and that the time span should be extended as far back as possible, that is, to Flamsteed's 1690 position.[69] This he did in his second attempt—a solution Gould would later characterize as "pre-eminently a discussion of limits, and a brilliant combination of ingenuity, of analytical skill, and of laborious calculation." Leverrier took January 1, 1800, as the epoch, and sought to derive a value for the unknown's mass by substituting in his equations forty different values for the longitude at epoch representing nine-degree intervals from 0° to 360°. After carrying out each computation, he rejected all values for the mass which were either negative or were large enough to cause perturbations of Saturn. He then analyzed what remained and concluded that the most probable longitude at epoch for the disturbing planet must lie between 108° and 162° or between 297° and 333°. But from further calculations, he found that neither of these ranges would satisfy the 1690 or 1747 positions of Uranus, so that "the consequence of which would seem to result from the discussion, thus conducted, would be, that it was impossible to represent the course of Uranus by means of the perturbative action of the new planet."[70]

At this point, Leverrier's investigation ground to a halt, and as Gould has noted, "many a good mathematician and experienced computer would, for less reason, have abandoned this apparently unprofitable labor in despair."[71]

But after turning the problem over in his mind for several months, the French geometer found that by neglecting two minor terms and going beyond the limits of his second solution, the theory of Uranus began to fall into place.

After additional calculations, he arrived at the conclusion that his unknown planet, to account for the perturbations of Uranus, must have had a mean longitude of between 243° and 252° on January 1, 1800. Further computations indicated that the most probable longitude for that date was 252° and from this he deduced that its heliocentric longitude—that is its position—for 1847.0 would be 325°, a figure less than two degrees away from Adams' unpublished result.[72]

Airy was not only impressed again by Leverrier's second memoir describing this work, but was now moved to action, writing Leverrier to congratulate him and to inquire whether his solution would also settle the matter of the error in Uranus tabular radius vector. Leverrier, like Adams, apparently did not consider the question particularly pertinent, but unlike Adams, he answered late in June 1846 that because of the nature of the solution and the data on which it was based, it would resolve the radius vector question. He also offered to send the Astronomer Royal the coordinates for his suspected planet as the basis for a telescopic search, but Airy did not take him up on the offer.[73]

Airy's interest, however, was now thoroughly aroused, for he had indeed noticed the close agreement between Adams's and Leverrier's final positions for the presumed planet, although he mentioned nothing of Adams's work in his letter to the French geometer. "To this time," his subsequent "Account" reveals, "I had considered that there was still room for doubt of the accuracy of Mr. Adams's investigation; for I think that the results of algebraic and numerical computations, so long and so complicated as those of an inverse problem of perturbations, are liable to many risks of error in the details of the process."[74]

On June 29, 1846, at a special meeting of the Board of Visitors at Greenwich Observatory attended by both Challis and Sir John F. W. Herschel, Adams' work was discussed, and Airy supported a proposal for a search for the postulated planet. On July 9, he wrote to Challis asking him to begin the project and a few days later he advised the Cambridge director that "the importance of this inquiry exceeds that of any current work, which is of such a nature as not to be totally lost by delay."[75] Airy also outlined a laborious and time-consuming procedure for the search, involving systematic sweeps of a large segment of the sky, which, as Grosser comments, "was better suited to the construction of a star map than the identification of a new planet."[76]

Challis advised the Astronomer Royal on July 18 that he intended to mount a search with Cambridge's 11.75-inch Northumberland refractor, and began observations on July 29, adopting Ariy's suggested procedure at first, but soon modifying it without making it any less cumbersome. Adams provided him with coordinates for the presumed planet, computed at twenty-day intervals from July 20 to October 28, and for every five degrees between longitudes 315° and 335°, bracketing the predicted position. But rather than narrowing the field of his search, Challis continued on his pedestrian way. The work was slowed by clouds and moonlight and over and above these problems does not seem to have have been marked by any sense of urgency.[77]

In August, parenthetically, after receiving Leverrier's second memior, the U.S. Coast Survey's Sears Cook Walker in Washington, D.C., suggested a search to Superintendent Matthew Fontaine Maury of the U.S. Naval Observatory, but existing projects in positional astronomy there held immediate priority. No one else, apparently, even considered making such a search.[78]

Leverrier, meanwhile, had further refined his work, and on August 31 presented his third memoir to the French Academy which contained these final elements for his postulated planet:

a' (mean distance) = 36.154 a.u.
e' (eccentricity) = 0.17061
ϖ' (longitude of perihelion) = 284° 45'
P' (period) = 217.387 years
m' (mass; sun = 1) = 0.0001075

He also reported that its mean longitude on January 1, 1847, would be 318° 47', that its true heliocentric longitude on that date—its actual position—would be 326° 32', and that its distance then would be 33.06 a.u. He wrote that this position "places the planet about 5° to the East of the star δ Capricorni," and he added an appeal to astronomers to launch a search for the body, noting that as it had come to opposition on August 19, the "advantages of its great angular distance from the sun will diminish steadily." He also estimated its telescopic diameter at 3.3 seconds of arc, and added pointedly that "a good refractor is perfectly capable of distinguishing such a disk...."[79]

Airy would later paraphrase the confident tone of this memoir: "Look in the place which I have indicated, and you will see the planet well."[80]

But despite such bold specificity, astronomers did not rush to their telescopes to seek Leverrier's predicted planet. Gould later remarked that "it must, indeed, be confessed that astronomers in general did not seem to consider the theoretical results, as published by Mr. Le Verrier, as necessarily indicating the *physical* existence of such an exterior planet.... Even in Paris, that focus of science, with its many and powerful telescopes, with its numerous astronomers, where Mr. Le Verrier was known and his brilliant genius appreciated ... we have no information that any attempts were made to test the physical accuracy of Le Verrier's results, or that the planet was even looked for on one single evening."[81] Challis, it may be added, later explained the plodding pace of his search for Adams' postulated planet in much the same way, declaring that "it was so novel a thing to undertake observations in reliance upon merely theoretical deductions; and that while much labor was certain, success appeared very doubtful."[82]

Meanwhile, Adams was having a run of bad luck. His September 2 letter to Airy reached Greenwich while Airy was in Wiesbaden, Germany, presumably enjoying the baths, and was answered by an assistant who merely offered Adams more data on Uranus for 1844 and 1845. Adams now also decided to prepare a brief summary of his work on the Uranian problem for the annual meeting of the British Association for the Advancement of Science scheduled in mid-September at Southhampton. He arrived there, paper in hand, on September 15, only to find that he was a day late, that the section on Mathematical and Physical Science had ended its session the day before.[83] Nor was there any good news from Challis, who was still sweeping the heavens and recording stars, having reported to Airy on September 2 that he had "lost no opportunity of searching for the planet," but that the work was going "very slowly."[84]

Things were going very slowly now for Leverrier, too, for he had been unable to persuade French astronomers to make a search for his suspected planet. He thus turned his attention beyond France and on September 18, taking an opportunity to acknowledge a paper he had received a year earlier, wrote to the German observer Johann Gottfried Galle at the Berlin Observatory. After belatedly thanking Galle for the paper, Leverrier confided:

Right now, I would like to find a persistent observer who would be willing to devote some time to an examination of a part of the sky in which there may be a planet to discover. I have been led to this conclusion by the theory of Uranus.... You will see, Sir, I demonstrate that it is impossible to satisfy the observations of Uranus without introducing the action of a new Planet, thus far unknown; and remarkably, there is only one single position in the ecliptic where the perturbing planet can be located.[85]

Leverrier then listed the elements contained in his August 31 memoir, along with the geocentric longitude of the planet for the latter part of September—324° 58'. "The actual position of this body shows that we are now, and will be for several months, in a favorable position for the discovery. Furthermore," he added, "the mass of the planet allows us to conclude that its apparent diameter is more than 3" of arc. This disk is perfectly distinguishable, in a good glass, from the spurious star diameters caused by various aberrations."[86]

Galle received Leverrier's letter on September 23, and immediately asked permission from his director, Johann Franz Encke, to search for the planet. A somewhat skeptical Encke acquiesced reluctantly and Galle, with student Heinrich Louis d'Arrest as a volunteer assistant, began the search that same night. He first directed the Berlin Observatory's 9-inch Frauenhofer refractor to the position indicated by Leverrier, but found nothing unusual. Then, using a newly-published star chart, Hora XXI, compiled by Carl Bremiker for the Berlin Academy, he began systematically observing stars in the area, describing them one by one to d'Arrest who checked them off on the chart. Very soon, he called out an 8th magnitude star that d'Arrest could not find.[87]

A careful observer, Galle notified Encke, who came to the telescope. The "star" was tracked until it set, and then observed again through the next night. Galle found that it was moving in a retrograde direction at about three seconds of arc per hour, and that its apparent diameter was 3.2 seconds—both in accord with Leverrier's estimates. Its discovery position was 325° 52′45″ of longitude, only about 55 minutes of arc from Leverrier's predicted position, and less than a degree and a half from Adams's assigned place.[88]

On September 25, Galle wrote Leverrier that "the planet whose position you have pointed out *actually exists.*"[89] Leverrier replied: "I thank you for the alacrity with which you applied my instructions.... We are thereby, thanks to you, definitely in possession of a new world."[90]

But nothing is ever so simple, and the discovery of Neptune is no exception to the rule. Even as the news went out from Berlin and Paris, a storm of controversies large and small broke out. The least of these, but the first in point of time, was over a name for the new planet.

Galle, in informing Leverrier of the discovery, had suggested the name "Janus," but Leverrier, in his reply, declared that the "Bureau of Longitudes has decided on *Neptune.* The symbol will be a trident. The name Janus would imply that this is the last planet of the solar system, which we have no reason at all to believe."[91] Grosser points out that Leverrier here may have been trying to forestall the use of other names, as in the ordinary course of events the Bureau of Longitudes would have nothing to do with naming the planet and, in fact, had not formally named it.[92] At the time, James Pillans, professor of Latin at Edinburgh University, pointed out that Leverrier had been apparently mistaken in thinking that Janus was the god of limits or boundaries, and that as the two-faced Roman deity of gates, doorways and beginnings, the name was quite appropriate, especially in view of Leverrier's own implication of still other planets remaining to be discovered.[93]

Within days, however, Leverrier changed his mind and on October 5, Arago advised the French Academy of Sciences that he had been given the honor of naming the new planet, forthwith proposing the name "Le Verrier."[94] Leverrier himself, in an effort to give this name both astronomical and political legitimacy, now began to call Uranus "Herschel's planet," and to argue that as comets were customarily named after their discoverers, it was only natural to extend the custom to newly discovered planets.[95] But most leading astronomers, including Encke and Gauss in Germany, F. G. W. Struve at Pulkowa Observatory in Russia, and Airy and Sir John Herschel in England, favored Neptune and that name has, of course, prevailed.[96]

Immediately, too, a far more important controversy, marked by misunderstandings, bitter recriminations and even patriotic chauvinism, exploded over the question of priorities in the prediction and discovery of the new planet. This began on October 3 when Herschel, in a brief letter to the London *Athenaeum,* first publicly revealed Adams's long investigation into the problem of Uranus's motion and his postulation of a trans-Uranian planet in almost precisely the same position assigned by Leverrier.[97] Astronomers and the public barely had time to react to this sensational revelation before Challis, in a series of seemingly contradictory private and published letters, managed to raise the question to the status of an international *cause célèbre.*

Challis, as it turned out, had not seen Leverrier's third and final memoir of August 31 until September 29, and that same night he had observed the night sky in the vicinity of Leverrier's predicted position, recording some three hundred stars and noting one in particular that seemed to have a disk. He did not check the object with a higher magnification, however, as the elder Herschel had done with Uranus, nor did he observe it the following night because, he later explained, of the moonlight. On October 1, he learned of Galle's discovery, and realized that he had seen the planet two nights before.[98]

This much Challis wrote on October 5 to François Arago of the French Academy, claiming his observation as a confirmation of Leverrier's prediction. He also remarked that he had been searching for a new planet in the course of making a star map, but he did not explain why, or refer to Adams or his predictions at all.[99]

Meanwhile, after learning of Galle's discovery, Challis had also begun to review the more than three thousand star positions he had recorded in his search since July 29, and soon found that he had also observed the planet on August 4, and again on August 12, without recognizing it for what it was. He later attributed this failure to the fact that, unlike Galle, he did not have an adequate chart of the region being searched, and that he had thus been forced to use the slow and tedious method of comparing observations over intervals of time.[100]

On October 12, Challis reported his two prediscovery observations to Airy, along with his September 29 sighting of the planet which, he advised the Astronomer Royal, he had made "strictly according to his [Leverrier's] suggestions."[101] On October 17, however, when Challis published the full details of his search, his prediscovery observations, and Adams's long work in the *Athenaeum*, he failed to mention his September 29 observation, Leverrier's memoir that had prompted it, or for that matter, Leverrier at all.[102]

Challis's selective sins of omission, and the seeming implications of his *Athenaeum* letter, caused an uproar in Paris and in the French Academy where Arago vehemently attacked not only the Cambridge director, but Airy, Herschel and Adams for what he and apparently most of his compatriots, along with not a few Englishmen, considered a blatant attempt to appropriate the exclusive credit for the prediction and discovery of Neptune for Adams. Challis and Airy, already under mounting criticism in their own country for their actions, or rather inactions, bore the brunt of Arago's wrathful accusations and denunciations. But even Adams, who had still not commented publicly concerning either the discovery or his work, was brutally excoriated. Arago, to shouts of approval from his Academy audience, flatly declared that Adams had "no right to figure in the history of the discovery of the planet Le Verrier ... by the slightest allusion."[103]

If Arago's Gallic emotions carried him too far in presenting what, *prima facie*, was a good case, the French press, ever alert to incursions from across the channel, immediately carried excess to extreme, pouring out torrents of abuse on England and everything English with particular reference to English astronomers. The vilification of Challis, Airy, Herschel and Adams reached such an intensity that within weeks Arago and Leverrier felt obliged to repudiate it.[104]

Airy, meanwhile, was not so unperceptive that he did not realize that serious trouble was brewing over his own sins of omission in regard to Adams's work, and that in view of Herschel's brief revelation, and as Astronomer Royal, some explanation must soon be forthcoming from him. Typically, he approached the prospect in an orderly manner, apparently hoping that through private correspondence he could at least forestall public controversy long enough to permit a full and dispassionate statement of the facts at the proper time and place. Airy, surely a man destined to be overtaken by events, found this hope shattered almost immediately, but he nonetheless pursued his intended course.

On October 14, three days before Challis's incendiary letter appeared in the *Athenaeum*, Airy wrote Leverrier to congratulate him on the "successful termination of your vast and skillfully directed labours" and to note that he had been "exceedingly struck by the completeness of your investigation." But then he added:

> I do not know whether you are aware that collateral researches had been going on in England and that they led to precisely the same results as yours. I think it probable that I shall be called on to give an account of these. If in this I shall give praise to others, I beg that you will not consider it as at all interfering with my acknowledgement of your claims. You are to be recognized beyond doubt as the real predictor of the planet's place. I may add that the English investigations, as I believe, were not quite so extensive as yours. They were known to me earlier than yours.[105]

Airy's assignment of credit to Leverrier is explicit enough in this, but the French geometer was already angered over reports arriving from England about the claims being made on Adams's behalf. On October 16, he wrote the Astronomer Royal to ask some pointed questions concerning their mutual correspondence the preceding June, following publication of his second memoir but preceding the start of Challis's search. Airy sought to reassure him two days later, but by this time Challis's *Athenaeum* letter had reached Paris to inflame the issue beyond the realm of reason. Airy wrote two subsequent letters to Leverrier later in October to explain his position, but, published in the French press to Airy's dismay, they only added fuel to the now largely emotional fire.[106]

The controversy reached a peak on November 13, 1846, at a historic meeting of the Royal Astronomical Society in London. Here, Airy presented his detailed "Account" of the events, as he knew them, leading up to Neptune's discovery, while Challis formally described his ineffective Cambridge search for Adams's postulated planet, and Adams himself gave his first public paper on his lengthy labors on the Uranian problem and their spectacular result. Airy and Challis came under lengthy questioning and strong criticism for their roles in the events they so candidly described. Only Adams drew applause and praise from the assembled members, but together the three presentations went a long way toward replacing the heat in the controversy over Neptune's discovery with light.

Airy, in presenting his "Account," justified his departure from the Society's unwritten rule against "mere historical statements" by declaring that "in the whole history of astronomy, I had almost said the whole history of science, there is nothing comparable to this." He was giving his "Account," he said, "because it will provide a valuable contribution to the history of science," and "because it may tend to do justice to some persons who otherwise would not receive in future times the credit they deserve." Certainly, it achieved both purposes.[107]

The Astronomer Royal's main point, however, was to contend that the discovery of Neptune was "a consequence of what may properly be called a movement of the age; that it has been urged by the feeling of the scientific world in general, and has been nearly perfected by the collateral but independent labors, of various persons possessing the talents and powers best suited to the different parts of the researches."[108] Thus, in his more or less chronological, step-by-step review, he began with the letter he had received in November 1834 from the Rev. Thomas J. Hussey concerning the possibility that an unknown exterior planet was perturbing Uranus. Before taking up Adams's and Leverrier's separate investigations, he briefly discussed his own observations of Uranus in the 1830s and his correspondence about the troubled planet with Eugene Bouvard. He did not neglect to confess his own skepticism about mathematical theories in general and Adams's theory of Uranus in particular, and he found occasion to mention his own role, following Leverrier's June 1846 memoir, in initiating Challis's search for Adams's planet. While fully detailing Adams's unpublished work, he had high praise for Leverrier's published results.[109]

Airy, in conclusion, drew two lessons from the events he described. The first concerned Bode's law and made the point that scientific ends may sometimes be served by unscientific means. "This history presents a remarkable instance of the importance, in doubtful cases, of using any received theory as far as it will go, even if that theory can claim no higher merit than that of being plausible. If the mathematicians whose labours I have described had not adopted Bode's law of distances (a law for which no physical theory of the rudest kind has ever been suggested), they would never have arrived at the elements of the orbit. At the same time, this assumption of the law is only an aid to calculation, and does not compel the computer to confine himself perpetually to the condition assigned by the law, as will have been remarked in the ultimate change of mean distance by both the mathematicians...."[110]

Secondly, Airy declared that the history of the discovery of Neptune "shews that, in certain cases, it is advantageous for the progress of science that the publication of theories, when so far matured as to leave no doubt of their general accuracy, should not be delayed till they are worked out to the highest imaginable perfection. It appears to be quite within probability," he conceded, "that a publication of the elements obtained by Adams in October 1845 might have led to the discovery of the planet in November 1845."[111]

Adams's paper, entitled "An Explanation of the observed Irregularities in the Motion of Uranus, on the Hypothesis of Disturbances caused by a more distant Planet; with a Determination of the Mass, Orbit and Position of the disturbing body," detailed his approach to, and solution of the Uranian problem. But of more immediate importance, it contained a statement of his own position in regard to the controversy over priorities in Neptune's prediction and discovery. After reviewing the chronology of his own and Leverrier's work, he declared:

I mention these dates merely to show that my results were arrived at independently, and previously to the publication of those of M. Le Verrier, and not with the intention of interfering with his just claims to the honours of the discovery; for there is no doubt that his researches were first published to the world, and led to the actual discovery of the planet by Dr. Galle, so that the facts stated above cannot detract, in the slightest degree, from the credit due M. Le Verrier.[112]

This disclaimer, along with Airy's candid statement of the events, did much to quiet the controversy without, however, ending it entirely. The Royal Astronomical Society, for example, unable to agree on whether Adams or Leverrier should receive its Gold Medal and precluded from giving it to both men, decided in its collective wisdom not to award the medal at all for the astronomically auspicious year of 1846. On the other hand, the Royal Society, recognizing Leverrier's effective publications, presented its prestigious Copley Medal to the French geometer. Subsequently, many honors accrued to both men, and their respective roles in the historic discovery have been well recognized.[113]

In more recent years, parenthetically, there has been a tendency to readvance the Adams claim to priority by arguing that his communication of his results to Challis and Airy in September and October 1845 constituted a formal publication of sorts. W. M. Smart, in his centennial essay on Adams and the discovery of Neptune, notes that at the time it was not uncommon for scientists to deposit the results of their work with scientific societies in "sealed packets" to protect the right of discovery.[114] Publication, however, serves other purposes than establishing priority. Gould, writing closer to the events than later commentators, makes still another point concerning Leverrier's publications that has been too often neglected. "Having fairly arrived at his results," he wrote, "he looked upon them as conclusive.... He gave them, therefore, fearlessly to the world, and staked his reputation upon their accuracy. This forms by no means the least part of his claims to the respect and admiration of scientists throughout the world. Had the planet not been found in the predicted place, Le Verrier would alone have borne the mortification."[115]

The strong criticism of Airy and Challis persisted for some time, but eventually subsided, to surface again occasionally in historical discussions of the discovery of Neptune. Adams and Leverrier themselves did not personally become embroiled in or embittered by the controversy over priorities, and gracefully accepted such shares of the credit as the world and history has been wont to give them. They met for the first time the following year in England, and immediately established a firm friendship that lasted until Leverrier's death in 1877. Many years after Neptune's discovery, incidentally, Leverrier received the Royal Astronomical Society's Gold Medal twice, the second time from the hand of John Couch Adams, then the Society's president, for his work on the orbit of Mercury.[116]

One other interesting event during the first weeks after Neptune's discovery was almost obscured by the larger controversy over priorities. By the time Challis's letter in the *Athenaeum* had precipitated the storm, the new planet was already suspected of having a satellite and more provocatively of being a ringed planet like Saturn.

On October 10, the English amateur astronomer William Lassell, who would discover the third and fourth satellites of Uranus five years hence, observed "a minute star, distant about 3 diameters of the planet.... The close situation of this star, and its minuteness," he later reported, "occasioned my strong suspicion that it may be a satellite, but I regret to say that a long interval of cloudy nights immediately succeeding, prevented my having an opportunity of verifying it."[117] Lassell did not obtain his desired verification until July 1847 and the satellite, which revolves around Neptune in a retrograde direction, was named Triton.[118]

But Lassell was, if anything, somewhat more certain that the new planet was girdled by a ring, reporting that "I have strong reason to think that it has such an appendage, though the weather in this very unfortunate climate has not afforded that absolute verification which I could have wished." In his observations of Neptune on October 3, he had been "struck with its shape, which was evidently not merely that of a round ball," and again on October 10, "I received many distinct impressions that the planet was surrounded by an obliquely situated ring ... having its major axis nearly at right angles to a parallel of declination...." With a better instrument and better atmosphere, he thought, the planet would "put on the appearance of Saturn, when his ring is nearly closed." To check these impressions, Lassell called in "several friends" to view the planet and to make drawings, which, he found, agreed well with his own observations. He also tried different sized specula on his 24-inch Newtonian reflector, but the impression of the ring was still there. Finally, for comparison, he "carefully scrutinized the planet Uranus without being able to persuade myself that any such appendage belonged to him."*[119]

Interestingly, it was Challis who three months later claimed confirmation

---

*It was common practice at this time for astronomers to refer to the planets anthropomorphically.

of Lassell's suspicion of a Neptunian ring. The Cambridge director declared that both he and an assistant named Morgan had observed such a ring on the nights of January 12, 14 and 15, 1847, with the Northumberland refractor. He estimated that the axis of the ring was at an angle of 65° to a parallel of declination, and that the ratio of its diameter to that of Neptune was 3:2. Observing conditions on January 12 and 14 had been "particularly good," he reported, adding that under similar conditions, he had "twice seen with the Northumberland Telescope the second [Encke's] division in Saturn's Ring."[120]

Six months later Lassell, in confirming his discovery of Triton, declared that he had "seen on two or three occasions the same appearance" of the ring, "but nothing more strongly confirmatory."[121] Confirmation of Neptune's "ring," of course, has yet to be forthcoming.[122]

CHAPTER 3

BEYOND NEPTUNE

Throughout the intense controversy over who should get what credit in the discovery of Neptune, it was widely assumed that Neptune was indeed the planet of Leverrier's and Adams' predictions. Yet very soon, as observations of the eighth planet began to accumulate, this assumption came under serious challenge, and a new debate erupted in which it was charged that Neptune's discovery was nothing more than "a happy accident."

The new controversy marked the entry of American astronomers and mathematicians into the history of planetary discovery. It began in the computations of Neptune's orbit by Sears Cook Walker of the U.S. Coast Survey and the provocative conclusions drawn therefrom by Harvard University's Benjamin Peirce. And it provides a direct link to Percival Lowell, one of Peirce's brightest students at Harvard, whose long and laborious analytical and photographic searches for a trans-Neptunian Planet X in the early years of the twentieth century led to the discovery of the ninth planet, Pluto. Lowell, in the course of his investigation, answered some of the questions raised by Peirce, his mathematical mentor, and others in the aftermath of Neptune's discovery.

The discovery of Neptune, parenthetically, also coincides roughly with the emergence of American astronomy on the world scientific scene. Harvard's William Cranch Bond, for example, was among the first to report observations of the new planet, made with a 4¼-inch refractor, in European journals.[1] And even as the post-discovery controversy over priorities raged, Harvard was installing America's first large telescope, a 15-inch refractor, at its observatory in Cambridge, Massachusetts.[2] Within three years, the first issue of the *Astronomical Journal* would be published by its founding editor, Benjamin Apthorp Gould, Jr.

The truly sensational aspect of Neptune's discovery, of course, was the fact that the planet was found almost precisely where Leverrier and Adams had said it would be found—the difference between forecast and fact being only 55 minutes of arc in the case of Leverrier, and less than a degree and a half in the case of Adams. But if these geometers had been incredibly right in predicting the planet's location, it soon appeared that they had been wrong on just about everything else.

The first casualty of the discovery, however, was Bode's law of planetary distances. With Neptune, as with Uranus, immediate attempts were made to calculate preliminary elements, and these at once showed that the new planet did not fit into the curious arithmetical progression that Bode had so assiduously promoted for so many years. Adams himself, eschewing controversy for computation, was the first to derive preliminary elements, using the August 4 and 12 prediscovery positions and Challis's subsequent observations of the planet up through October 13. He found, Challis reported on October 21, that the new planet's mean distance from the sun was only 30.05 astronomical units, far short of the 38.4 a.u. to be expected from Bode's law and, it should be noted, considerably less than Adams's and Leverrier's estimates for their postulated planets.[3] Just one day later, Encke reported that Galle's computations indicated a mean distance of 30.03885 a.u., and a period of about 165 years.[4] Other computers subsequently obtained similar non-Bodean values and, as Challis would soon declare, "we are compelled to conclude that this singular law fails in this instance."[5] Henceforth, Bode's law does not figure seriously in the history of planetary discovery, and later planet-hunters, as will be seen, will rely on other indications in making the crucial assumption of mean distance.

The time span of the observations available for these preliminary calculations was very short and represented less than a degree of arc of the planet's full orbital motion. But it was long enough to provide a basis for searches in star catalogs and old observation records for possible prediscovery sightings that would extend the orbital arc and thus permit the immediate computation of a relatively accurate orbit and ephemeris for the new planet.

The unimaginative Airy, incidentally, felt such efforts were premature, and in a letter on February 1, 1847, as the planet temporarily disappeared behind the sun, he sought to discourage Adams from further attempts to refine its orbit until its reappearance could be observed two months hence.[6] Within days of this letter, ironically, a prediscovery observation of Neptune was found by Walker, and within weeks, the same observation was discovered independently by astronomer Adolf Cornelius Petersen of Germany's Altona Observatory.[7]

Walker had computed circular elements for Neptune by the end of December 1846, and ellipitical elements by late January 1847. But the probable error left by this latter theory still averaged ±48 seconds of arc. The orbit, he realized, could be improved if a prediscovery observation could be found in "some ancient Catalogues" and, he concluded, Joseph J. F. Lalande's *Histoire Céleste Français* "was the only one that afforded a present prospect of success. A preliminary ephemeris indicated the nights of May 8 and 10, 1795, as the only ones that probably comprised the place of Neptune."[8]

Walker then computed Neptune's probable position on May 10, 1795, from his elliptical elements, drew up a list of nine stars in Lalande's catalog that were within 15 minutes of arc of this position, and then compared them with Bessel's later star charts of the region. "The stars seen afterwards by Bessel were at once excluded," he later wrote, "leaving Nos. 1, 2, and 8, not in the Zone of Bessel. Of these, No. 1 was excluded by its magnitude (9,10), No. 2 in my opinion by being 17' south of the geocentric path. There remained then in all the Histoire Céleste only one star No. 8 that was a candidate for having been the planet Neptune.'"[9]

Walker reached this conclusion on February 2, a date he considered "the true date of discovery." But that night "was one of continuous storm" and so he drew up a report of his findings which he immediately sent to the U.S. Naval Observatory's superintendent, Lieutenant Matthew Fontaine Maury. The following day he communicated his results to other Naval Observatory astronomers including Professor Joseph Hubbard who, on the night of February 4, the first clear night after the discovery, made a telescopic search and determined that Lalande's star was now indeed missing.[10] Walker's discovery was published in the *Washington Union* for February 9, 1847.

Shortly after this, Petersen at Altona was also led to Lalande's *Histoire Céleste*. He, however, compared Lalande's observations directly with the heavens in observations beginning March 17, and found that three of Lalande's stars observed on May 10, 1795, were missing. Probable typographical error, he concluded, eliminated two of the suspects, leaving the third unaccounted for. From Galle's circular elements for Neptune, he then computed the planet's probable position on May 10, 1795, and found it agreed closely with the missing "star."[11]

Both Walker and Petersen advised Leverrier of the results of their independent labors and Leverrier, in turn, reported their findings simultaneously to the French Academy of Sciences on March 29. A brief announcement of Petersen's work was published in *Astronomische Nachrichten* for April 3, 1847, followed by a full report on April 15. Walker's discovery was not formally published in Europe until May 20. Thus the occasion marks probably the first time important astronomical knowledge was available to American astronomers before it was available in Europe.

Yet despite Walker's and Petersen's work, the status of Lalande's May 10, 1795, observation as a prediscovery position of Neptune remained cloudy, for the listing of its coordinates in the *Histoire Céleste* was followed by a colon (:), Lalande's method of indicating a doubtful position. This cloud was soon dissipated, however, when Félix Victor Mauvais went back to Lalande's original manuscripts, preserved in the Paris Observatory, and found that Lalande had observed the "star" not once, but twice, on both May 8 and 10, 1795. Because the two positions were slightly different, Lalande had suppressed the first and marked the second as doubtful in printing it in his catalog. Mauvais and Walker quickly determined that the difference in the two positions was exactly that of Neptune's angular motion over a two-day period, thus providing full confirmation of the identity of Lalande's "star" and Neptune.[12]

Once convinced that Lalande had observed the planet in May 1795, Walker had immediately used the observation to compute a third set of elements for Neptune which he completed on February 6, and which, like his earlier ones and those of various European computers, agreed in showing that the planet's distance from the sun was of the order of 30 a.u. In addition, it indicated that the planet's other orbital elements also differed widely from those of both Leverrier's and Adams's hypothetical trans-Uranian planets. Neptune's mass, for example, was reduced by nearly half, and its eccentricity now indicated a nearly circular orbit.

Walker's third orbit for Neptune, published Feburary 9, 1847, is compared below with those of Leverrier's and Adams's hypothetical planets in a table which has been adapted from one published in 1949 by Smart:[13]

Elements	*Walker*	*Leverrier*	*Adams*
a (mean distance)	30.25	36.15	37.25
e (eccentricity)	0.00884	0.10761	0.12062
i (inclination)	1°54'54"	*	*
Ω (longitude of node)	131°17'38"	*	*
ω (longitude of perihelion)	0°12'25"	284°45'	299°11'
P (period, in years)	166.381	217.387	227.3
m (mass, sun = 1)	0.0000666	0.0001073	0.000150
(longitude 1847.0)	328°7'57"	326°32'	329°57'

*Not predicted.

If Walker's orbit surprised astronomers, however, Harvard's Benjamin Peirce promptly turned surprise into sensation. From the mean distance alone, verified by his own observations and those of Harvard's George P. Bond, he now announced that "the planet Neptune is not the planet to which geometrical analysis had directed the telescope; its orbit is not contained within the limits of space which have been explored by geometers searching for the source of the disturbance of Uranus; and its discovery by Galle must be regarded as a happy accident."[14]

Peirce's pronouncement touched off a new controversy which, despite the fact that it raised pertinent mathematical questions concerning Neptune's

Harvard University Archives

Benjamin Peirce of Harvard University.

discovery, has been described only briefly and cursorily in the subsequent literature on the subject, perhaps because it is at once more complex and less emotional than the broader controversy over priorities.

Initially, Peirce based his argument on two propositions in Leverrier's solution to the Uranian problem: (1) that his hypothetical planet's distance from the sun must be between 35 and 37.9 a.u., and (2) that its mean longitude on January 1, 1800, must have been between 243° and 252°.

"Neither of these propositions is of itself necessarily opposed to the observations which have been made upon Neptune," he declared, "but the two combined are decidedly inconsistent with observation. It is impossible to find an orbit, which, satisfying the observed distance and motion, is subject at the same time to both of these propositions, or even approximately subject to them. If, for instance, a mean longitude and time of revolution are adopted according with the first, the corresponding mean longitude in 1800 must have been at least 40° distant from the limits of the second proposition. And again, if the planet is assumed to have had in 1800 a mean longitude near the limits of the second proposition, the corresponding time of revolution with which its motions satisfy the present observations cannot exceed 170 years, and must therefore be about 40 years less than the limits of the first proposition. Neptune cannot, then, be the planet of M. Le Verrier's theory, and cannot account for the observed perturbations of Uranus under the form of inequalities involved in his analysis."[15]

This was not to say that Neptune itself could not account for the perturbations of Uranus, he quickly added, "for its probable mean distance of about 30 is so much less than the limits of the previous researches, that no inference from them can be extended to it. An important change, indeed, in the character of the perturbations, takes place near the distance of 35.3; so that the continuous law by which such inferences are justified is abruptly broken at this point, and it was hence an oversight in M. Le Verrier to extend his inner limit to 35." Peirce here was referring to the idea of commensurable periods, then called "Laplacian librations" from the French astronomer's earlier theoretical investigations of the motions of the inner satellites of Jupiter and Saturn. "A planet at the distance of 35.3 would revolve about the sun in 210 years, which is exactly two and a half times the period of revolution of Uranus," he explained. "Now if the times of revolution of the two planets were exactly as 2 to 5, the effects of their mutual influence would be peculiar and complicated, and even a near approach to this ratio would give rise to those remarkable irregularities of motion which are exhibited in Jupiter and Saturn, and which greatly perplexed geometers until they were traced to their source by Laplace. This distance of 35.3, then, is a complete barrier to any logical deduction, and the investigations with regard to the outer space cannot be extended to the interior."[16] Lowell, as will be seen, provided a mathematical refutation of Peirce's position here many years later in the course of his search for a trans-Neptunian planet.

But Peirce found in Walker's elements for Neptune an even more remarkable commensurability. "The observed distance 30, which is probably not very far from the mean distance, belongs to a region which is even more interesting in reference to Uranus than that of 35.3. The time of revolution which corresponds to the mean distance 30.4 is 168 years, being exactly double the year of Uranus, and the influence of a mass revolving in this time would give rise to very singular and marked irregularities in the motions of this planet.... But it is highly probable that the case of Neptune and Uranus is not merely that of a near approach to the ratio of 2 to 1 in their times of revolution; for it may be shown, as was shown by Laplace for the ratio of two-fifths, that a sufficiently near approach to it must, on account of the mutual action of the planets, result in the permanent establishment of this remarkable ratio." Thus, after demonstrating the point mathematically,

Peirce predicted that, with further refinements of Neptune's orbit, its period would be found to be "precisely twice as long as that of Uranus" of 168 years.[17] Neptune, of course, is now known to have a revolution period of very nearly 164.8 years, somewhat short of precise commensurability, and again this fact will be later considered by Percival Lowell in the course of his trans-Neptunian planet investigation.

Peirce communicated these conclusions to the American Academy of Arts and Sciences on March 16, 1847. More than a month later, he presented another paper to the Academy, now showing that the Uranian problem, as considered by Leverrier and Adams, had a number of solutions, each different from the other and from those of Leverrier and Adams, and each equally complete. "The present place of the theoretical planet which might have caused the observed irregularities in the motions of Uranus, would, in two of them, be about one hundred and twenty degrees from that of Neptune, the one being behind, and the other before, this planet," he declared. "If Le Verrier or Adams had fallen upon either of the above solutions instead of that which was obtained, Neptune would not have been discovered in consequence of geometrical prediction."[18]

Peirce also now reported that he was having difficulty explaining the motions of Uranus by the action of Neptune itself, a failure apparently stemming from a deficiency of degree in his method, that of Leverrier, and the value he used for Neptune's mass.[19] For subsequently, he succeeded by adopting a mass of 1/19840 (0.0000504) for the planet, a value considerably smaller than even Walker's estimate. Neptune's mass was then, and for many years, a serious problem for celestial mechanicians. Observations of Lassell's satellite by observatories in Europe had yielded widely varied values all of which, however, were well below those assigned by Adams and Leverrier. Peirce's value, incidentally, compares well with Gaillot's figure of 1/19094 (0.0000523), derived more than half a century later and used by Lowell in his Planet X computations.[20]

Peirce continued his work on Neptune through the summer of 1847, determining the approximate perturbations of the planet by other planets and communicating his results to Walker. Walker, in turn, used them to calculate a pure elliptical orbit which he presented to the American Academy on December 7, and then a second such orbit presented to the Academy on March 6, 1848. The latter, as B. A. Gould noted in 1850, "has represented the course of Neptune so well up to the present time, as to render a nearer approximation unnecessary, if indeed it were possible."[21]

Peirce's startling deductions from Walker's orbits were, of course, disputed in Europe, primarily by Leverrier and Sir John F. W. Herschel. Adams, perhaps typically, again eschewed controversy. "Mr. Adams has taken no personal part in the controversies which have arisen since the discovery of Neptune, but has continued to devote himself to the pursuit of science," Gould noted in summarizing the arguments against Peirce's thesis. "Mr. Le Verrier has published several articles in the *Comptes Rendus* in order

Harvard University Archives

Benjamin Apthorp Gould.

to defend his claim to be considered the actual discoverer of Neptune, by showing that this planet might have been brought within the limits of his theory. In England, Sir John Herschel has taken similar ground in favor of Mr. Adams."[22]

Gould, in presenting and assessing these arguments, computed the true longitude and radius vector for Neptune and for Leverrier's and Adams's hypothetical planets for every tenth year from 1690 through 1890 "to show as clearly as possible the relative positions" over the two-hundred-year period. The table is reproduced below because the longitude portion of it for the years 1800 and 1860 has since been refined and used as a point against Peirce's "happy accident" claim.[23] Gould himself noted that "the longitude of the planet of Le Verrier's theory coincided with that of Neptune in 1840; and that Neptune would be in conjunction with Adams's planet about the year 1856," and that "the closest agreement of the radius-vectors was not far from the year 1830...." But he also pointed out the great discordance in the radius-vectors, of the order of ten astronomical units, in the years 1710 and 1720, and he might well have called attention to the large discrepancies in the longitudes for these years as well.[24]

Beyond Neptune   67

	True Longitude			Radius-Vector		
Date	Neptune	Le Verrier	Adams	Neptune	Le Verrier	Adams
1690	341.1	65.1	81.1	29.92	39.09	40.69
1700	4.4	79.2	94.5	29.84	39.63	41.30
1710	26.7	93.0	107.3	29.79	40.89	41.66
1720	49.1	106.5	119.8	29.77	40.04	41.74
1730	70.8	120.2	132.0	29.80	39.85	41.48
1740	93.8	134.0	145.5	29.86	39.40	41.09
1750	116.0	148.2	158.9	29.95	38.71	40.39
1760	138.0	163.0	172.6	30.04	37.84	39.46
1770	160.1	178.9	187.3	30.10	36.93	38.37
1780	181.8	195.1	202.7	30.22	35.68	37.15
1790	203.5	212.7	219.2	30.28	34.58	35.91
1800	225.9	231.4	236.8	30.30	33.57	34.74
1810	246.8	251.2	255.5	30.28	32.78	33.75
1820	268.5	270.9	275.2	30.23	32.64	33.06
1830	290.2	291.4	295.5	30.15	32.29	32.77
1840	312.0	312.0	315.9	30.02	32.63	32.91
1850	334.2	332.0	335.9	29.96	33.32	33.47
1860	356.4	351.0	355.0	29.87	34.26	34.37
1870	18.7	8.9	14.1	29.81	35.43	35.49
1880	41.1	25.8	29.9	29.77	36.48	36.73
1890	63.4	39.4	45.7	29.78	37.99	37.97

Gould first considered Leverrier's argument that if the observations of Uranus were assigned an error factor of 10 percent, then the deductions made from them could be expected to have a 10 percent error. "I should not have alluded to this reasoning had not Mr. Le Verrier published it," Gould declared. "According to this argument, an error of 3.0 [astronomical units] would be allowable in the mean distance of 30,—of 4.0 if the mean distance were 40, etc.—errors which would make the attraction of the planet to be exerted in a direction totally different from the true one. But even this allowance would not correct the error in the radius-vector in 1710 and 1890."[25]

Gould next noted Leverrier's claim that "when there are perturbations, he can tell where Neptune is." Nonetheless, Gould pointed out, "at the time of the early observations, when the radius-vector of Neptune differed from that of the theoretical planet by ten times the radius of the earth's orbit, Uranus was, according to Mr. Le Verrier's theory, undergoing a perturbation by Neptune." And here he also cited Leverrier's remark, made to the French Academy, that the small eccentricity of Walker's orbit for Neptune computed with the 1795 Lalande sighting, would be "imcompatible with the nature of the perturbations of Uranus."[26]

Leverrier also sought to explain the indications that Neptune's mass was only about half what his theory required by showing that the error corresponded to only 20 percent in Neptune's diameter. "This," Gould commented dryly, "is true,— no schoolboy will deny it,— but it was the mass not the diameter, which he sought." Gould conceded, however, that as the mass of Uranus was still in question, some discordance in the mass of Neptune as deduced from that of Uranus should be expected. Whether the discrepancies in the values for Uranus' mass might be reconciled by investigations of the influence of Neptune, he felt, was "one of the most important questions, connected with Neptune," which remained undecided.[27]

In defending his theory against Peirce's "happy accident" thesis, Leverrier also insinuated that Walker's orbit, on which the thesis was based, might be grossly in error, and here Gould rose to the defense of his colleague and compatriot. "The only other point of Mr. Le Verrier's argument to which I will allude," he wrote, "is that in which he says,—'The orbit calculated by Mr. Walker, from a position in 1795, and the small arc observed since the discovery, can very well be erroneous by many degrees either in 1887 or in 1757, and if I have admitted the positions which it has given for these epochs, it is solely by courtesy, and because it presents me no inconvenience.'

"Of this I may be permitted to say," Gould replied, "that Mr. Walker's laborious and accurate investigations have given us the orbit of Neptune to a very high degree of precision, and deserve the gratitude and admiration of astronomers,—not such an imputation as this. It would be contrary to all probability should the place given by Mr. Walker's orbit for those years be false by two minutes." And in a footnote, he declared that "the error in 1887 and 1757 would, according to the doctrine of chance, be to that in either of the years above named in the ratio of $65^2:27^2 = 5''.8:1''.0$, and the resulting error, therefore, less than six seconds."[28]

Herschel's objections to Peirce's "happy accident" claims were more general, and rested primarily on the fact that both Adams and Leverrier had placed their hypothetical planets in positions nearly identical with Neptune's discovery position. Herschel's argument, summarized in his textbook, *Outlines of Astronomy,* in 1849, and thereafter, was based on the contention that Uranus and Neptune had been in conjunction in 1822, producing a maximum perturbation which led to Adams's and Leverrier's success in predicting the disturbing planet's longitude and, incidentally, to the failure of Bouvard's ill-timed 1821 tables. He also argued that the radius-vectors of Adams's and Leverrier's planets were really not so very different from Neptune's at the time of that planet's discovery, which led Gould to remark that "surely it cannot be considered as an analogy between two orbits, that the perihelion of one was so near the aphelion of the other."[29]

Gould, it is clear, was not impressed by such arguments. "Reasoning like this seems, however, utterly inapplicable to researches of such nicety and analytical refinement as characterized those upon Uranus. It would allow these investigations no other merit than the success with which Neptune's apparent place was approximately predicted. It is an effort to show that the uncertainty of the calculations was so great, that Neptune's perturbative influence may be included within their limits."[30]

Gould's own conclusion is perhaps worth stating here. "The arguments which tend to prove that Neptune is the planet of their theory," he wrote, "can only be based upon the supposition of error in that theory, a supposition which I am unwilling to admit. Investigations conducted with the care and precision which characterized these must not be so lightly dealt with. The combined labors of Leverrier and Peirce have incontrovertibly proved, that, by reducing the limits of error assumed by the modern observations to 3", there can be but two possible solutions to the problem. There are two different mean distances of least possible error,—one of which is 36, and the other 30. The one is included within the theory and limits of Leverrier, and corresponds with Adams's solution; the other is the orbit of Neptune.

"This simple view of the case," he added, "... reconciles all the computations and observations, as well as the discords and contentions. It does not detract in the slightest degree from the well-earned fame of the illustrious geometers who had arrived at a solution of the problem, and I am not aware that it has ever been opposed by mathematical reasoning."[31]

Gould's last statement apparently still holds, although Sir John Herschel's arguments, perhaps because of their relative simplicity and a residue of emotion, have tended to prevail in discussions of the "happy accident" controversy in later years. More recently, for instance, W. M. Smart, the first John Couch Adams Astronomer at Cambridge University, expanded on Herschel's position in his lengthy summary of the history of the discovery of Neptune written on the occasion of the centennial of that historic event.[32]

Smart responded to Peirce's century-old challenge by asking: "Was the discovery of Neptune a happy accident? The answer is emphatically 'no.'" Smart then presented a diagram to illustrate the maximum perturbative effect of Neptune on Uranus near the 1822 conjunction of the two planets, and to show that "the disturbance of Neptune on Uranus between 1693 and 1779 must be small compared with the disturbance between 1779 and 1843." Thus, he noted, the ancient observations of Uranus, if reasonably accurate, ought to provide substantially accurate elements of Uranus's orbit; whereas Bouvard, in his 1821 tables, had ignored these ancient observations in favor of the modern ones, which were subject to greater perturbations, to his undoing.[33]

The maximum perturbations of Uranus by Neptune, Smart pointed out, would come as Uranus passed conjunction, and were dependent on the mass of Neptune, the distance between the two planets, and the differences in their heliocentric longitudes. "These perturbations would not be greatly different if we increased the heliocentric distance of Neptune and counterbalanced the diminution of the attractive force by increasing the mass of Neptune. In other words, the assumption of any value of the semi-major axis of the unknown planet's orbit, not differing too wildly from the true value, would lead to values of Neptune's heliocentric longitudes, at any time between 1801 and 1843, within a few degrees of the true values, thus providing adequate information for the telescopic detection of the unknown planet."

Smart then gave the comparative values of the heliocentric longitudes of Neptune and Adams's and Leverrier's hypothetical planets for ten-year intervals from 1800 to 1860, when prediction and fact came closest to agreement, to illustrate his point. His values compare with those listed by Gould for these same years in his more comprehensive table. Thus for Smart: "If there was any element of 'accident' in the discovery of Neptune it arises simply from the fact that the time of the planet's nearest approach to Uranus occurred as it did, that is, in 1822." Had the conjunction of Uranus and Neptune occurred some years earlier, he added, "the discovery of Neptune by mathematical process—if the planet had succeeded in escaping telescopic discovery meanwhile—might have been delayed to the beginning of the present century."[34]

The "happy accident" controversy, interestingly, is today little known, is seldom discussed in modern astronomy textbooks, and when mentioned in historical reviews of the discovery of Neptune is usually dismissed briefly, as if it were of no importance. Grosser, in his excellent *The Discovery of Neptune* (1962) refers to it, but only in passing, declaring that "although supported by mathematical arguments, Peirce's conclusion was drastically overstated," and then citing Smart's statements quoted above to his point.[35] Peirce's work, however, is important in the history of planetary discovery, and not in the least because of its impact on Percival Lowell. Lowell's interest in mathematical astronomy was certainly nurtured in part by it, and he used it selectively and critically in his analytical search for a planet beyond Neptune.

That there was a planet beyond Neptune seems to have been casually accepted by most astronomers in the years following Neptune's discovery, although few thought it would be discovered as a practical matter, and even fewer actively concerned themselves with the possibility through the remainder of the nineteenth and into the early years of the twentieth century. Leverrier, as noted, did not think Neptune should be considered the last of the major planets, and soon after its discovery expressed the hope that "after thirty or forty years of observing the new planet we will be able to use it in turn for the discovery of the one that follows it in order of distance from the Sun."[36] He did not think it feasible, however, to make such an attempt at the time, nor did he ever undertake a serious mathematical exploration of trans-Neptunian space.

The great French geometer, nonetheless, appears again in the history of planetary discovery in another of its more bizarre incidents. Once the Neptunian controversies had subsided, Leverrier, now director of the Paris Observatory, began the herculean task of reconstructing the theories of all the planets, publishing his results in fourteen volumes of the *Annales de L'Observatoire Impérial de Paris* between 1855 and 1877, the year of his death. In 1859, he completed his "Theorie et Tables du mouvement de Mercure," as Volume V, in which he declared that "Mercury is without doubt perturbed in its path by some planet or by a group of asteroids as yet unknown."[37]

Leverrier's new investigation of Mercury's orbit was, in effect, a continuation of the inconclusive studies that he had made of that planet in 1841–43, and his postulation of a disturbing mass between Mercury and the sun now was designed to account for an anomaly in the precession of the apsides of the planet, amounting to 38 seconds of arc per century, that neither he nor anyone else had been able to explain by known gravitational factors.

Leverrier presented his Mercury memoir to the French Academy on September 12, 1859, and communicated his major findings that same day in a letter to Hervé Faye which was immediately published in the *Comptes Rendus* and subsequently in the *Monthly Notices of the Royal Astronomical Society* under the title: "Suspected Existence of a Zone of Asteroids Revolving Between Mercury and the Sun." The latter journal, incidentally, noted that the letter should be read in connection with a suggestion of "a zone of meteors in the neighborhood of the sun" which physicist William Thompson (later Lord Kelvin) in 1854 had "thrown out as a possible source of supply of the solar heat and energy...."[38]

Leverrier, it is clear, favored a multiplicity of asteroidal bodies over a single planet as the disturbing influence on Mercury, orbiting the sun at a mean distance of no more than 0.3 astronomical unit, and thus inside Mercury's orbit.[39] Three months later, however, while his memoir was still in press, he received a startling report which changed his thinking. When his memoir was finally published, it contained an addendum in the form of a letter from a Dr. Lescarbault, written December 22, 1859, under the heading: "Passage d'une planète sur le disque du soleil, observé à Orgéres (Eure-et-Loir) par M. Lescarbault."[40]

Lescarbault, a rural physican and amateur astronomer, wrote that on March 26, 1859, he had observed what he believed was an intra-mercurial planet in transit across the sun's disk, and explained that he had delayed reporting the sighting "in the hope of seeing the little heavenly body again." It had appeared as a "very small" and "perfectly round" black dot, he said, and expressing his conviction that it would be seen again, he estimated that its orbital inclination would be between 5°20″ and 7°20″, its longitude of node about 183°, its eccentricity "enormous," and its transit time across the diameter of the sun 4 hours and 30 minutes.[41]

Almost at once, Leverrier set out for Orgéres to check Lescarbault's claim, arriving unannounced on New Year's Eve. Without at first revealing his identity, he questioned the good doctor closely, pressing "hard from step to step till he had obtained such material and verbal evidence as no longer permitted him to doubt the reality of the observation or the good faith of the observer," as the *Monthly Notices* reported only two weeks later. He also obtained collateral evidence "of the high character and worth" of Lescarbault "from others of station in the neighborhood." The anonymous report concluded that "astronomers in all countries will unite in applauding this second triumphal conclusion to the theoretical inquiries of M. Le Verrier...."[42]

Using Lescarbault's observations, Leverrier calculated a circular orbit for the planet he named "Vulcan," with a mean distance of 0.1427 a.u.; a period of 19 days, 7 hours; an inclination of 12°; a longitude of node of 13°; and a mass 1/17 that of Mercury—"extremely small" but still sufficient to "produce all the anomalies in the motion of the perihelion of Mercury."[43]

For the next half-century and more, hundreds of astronomers—some of them very eminent—searched sporadically but vainly for another glimpse of "Vulcan," usually looking for Lescarbault's elusive planet close to the sun's disk during solar eclipses. There were, indeed, a few sightings of such a body reported, but these were subsequently found to be faint stars seen close to the sun, or simply sunspots. Almost immediately, too, it was realized that Lescarbault's observation was not unique. "The investigations of Professor Le Verrier on the theory of Mercury, and the procrastinated publication of Dr. Lescarbault's remarkable observations, have more than ever called the attention of the learned to infra-Mercurial planets," one astronomer wrote in the *Monthly Notices* soon after Lescarbault's discovery became known. "Since 1761, mention has frequently been made of one or more dark spots passing before the sun," he reported, adding that a list of "upwards of twenty such occurrances" had recently been published, two of which "have a very probable relation with the planet of 1859...."[44]

Over the next few months, astronomers not only in England and Europe, but in India and Australia launched searches for the supposed new planet.[45] But they might have better heeded a brief notice that appeared in April, first in *Astronomische Nachrichten* and then in the *Monthly Notices* regarding the activities of the French astronomer E. Liais of the Brazilian Coast Survey. Liais reported that he had been observing the sun on March 26, 1859, at the same time as had Lescarbault, and "positively denied" that any planet had transited its disk.[46]

Interest in Lescarbault's planet waned somewhat over the next fifteen years, but it was revived again early in 1876. In February, then president John Couch Adams presented the Royal Astronomical Society's Gold Medal to Leverrier for his work on the theory of Mercury and the problem of the anomaly in the motion of its apsides. "The theory of Mercury," Adams declared, "has been established with so much care, and the transits of the planet furnish such accurate observations, as to leave no doubt of the reality of the phenomenon in question; the only way of accounting for it appears to suppose, with M. Le Verrier, the existence of several minute planets, or of a certain quantity of diffused matter, circulating about the sun within the orbit of Mercury."[47]

Less than two months later, a transit of "Vulcan" was reported as having been detected on photographic plates made at the Greenwich Observatory, but again the object was shown to have been a sunspot. Leverrier, meanwhile, had been studying twenty earlier observations of dark spots transiting the sun, and had concluded that five of them referred to a single body. He calculated probable transits of this body would occur about March 22, 1877, and October 15, 1882, and urged astronomers to look for the planet at these times.[48]

Leverrier's colleague, Pierre Jule César Janssen immediately called for worldwide observations, particularly by photography, and Sir George Airy, still Astronomer Royal, quickly echoed his call. Recognizing that observations should be made from widely separated points, and that the actual time of transit was only an approximation, Airy sent telegrams to observatories in India, Australia, New Zealand, the United States and Chile requesting observations, and voiced the hope that the predicted transit of "Vulcan" would also be observed in eastern Siberia and in Japan. And in urging photographic, rather than visual observations, he pointed out that this newly developed technique, "admitting of subsequent long and continuous study, would be preferable to eyeviews in a telescope in which the power of deciding on the visibility of a spot must terminate with the actual view."[49] The sun was duly and well observed during these predicted transits but, as the noted historian of astronomy, Agnes M. Clerke, later commented: "Widespread watchfulness notwithstanding, no suspicious object came into view at either epoch."[50]

"Vulcan" was also reported by two American observers on July 29, 1878, during a solar eclipse "as a ruddy star with a minute disc southwest of the obscured sun." The observers were James C. Watson at Rawlins, Wyoming, and Lewis Swift at Denver, Colorado, and they later claimed to have seen a pair of planets each. Their observations disagreed, however, nor could they be reconciled with Leverrier's and Lescarbault's "Vulcan." The "most feasible" explanation, Miss Clerke felt, was that both men had seen two faint stars in Cancer, "haste and excitement doing the rest."[51]

Solar eclipses subsequently provided opportunities for systematic searches for "Vulcan" through the nineteenth and into the twentieth century, but all such efforts gave negative results. The Smithsonian Institution mounted eclipse expeditions in 1900, 1901 and 1908, in which the search for intra-mercurial planets was a major part of the observing programs, and which included such eminent solar scientists as Samuel Pierpont Langley, then the Smithsonian's secretary, and his successor, Charles Greeley Abbot, among the observers. Of the 1900 expedition, Abbot reported that they "were able to photograph many stars" and that there were "several starlike objects among them not to be found in star maps. But neither at that eclipse, nor at subsequent ones, have these objects been confirmed by other observers. It is likely that they were no more than starlike defects of the photographic plates."[52]

Lick Observatory astronomers also made special efforts to find an intra-mercurial planet during these and other eclipses in the first decade of the twentieth century. Lick Observatory's W. W. Campbell concluded after observations of the January 3, 1908, eclipse from the South Seas that the search for such a body was futile.[53] Abbot agreed: "Their results seem to prove definitely that the supposed planets nearer the sun than Mercury do not exist."[54]

The possibility of finding a ninth planet beyond Neptune does not seem to have been as popular with the late-nineteenth-century astronomers as that of an intra-mercurial planet, or planets, and in fact a general skepticism seems to have prevailed on the subject right up to the discovery of Pluto. But nonetheless, a dozen or so trans-Neptunian bodies were postulated from time to time in the years following Neptune's discovery, and several visual and photographic searches were undertaken.

Amherst College Observatory
David Peck Todd.

Morton Grosser, in an excellent 1964 paper in *Isis*, has summarized these early trans-Neptunian speculations, for they were little more than that, and thus they can be treated briefly here. In all of them, Bode's law of planetary distances is conspicuous by its absence, and two alternative methods of making this crucial assumption now dominate. The first of these, developed in 1877 by an American, David Peck Todd, involved the analysis of the residuals of Uranus by a graphical method similar to that used in Sir John Herschel's *Outlines of Astronomy* (1849) to illustrate his discussion of Neptune's discovery. The idea here was to determine the hypothetical planet's period, and thus its distance, from the curve of the perturbations of Uranus, assuming the peaks of this curve to represent conjunctions of the presumed planet with Uranus, and to estimate its longitude at a given date from the peaks. The second method, which grew out of a suggestion by French astronomer Camille Flammarion in 1879, involved extrapolations from the observed clustering of the aphelion points of periodic comets and meteor streams near the orbits of the outer planets, the assumption being that any such clustering beyond Neptune might be an indication that an unknown planet was in orbit at that distance.

The earliest prediction of a trans-Neptunian planet, if it may properly be called a prediction at all, seems to have been made by a French physicist, Jacques Babinet, in 1848. Babinet used simple arithmetic to derive differences between Leverrier's prediction and the actual orbit of Neptune, concluding therefrom that a planet, which he named Hyperion, was in orbit at a mean distance of 47–48 a.u., with a period of 336 years, a mass of 1/25900 (0.0000386), and a magnitude of 10–11. Leverrier himself ridiculed this result, and no one else apparently gave it any serious attention.[55]

Two years later, a curious incident gave rise to the short-lived suspicion that a trans-Neptunian planet actually had been seen. In 1850, the U.S. Naval Observatory's James Ferguson was pursuing a series of observations of newly-discovered asteroids, the results of which he duly published in the newly-established *Astronomical Journal*. The following year, the English observer John Russell Hind, in the course of mapping stars, found that the 9th magnitude star Ferguson had used as a comparison star for Hygea was missing. Subsequent searches seemed to confirm the disappearance, and Hind soon suggested that the "star" had been a planet, at a distance of about 137 a.u. and with a period of 1600 years—very trans-Neptunian indeed. The missing star remained elusive, however, and interest in it gradually waned. The mystery was not solved until more than a quarter of a century later when C. H. F. Peters, the director of the Hamilton College Observatory at Clinton, New York, demonstrated that the star had never really been missing at all. Ferguson, as it turned out, apparently had mislabeled a micrometer wire in his initial Hygea observations, and thus recorded the star in an incorrect position.[56]

The first serious attempt to find a trans-Neptunian planet came, as noted, in 1877, with Todd's work.[57] His graphical method, he later wrote, had "suggested itself" to him as early as 1874, and he had not realized until later that it was "quite similar" to Herschel's treatment of the discovery of Neptune. He started work on the problem in August 1877, and in spite of the admitted inconclusiveness of the residuals of Uranus, by late October he had derived elements for a planet at a mean distance of 52 a.u., with a period of 375 years, and a magnitude of 13+. Its longitude for 1877.84 was $170° \pm 10°$, and he estimated that its angular diameter would be 2.1 seconds of arc. He also found that its inclination would be 1°24' and its longitude of node 103°.[58]

Todd searched for this planet on thirty nights between November 3, 1877, and March 6, 1878, with the Naval Observatory's 26-inch refractor at Washington, D.C., but in vain. His failure, along with the skepticism of most astronomers concerning new planets, probably accounts for the fact that he published nothing about his search for three years and then only after another planet-hunter, Professor George Forbes of Glasgow, announced predictions of a pair of trans-Neptunian planets at distances of about 100 and 300 a.u. with periods of about 1000 and 5000 years respectively.[59]

Forbes based his predictions on cometary aphelia, following an almost casual suggestion made a year earlier by Flammarion. In 1879, the French astronomer, in the first edition of his *Astronomie populaire*, had noted that the existence of the aphelion points of two periodic comets and a periodic meteor stream at about 48 a.u. from the sun might indicate an unknown planet at that distance. His suggestion, expanded but not further supported in subsequent editions and translations of his book, received some theoretical backing that same year from studies of comets reported by Hubert Newton, a mathematician at Yale University.[60]

Forbes proposed his two remote planets beginning in February 1880 in a series of papers presented to the Royal Society of Edinburgh, basing his hypothetical inner planet on the orbits of four periodic comets, and his outer one on six. He also claimed that the motion of Uranus was not inconsistent with his deductions from the cometary aphelia.[61] Perhaps the most interesting feature of Forbes' predictions, as Todd himself pointed out, was that the longitude he had assigned the inner planet of his pair was only four degrees away from that assigned by Todd to his single planet.[62]

Some years later, Forbes sent coordinates for his inner planet to British amateur astronomer Isaac Roberts, a pioneer in astronomical photography, who then made a careful photographic search in the indicated region out to the fifteenth magnitude, exposing pairs of plates at approximately 7-day intervals and comparing them by superposing one over the other. In 1892, Roberts reported that no new planet had been found.[53]

Around the turn of the century, predictions of trans-Neptunian planets seem to have become at once more serious and more superficial. In the first category, certainly, belongs the work of Hans-Emil Lau, of Copenhagen, who in 1899 began to study the problem with classic analytical methods, reviewing the observations of Uranus from 1690 through 1895. In 1900 he published his conclusion that the nature of such Uranian perturbations as seemed to exist indicated not one, but two trans-Neptunian bodies, and these he assigned distances of 46.6 and 70.7 a.u., with masses of 9 and 47.2 times that of the earth, and a magnitude for the nearer planet of between 10 and 10.5. Recognizing the duality of the solution, he listed the 1900 longitudes of these hypothetical bodies as $274° \pm 180°$ and $343° \pm 180°$.[64] Brief searches to find Lau's inner planet, however, were unsuccessful.

In 1901, Gabriel Dallet, also working from observations of Uranus, deduced a hypothetical planet at a distance of 47 a.u., with a magnitude of between 9.5 and 10.5, and a longitude in 1900 of 358°.[65] Inspired by Dallet's work, Theodore Grigull later the same year used the motions of three comets to derive a longitude for a trans-Neptunian planet less than six degrees away from Dallet's position, and the following year, from observations of twenty comets spanning almost five hundred years, he brought the differences in longitudes to within 2.5 degrees.[66] The degree of sophistication of Grigull's work is perhaps indicated by the fact that he noted that his planet's presumed distance of 50.6 a.u. agreed quite well with the distance of 53.3 that he had extrapolated from a curious scheme devised a year earlier by the Vicomte du Ligondès which presumed to relate the distances of the known planets to a conical structure with its vertex at the sun. In 1903, Ligondès obligingly modified his scheme, which made the agreement with Grigull's planet even closer.[67]

But even more superficial predictions of trans-Neptunian planets were to come. Beginning in 1904, the controversial Dr. Thomas Jefferson Jackson See, former assistant both to George Ellery Hale and Percival Lowell and now a U.S. Navy astronomer at Mare Island, California, took up the problem in connection with his speculative theories in cosmology. Subsequently, on the basis of his "Capture Theory" and something he called "the Theory of the Resisting Medium," he proclaimed "certainly one, most likely two, and probably three planets beyond Neptune." As he proposed that all the planets had been captured by the sun in its course through space, his rationale here seems to have been that a resisting medium surrounding the sun tended to force the originally eccentric inner planets into circular or nearly circular orbits, and that planets very far from the sun, where the resisting medium was less dense, should have more eccentric orbits. "To suppose the planetary system to terminate with an orbit so round as that of Neptune is as absurd as to suppose that Jupiter's system terminates with the orbit of the fourth satellite," See declared. See placed his three supposed planets at distances of 42.25, 56 and 72 a.u., assigning the inner one a period of 272.2 years and a longitude in 1904 of 200°.[68]

See's three trans-Neptunian planets do not represent the limit of the imagination of planetary prognosticators, however. As Grosser has noted, among other predictions advanced at this time was one in 1902 by a Russian general named Alexander Garnowsky, who sent a brief communication, in Russian, to the French Astronomical Society concerning four hypothetical planets beyond Neptune. The Society requested more detail, but this apparently was not forthcoming.[69]

Despite the plenitude of postulated trans-Neptunian planets, the prevailing attitude among almost all astronomers remained one of skepticism, and this was undoubtedly reinforced by the scientifically superficial nature of many of these predictions. In 1907, moreover, French astronomer Jean Baptiste Aimable Gaillot began a revision of the theories and tables of both

Uranus and Neptune, publishing his work in 1909 as a volume of the *Annales de L'Observatoire de Paris* and finding, in the course of his computations, no particular indication of any exterior bodies in the motions of either planet.[70]

Late in 1908, however, Harvard's William H. Pickering announced a prediction, based on a graphical analysis of residuals of Uranus, of a "planet O" at a distance of 51.9 a.u., with a period of 373.5 years, a mass twice the

Harvard University Archives

Lowell Observatory

Thomas Jefferson Jackson See, *right above*, an early planet hunter, at the Lowell Observatory's 24-inch refractor with Wilbur A. Cogshall in 1896. Both were then members of the Lowell staff. *At left*, a rare photo of William Henry Pickering of the Harvard College Observatory.

earth's, and a magnitude of 11.5 or 13.4.[71] This prediction, along with at least six and probably eight others Pickering would make over the next twenty-four years, will be discussed in greater detail in subsequent chapters. Here, it is necessary to note that Pickering's prediction had two important effects.

First, the fact that he had found anything significant in the residuals of Uranus led Gaillot to review his own work on that planet and to now concede, though not very confidently, that two planets might possibly be orbiting the sun beyond Neptune. On the assumption of circular orbits, he placed these at distances of 44 and 66 a.u. from the sun and gave them masses of five and 24 times that of the earth.[72]

But far more importantly, Pickering's O prediction gave strong impetus and eventually new direction to a quiet, but intensive, exploration of trans-Neptunian space that had already been underway for several years in Boston, Massachusetts, and Flagstaff, Arizona, under the active direction of Percival Lowell.

Lowell was one of the most famous astronomers of his day, and certainly the most controversial, primarily because of his much-publicized observations of the so-called "canals" of Mars and his sensational deduction therefrom that the red planet was the abode of intelligent beings. Lowell's provocative theory of the probable existence of sapient martians was discussed and debated throughout the world during his lifetime and, though now discredited, influenced the subsequent development and direction of planetary astronomy, as well as of that sizeable body of modern literature known as science fiction.

But the world, strange to say, knew little of Lowell's long preoccupation with the trans-Neptunian planet problem until March 13, 1930, fourteen years after his death, and the announcement of the discovery of a trans-Neptunian planet at the observatory he had founded at Flagstaff in 1894. Although he had published the results of his tedious analytical labors in 1915, along with two possible orbits for a trans-Neptunian "planet X," this work attracted little attention, even within astronomy, and remained obscure until it was finally revealed that it had been instrumental in the discovery of the ninth planet, Pluto.

Percival Lowell was born on March 13, 1855, in Boston, the eldest of an exceptionally gifted generation of one of New England's oldest and most affluent families. His brother, A. Lawrence Lowell, became a distinguished educator, serving as president of Harvard University from 1909 to 1933. His youngest sister Amy, his junior by twenty years, won renown in the early twentieth century as a cigar-smoking, avant-garde poetess and author. His other two sisters, Katherine and Elizabeth, lived quieter, less public lives, but it may be noted that Elizabeth's son, Roger Lowell Putnam played a key role in mounting the final search for Lowell's Planet X that led to Pluto's discovery.[73]

Lowell was educated at private schools in Boston and in Europe, acquiring fluency in several languages by the age of 12, a sound background in the classics, and what proved to be a lifelong taste for travel. Even as a youth, his brother has recalled, he displayed an interest in astronomy and observed the stars and planets with a 2¼-inch refractor from the roof of his family home in Boston. In 1872 he entered Harvard, studying mathematics under Benjamin Peirce, who had precipitated the "happy accident" controversy over Neptune, taking second year honors in the subject. At his graduation in 1876, he gave a talk on "The Nebular Hypothesis" as developed by the great French mathematical astronomer, Pierre Simon Laplace. He would later, incidentally, find Laplace's hypothesis wanting.[74]

After leaving Harvard, he traveled throughout Europe and to the Middle East, and then returned to Boston and his grandfather's office where, as a man of business for the next seven years, he amassed considerable financial resources of his own.

In 1883, Lowell abruptly embarked on a new career, traveling to the Far East, and specifically Korea, where he lived for a while and served briefly as secretary to a Korean diplomatic mission to the United States. Over the next ten years, Lowell made three other extended visits to this then remote and little-known part of the world, writing no less than four books and a number of articles for *The Atlantic Monthly* about his experiences and impressions of the land and the peoples there. These brought him a reputation both as a knowledgeable Orientalist and as a colorful, authoritative writer.[75]

During his final trip to the Far East in 1893, made to Japan in the company of his friend, George Russell Agassiz, Lowell learned that Giovanni Virginio Schiaparelli, who had discovered the "canals" of Mars in 1877 as a geometric network of faint, fine lines criss-crossing the planet's disk, was losing his eyesight, and he now determined to take up the respected Italian astronomer's studies of the planets. To this end, on his return to Boston late in 1893, he set to work to establish an astronomical observatory, with the declared purpose of studying Mars, and with the help of Harvard astronomers William H. Pickering and Andrew Ellicott Douglass, both experienced Mars observers.

On the basis of atmospheric tests by Douglass, Lowell selected Flagstaff, a pine-forested oasis amid the Arizona desert at an elevation of 7,000 feet, as the site. By June 1, 1894, his new observatory, equipped with borrowed 18-inch and 12-inch refracting telescopes, was ready and observations of Mars began.[76] Pickering, incidentally, remained only a few months before returning to Harvard, and would later become a persistent critic of Lowell's martian ideas as well as a rival of sorts in the search for a trans-Neptunian planet.[77]

Thus Lowell launched still another career, and one which would now occupy his active mind to the exclusion of all other interests for the remaining twenty-two years of his life. Lowell promulgated his sensational conclusions of the probable existence of canal-building martians in 1894–95 in serialized articles in both professional and popular journals, in public lectures, and in his first astronomical book, *Mars*, which appeared late in 1895. Over the ensuing years, he and various assistants, observing ten successive oppositions of

Mars, amassed an unprecedented body of systematic data about the red planet which Lowell used to refine and bolster his logical arguments for the presence of martian intelligence. In hundreds of articles in the popular press, in scores of crowded lecture rooms, and in two other popular books, he repeatedly proclaimed his theory and the "proofs" he had been accumulating for it at his Flagstaff observatory. Newspapers around the world, from Budapest, Hungary, to Sydney, Australia, published lengthy pieces about his work, and chronicled the martian controversy for their readers. For the 1907 opposition of Mars, Lowell dispatched an expedition to South America to observe the planet, headed, incidentally, by planet-hunter David Todd. In 1909, he even claimed that his martians had built two new "canals" on their planet in just the past two years. Only months before his death on November 12, 1916, he was still urging his theory in much the same terms he had used in 1894-95 and contending that it had been fully confirmed by his long observations from Flagstaff.[78]

This massive, sustained effort, however, was in vain. For although a credulous public remained fascinated by his ideas and observations for many years, initial skepticism among astronomers soon turned into sharp criticism of his work and the conclusions he had drawn from it. This criticism steadily mounted in intensity until, by 1909, scientists in other disciplines, from which Lowell had eclectically borrowed concepts to support his now embattled theory, joined the attack, decrying not only his theory, but his methods and even his personal integrity. It is just after this bitter climax to the worldwide furor over his postulation of intelligent martian life that Lowell launched the most intensive phase of his analytical Planet X search.[79]

Mars, of course, was not the only planet to which Lowell turned his attention during these years. Beginning in 1896, when his fine 24-inch refractor was completed by Alvan Clark & Sons, and thereafter over the years, Lowell and his assistants made systematic observations of Mercury, Venus, Jupiter, Saturn and Uranus. Some of this work proved to be almost as controversial as his Mars studies, at least as far as astronomers were concerned, but much of it also was sound.[80]

Lowell's many-faceted astronomical interests have been detailed at considerable length elsewhere and need not be summarized further here.[81]

One other important point, however, should be made. Lowell concerned himself largely with what might be called the "big" problems in astronomy —problems that must inevitably bring prestige and fame to the person who solved them—and his concentration on the enigma of the martian "canals" is an obvious case in point. Thus it is not surprising that the problem of finding a trans-Neptunian planet was among them.

CHAPTER 4

THE FIRST SEARCH FOR PLANET X

Percival Lowell's interest in the problem of a trans-Neptunian planet first surfaces in December 1902, when in a series of lectures as newly appointed nonresident professor of astronomy at the Massachusetts Institute of Technology, he confidently, if quite casually, proclaimed the existence and eventual detection of such a body.

His pronouncement, however, attracted almost no notice, even after his lectures were published the following spring as a small book entitled *The Solar System*. For his references to an unknown planet beyond Neptune were brief and incidental to his purpose, which was a broad review of the state of contemporary cosmology and his own views thereon. He was not, of course, suggesting anything new and in fact he based his prediction on extrapolations from the observed association of the orbits of comets and meteor streams with the orbits of the outer planets, as Flammarion and Forbes had done more than twenty years before.

Lowell touched on the subject only twice—and only in passing during his lectures—first in his discussion of meteor streams. "Each of them is associated with the orbit of a particular planet," he noted, explaining that the "stream's perihelion remains at the sun, but its aphelion becomes its periplaneta. It sweeps around the planet at one end of its path somewhat as it sweeps around the sun at the other."

The Andromedes are thus dependent on Jupiter, the Leonids on Uranus; while the Perseids and Lyrids go out to meet the unknown planet which circles at a distance of about forty-five astronomical units from the sun. It may to you seem strange to speak thus confidently of what no mortal eye has seen, but the finger of the sign-board of phenomena points so clearly as to justify the definite article. The eye of analysis has already suspected the invisible.[1]

And in a later lecture, he made a similarly offhand reference while noting the clustering of the outermost points of the orbits of comets near the orbits of Jupiter, Saturn, Uranus and Neptune. "But the clusters do not stop with Neptune," he pointed out. "Beyond that planet is a gap, and then at 49 and 50 units we find two more aphelia, and then nothing again until we reach 75 units out."

This can hardly be an accident; and if not chance, it means a planet out there as yet unseen by man, but certain sometime to be detected and added to the others. Thus not only are comets a part of our system now recognized, but they act as finger-posts to planets not yet known.[2]

These short, parenthetical passages hardly qualify as a serious attempt at planetary prediction, and surely Lowell did not consider them as such. For despite the confidence of his tone, he apparently did not have the same confidence in the predictions themselves. Seven years later he would record that he had reached his 1902 conclusions as the result of a study of the orbits of all observed comets as cataloged by J. G. Galle, noting that a "like diagnosis and deduction has occurred independently to several astronomers: first I believe in 1879 to Flammarion and since to Forbes and others." But he would then add that while the "presence of these gaps and groupings is very significant ... it cannot be said to be conclusive of the presence of an unknown planet because the groupings might be due to chance—there being so few comets there at all."[3]

Lowell's 1902 predictions nonetheless reveal that he was generally familiar with the work of earlier planet-hunters, and that he accepted their basic premise that one or more discoverable planets were in orbit in the farther reaches of the solar system. This, in turn, reflects his extensive reading in the astronomical literature. It is possible, too, that he had discussed the question of trans-Neptunian planets informally with Flammarion or Todd, both of whom he knew personally. Certainly, he was fully conversant with the circumstances surrounding Adams' and Leverrier's analytical search for Neptune and the controversy that followed its discovery in 1846. For during his years at Harvard, his mathematical mentor had been Benjamin Peirce, a key figure in that controversy, "to whom as my master," he later wrote, "I owe my first knowledge of the facts."[4]

Lowell's search for what he soon came to call Planet X, however, more properly begins early in 1905, and it was carried on, sporadically as his other astronomical interests permitted and with varying degrees of urgency and intensity, over the next eleven years. From the first, it was both observational, through the comparative photography of what he considered to be likely regions of the sky, and theoretical, through the progressively rigorous application of the methods of celestial mechanics.

Actually, Lowell made two searches. In the first of these, from 1905 to mid-1909, the photographic and mathematical phases were pursued more or less independently. The photographic phase involved little more than a systematic survey of the sky along a narrow band around the ecliptic, while mathematically it called for the painstaking gathering of data on the perturbations of Uranus and Neptune to which he eventually applied essentially empirical techniques of analysis. In the second search from 1910 to 1915, in contrast, the mathematical and photographic phases were closely correlated, the "eye of analysis" now being used to guide the telescope. And in this second search, Lowell rejected all empirical attempts at a solution as useless, applying the full power of the analytical methods of Adams and Leverrier and carrying his calculations far beyond the point to which these two geometers had ventured or, indeed, had been obliged to go.[5]

Lowell expended large amounts of time and money on the search. In all, more than a dozen assistants participated at one time or another, working either at the observatory he had established in 1894 at Flagstaff in the Arizona Territory, or in his Boston office. Initially, he employed only a single, part-time computer, recruited from the Naval Observatory's Nautical Almanac Office in Washington, D.C. In the final stages, he kept a staff of five young mathematicians busy in Boston computing possible positions for his Planet X, while up to three observers pursued the hunt photographically at Flagstaff. Lowell himself devoted long hours for extended periods to the mathematical investigation and, indeed, worked so intensively through the summer and fall of 1912 that he collapsed from nervous exhaustion and was forced temporarily to direct the search from the sickbed of his Boston home.

One of the most unusual aspects of this unprecedented effort is the fact that in sharp contrast to most of Lowell's other well-publicized astronomical activities, it was conducted quietly and, initially at least, in virtual secrecy. Only a few of his closest associates and most trusted friends were aware of what he was up to and apparently they were under some interdiction not to reveal his purpose. In the earlier years, indeed, Lowell and his assistants studiously avoided specific mention of this purpose even in their mutual correspondence, referring to the search only vaguely as "the invariable plane work." When Lowell allowed himself a rare exception to this rule, he did so only to warn against inadvertent disclosure of "the trans-Neptunian exploration."[6]

Later, after he realized that the extreme difficulty of the problem alone was enough to discourage opportunistic competition, the search was carried on with somewhat greater candor among the searchers themselves. But the fact that such a search had even been undertaken did not become public knowledge until January 13, 1915, when he presented the theoretical results of his work to the American Academy of Arts and Sciences, of which he was a Fellow. His rigorous computations and carefully qualified conclusions then, however, caused no more stir than had his facile forecasts at M.I.T. in 1902. A final frustration, perhaps, came when the Academy balked at publishing his lengthy mathematical analysis of the problem and Lowell wound up paying for the publication of his now-famous "Memoir of a Trans-Neptunian Planet."

The supreme irony of this massive search for a new planet is the fact that it was successful even in Lowell's lifetime, although he died without knowing this; for in the spring of 1915, as he readied his Planet X memoir for the printer, faint images of a new solar system body beyond Neptune were recorded on two of the X search plates exposed at Flagstaff. The images were not found at the time, and it was not until fifteen years later—after Lowell Observatory's resumption of the X search in 1929 had led to the discovery of Pluto in 1930—that they were recognized for what they were.[7] As he was at his observatory when these plates were made, Lowell may have handled them himself and even scanned them, although perhaps cursorily, for a glimpse of his elusive planet X blinking balefully back and forth from among the unblinking fixed stars. That Lowell failed to find X, his brother and biographer A. Lawrence Lowell would later write, "was the sharpest disappointment of his life."[8] Lowell's lifelong friend George Russell Agassiz felt this failure "virtually killed him."[9]

Lowell launched his search for a trans-Neptunian planet at a particularly busy time in his astronomical career. In May 1905, Mars would reach its first really favorable opposition in eleven years, and Lowell was hoping that this relatively near approach to the earth would provide new and irrefutable proofs of the reality of the so-called "canals," and of the planet's general habitability, thus confirming the key points in his controversial theory of the probable existence of intelligent beings on Mars. This theory, first promulgated in 1894, had been drawing increasingly strident criticisms from a growing number of astronomers who, unlike a more credulous public, refused to be persuaded by Lowell's provocative logic that mankind might have sapient neighbors in space[10]

To reinforce this embattled theory then, Lowell was planning comprehensive observations of Mars before, during and after the opposition. He himself intended to observe the planet visually for what he, as well as most other astronomers, believed were "seasonal" changes on the planet, while his two assistants at Flagstaff, Vesto Melvin Slipher and Carl Otto Lampland, were preparing spectrographic and photographic studies respectively which Lowell hoped would first, reveal the spectral lines of life-essential water vapor and oxygen in the martian atmosphere; and second, confirm with the camera his widely disputed descriptions of a vast and geometrical network of "canals" on Mars. Slipher also was to begin a spectrographic study of the dark markings on Mars to determine whether chlorophyll, the ubiquitous green stuff in plants which carries on the vital process of photosynthesis, might be present there.[11] "We will get out something to make them sit up!" Lowell vowed of this ambitious observational campaign.[12]

Despite these preoccupations, however, Lowell early in February 1905 began to lay the foundation for a photographic search for a trans-Neptunian planet. His first step was to contact E. C. Pickering, director of Harvard College Observatory, and Flammarion in France, to request photographs and astrographic charts of the star fields along the invariable plane—the mean of the orbital planes of the known planets, and thus the most reasonable place to start looking for some unknown body that might also be orbiting the sun.[13] Such photographs and charts were necessary for comparison purposes, at least in the initial stages of the search and until adequate survey plates of the pertinent areas of the sky could be made at his own observatory.

Lowell also had to find a competent observer to carry on the search, for he planned to keep his regular assistants far too busy on Mars over the next few years to take on the added task of systematically photographing a sizeable segment of the heavens. Thus the Lawrence Fellowship program of the Lowell Observatory was founded in April 1905, in cooperation with Indiana University, ostensibly to give selected Indiana astronomy graduates a year's observing experience and a chance to pursue their thesis research for a modest stipend of $50 a month. In point of fact, the experience gained by the three young men successively appointed as Lawrence Fellows between 1905 and 1907 was almost entirely related to "the invariable plane work" and when that work ended in September 1907, the fellowship program ended with it.[14]

Parenthetically, Lowell's choice of Indiana's astronomy department as beneficiary of this short-lived fellowship program undoubtedly was influenced by Slipher and Lampland, both Hoosier graduates who had maintained close contact with their alma mater and with their astronomy professors there—John A. Miller and Wilbur A. Cogshall, who had been a Lowell assistant himself in 1896–97. Indeed, it was not Lowell but Slipher who negotiated with Miller about establishing the fellowship.[15] Miller, who later became director of Sproul Observatory at Swarthmore College, and Cogshall both subsequently contributed to the long Planet X search and to its culmination in the discovery of Pluto.

The three Lawrence Fellows who carried out the first photographic search for Planet X were John C. Duncan, 1905–06, later a professor of astronomy at Harvard and at Wellesley College who distinguished himself in stellar astronomy and as a textbook author; Earl C. Slipher, 1906–07, V. M. Slipher's younger brother who joined the Lowell staff as a permanent assistant in 1907 and later became a leading expert on Mars and planetary photography; and Kenneth P. Williams, 1907, who returned to Indiana when the fellowship program ended to teach celestial mechanics and eventually to become an authority on Civil War history as well.

The first phase of the photographic "trans-Neptunian exploration" was one of trial and error as Lowell's assistants experimented to find the right

Lowell Observatory

The Lowell Observatory staff in 1905 outside the 24-inch telescope dome. From left: Dome porter Harry Hussey, Vesto Melvin Slipher, John Charles Duncan, Percival Lowell, Lowell's secretary Wrexie Louise Leonard, and Carl Otto Lampland.

Lowell Observatory Indiana University

Two of the early searchers for Percival Lowell's Planet X in their later years: Earl Carl Slipher, *left,* and Kenneth P. Williams, *right.*

lenses, exposure times and observing techniques. Duncan arrived in Flagstaff early in July and on August 3 began making trial plates to determine exposures needed to reach thirteenth magnitude stars, using the 24-inch refractor with a camera rigged up by Lampland from equipment on hand. On August 22 he began photographing along the invariable plane, and it quickly became apparent that the small field and slow speed of the instrument were not suited to the problem. Consequently in September, Slipher obtained a 5-inch Voigtlander photographic lens on approval, and on September 23 Duncan began testing it, finding that while Lampland's makeshift apparatus on the 24-inch gave fields only about a degree in width, the Voigtlander yielded a field some 15° wide, with sharp definition over neary 10°.[16]

Duncan's initial invariable plane plates were shipped to Boston for Lowell's examination. But Lowell, in England to triumphantly proclaim Lampland's success in photographing the martian "canals," did not inspect them until October and then he found them wanting.

"The star images are so far from perfect in the Voigtlander," he wrote to Slipher on Oct. 17, "that I am really unable at present to decide whether it would ever be made a useful tool or not."[17] However, the following day, he advised Slipher to continue working with the Voigtlander temporarily at least "as it strikes me that it might be useful to have the whole region I want gone

Lowell Observatory

The 5-inch Brashear camera used by Lowell Observatory in the first search for Percival Lowell's postulated trans-Neptunian Planet X.

over with it twice, provided the proper star images can be got.... On further consideration let Mr. Duncan devote himself to the perfection of the Vorgtlander [sic] plates, giving me the stellar magnitude on one of the plates and a scale in a corner and at the edges the trace of the invariable plane."[18] But two weeks later he decided against the Voigtlander, writing Slipher: "I think it would be better to get the best first-hand lens of the sort that you can instead of taking this Voigtlander one."[19] And in another week, he sent instructions for Duncan "to take at the last possible minute another set of Voigtlander plates covering the whole region with it a second time."[20] This Duncan did, starting in mid-November. By December, he was testing a Roettger 6⅜-inch lens.[21]

These initial photographic probings of selected areas along the invariable plane did not, of course, produce a new planet, but they were not wholly without incident. On December 2, Slipher found an image "which I take to be that of a comet" on one of Duncan's Voigtlander plates exposed November 29 and after an unsuccessful search for the object that same night, he sent the plate to Lowell a few days later.[22] Lowell promptly reported the discovery of a "comet" to Harvard College Observatory which telegraphed the report to other observatories. Then on December 29, a second Harvard telegram advised astronomers that Lowell himself had discovered a second comet on the same plate and one which, moreover, "had two tails." As the delay in reporting these discoveries, if such they were, precluded their confirmation by other astronomers, astronomer Robert G. Aitken at California's Lick Observatory, where much of the criticism of Lowell's Mars work was centered, was moved to satiric comment about them in print. "The discovery of a comet by photography is unusual enough to be noteworthy," he wrote, "but to find two on a single plate is a unique achievement.... It might be suggested that photographic defects often look wonderfully like comet tails; but it is of course assumed that Professor Lowell took precautions to guard against such deception before announcing the discoveries."[23] Later the elder Slipher carefully explained the delay in reporting the "comets" to Aitken,[24] and Lowell privately expressed "considerable doubt about No. 2," as his longtime secretary Miss Wrexie Louise Leonard noted in answering an inquiry about the discoveries.[25] No other astronomers seemed to have sighted the purported comets.

There were other problems in the invariable plane work. In December, Lowell complained to Slipher that Duncan should "give me the centering of his plates a little more carefully so that I can compare one with the other which I find it now very difficult to do."[26] A few days later, he instructed Duncan to extend his exposures from 60 to 90 minutes to "bring out the fainter stars and any trails that may exist."[27] The Roettger lens, like the Voigtlander, proved to be less than satisfactory, too, and thus in January Lowell informed Slipher that he had ordered a new 5-inch photographic doublet lens from John A. Brashear of Pittsburgh's Allegheny Observatory, who had made Lowell's fine spectrograph in 1901.[28] The lens duly arrived in Flagstaff and Duncan began making test plates of the Neptune region on January 23.[29] These, sent to Boston, were "just what I want," Lowell enthused to Slipher. "I am awaiting with impatience more of his work with the new lens."[30]

But when more plates were forthcoming, his enthusiasm cooled somewhat. "The field of definition of the Brashear lens does not seem to be as large as the original Voigtlander although the lens itself seems to be more powerful," he noted in a letter to Duncan.[31] Slipher also felt that "the size of the well-defined field is less than I thought such a lens should give" and sought the advice of Edward Emerson Barnard, an authority on astronomical photography, at Yerkes Observatory.[32] Lowell now briefly considered getting a still larger lens and Slipher advised him that "it depends on what work you have in mind of the nature of the Invariable plane work whether it would be wise or necessary to have a larger lens of this kind made."[33] These doubts vanished late in April, however, when Lampland sent Lowell an enlarged print of a three-hour exposure with the Brashear that recorded stars down to the sixteenth magnitude, outdistancing by two magnitudes the *Carte Photographique du Ciel,* one of the great sky charts of the day.[34] Lowell, enthused again, promptly dispatched a $350 check to Brashear as payment in full for the lens.[35]

Duncan continued to make plates, now using three-hour exposures and duplicating them at about two-week intervals, but nothing unusual appeared on them except, as Lowell noted dryly early in May 1906, "a great many specks. Possibly they needed to be more carefully brushed before being used," he suggested.[36] On May 30–31, an asteroid was recorded and Lowell forwarded its position to Max Wolf and Anton Berberich in Germany, who numbered observations of the minor planets among their special interests, explaining to Wolf that "I am not on the hunt for these bodies; they are merely bye-products [sic]. ..." He did not say of what, however.[37]

Duncan's year as a Lawrence Fellow was up in June and he exposed his last invariable plane plate—No. 209—on June 20. Lampland, who had joined the search in mid-May, kept making plates until Slipher's brother Earl, whose appointment had been suggested by Slipher in March but not approved by Lowell until the end of June, was ready to take over.[38] Lowell was eager to press on with the invariable plane photography and, from Europe where he was spending the summer, urged Slipher to set his brother to the task as soon as he arrived at the observatory.[39] "He came last week," Slipher replied in mid-July, "and will carry on the Invariable plane work as rapidly as possible. The plates will be sent to you in Boston without examination, and I hope when you return you may find their examination better than you anticipate."[40] Young Slipher made his first plate—No. 228—July 28, 1906.[41]

On his return in September, however, Lowell plunged into a busy regimen of other work. Planning for the highly favorable opposition of Mars due in July 1907 now held top priority, for as the evanescent "canals" in Lampland's 1905 photographs had failed to convince astronomers in general of their reality, he was determined to get new and better photographs of Mars that would confound his critics once and for all. To this end, he now ordered that the objective of the 24-inch refractor at Flagstaff be brought back to Boston for refiguring by the well-known telescope-making firm of Alvan Clark and Sons, which had made the instrument for him in 1896.[42]

He also now set to work preparing what proved to be a highly popular series of lectures on Mars and planetary evolution for the Lowell Institute which he delivered to capacity audiences in Boston's Huntington Hall late in October and which were later serialized in *Century Magazine* and then published as a book, *Mars as the Abode of Life,* in 1908. Throughout the autumn, he was also occupied with the publication of his magnum opus on Mars, *Mars and Its Canals,* which appeared in late November. There were other matters, too. As will be seen, computations of Uranian residuals for the mathematical phase of his trans-Neptunian planet search were awaiting his attention at this time. Thus it was not until late December that he got around to examining young Slipher's plates.

Then he found them lacking in quantity if not quality. On New Year's Eve he telegraphed Flagstaff asking for more plates, and followed this up with a letter to the elder Slipher in which he praised the younger Slipher's work but complained: "I have received no plates of the invariable plane region since those dated Oct. 26, except one dated Nov. 14. I am at a loss to know what this means...."[43] And two days later he asked: "Can you tell me, too, the reason why there were such gaps? I have no plates between 298 and 306 and am lacking also 309 and 325 and 327—are these poor plates or what?"[44]

Slipher replied immediately that several dozen plates had been sent to Boston a few days before. "I suppose these have reached you now," he wrote, explaining: that his brother had been "keeping the first plate until he [Slipher's brother] can get the duplicate since it simplifies the work of orienting and marking the plates if both can be done at once."[45] And in a subsequent letter, he outlined another problem that his brother had encountered. "Heretofore," he wrote, "he has been trying to do the invariable plane as it went along, without gaps, and, as nearly as could conveniently be done he took the regions when opposite the sun. But of course with the Moon and unfavorable weather interfering, one's chances of doing a region at opposition are very seriously curtailed. The weather for the Moonless part of December was good and my brother accomplished a good deal.... But with three hour exposures plates do not multiply rapidly even in good weather. How nice it would be for him and for that work if the clouds and the Moon could keep close company the one with the other!"[46]

Lowell now also made an important change in the observing plan. Rather than pairs of plates made at about two-week intervals, he advised the elder Slipher: "I would like now plates taken exactly opposite the sun ... and repeated three days apart in triplicate."[47] And in another letter, he explained that "as the effective field of each [plate] is at least four degrees across and the sun moves a degree a day, there should be time to sweep the whole Invariable Plane region in this way."[48] But neither the clouds nor the moon cooperated in this rigid schedule,[49] and within a few weeks he was forced to modify his instructions. "The plates should be taken opposite the sun so nearly as possible, within five minutes R.A. [right ascension]," he now directed. "Duplicate the region in three or four days. Triplicate it only when you have time and then a trifle eastward say ten minutes of R.A. and within about a week of the first. This third is not so important and may be omitted when it is too much work."[50]

Young Slipher hardly settled into this new observing pattern, however, before Lowell's martian priorities again intruded on the search. Early in February 1907, Lowell decided to sponsor an expedition to the west coast of South America to observe the favorable July opposition of Mars, writing to Lampland that "if young Mr. Slipher would like to go I propose that you should instruct him as minutely as possible in the photography of Mars."[51] The expedition was to be directed by one-time planet-hunter David P. Todd of Amherst College, who agreed to undertake "every cost ... for the sum of $3,500," and to be equipped with Amherst's 18-inch refractor.[52] Young Mr. Slipher wanted to go, as it turned out, and consequently late in March Lowell wrote Cogshall at Indiana to ask if Williams, then an astronomy student there, would "substitute" as Lawrence Fellow while young Slipher was away.[53] Slipher exposed his last invariable plane plate—No. 386—on April 21 and Williams arrived in Flagstaff a few days later to begin his brief service in the trans-Neptunian exploration on May 1.[54]

Young Slipher's abbreviated labors on "the invariable plane work" added more than 150 plates to Lowell's accumulating store. Moreover, he had turned up "asteroids *en masse*," as Lowell noted in April.[55] Indeed, Lowell found the trails of these tiny bodies on no less than eight pairs of Slipher's plates exposed in a single three-week period. Again he sent positions to Wolf and Berberich.[56]

Lowell's first photographic trans-Neptunian exploration was now coming to a close, however. His preoccupation with the work of his South American expedition, with his own and Lampland's martian observations at Flagstaff in the summer of 1907, and with the publication of the results, left him little time to concern himself with problematical planets. From May 1 to September 6, Williams made only fifty-four plates. With his September 6 plate—No. 440—the "invariable plane work" ended.[57]

It is clear in retrospect that the reasons for its failure were that, granting a lack of established precedent, both the methods and the instruments Lowell and his assistants used were crude and inadequate to the exacting demands of such a search. The Brashear lens, which bore the brunt of the work, yielded a sharply defined field of only about 5°.[58] Yet in 1905–07, Pluto's highly inclined orbit placed that planet far from the ecliptic and, as its eventual discoverer Clyde W. Tombaugh has since pointed out, "outside the belt covered by the plates with the 5-inch camera." Pluto then was also only a sixteenth magnitude object, as Tombaugh also noted, and thus was at the very limit of the search plates.[59] In addition, many of the earlier plates were not taken at opposition.

Moreover, Lowell's method of examining and comparing plates was at best cumbersome and at worst unsystematic and imprecise. For he simply superimposed, star for star, pairs of plates of particular regions of the sky and then scanned them painstakingly with a hand magnifying glass for evidence of some previously unobserved object. When Lampland in March 1906 suggested that he use a Zeiss stereo-comparator for this tedious work and sent him a prospectus for the instrument, Lowell wrote: "I question whether the advantage of it is worth the price but I shall look into the matter."[60] This he did apparently during his trip to Europe that summer but he found his experience with the comparator "not reassuring."[61]

Yet while this initial photographic search failed to find a new planet, it provided some sound lessons on how to conduct such a search, lessons that would prove valuable in the future—to his assistants in particular. It also showed the practical futility of searching at random through trans-Neptunian space for a body whose presence was merely presumed and about which, if indeed it did exist, nothing whatever was known. What was needed, Lowell was well aware, was some indication of where to look, however general in nature, some "finger-post" that would point out the more probable places where a hitherto unseen planet might be hiding among the millions of stars in the sky. Thus in March 1905, he launched a parallel mathematical investigation which he hoped would eventually narrow the field of the search, and thus the odds against the searcher.

For such a mathematical search there were, of course, precedents of method, ranging from the empirical approaches used by Todd, Flammarion, Forbes and others to the rigorous analytical methods developed by Lagrange, Laplace and Gauss in the late eighteenth and early nineteenth centuries and later applied with dramatic success by Adams and Leverrier to the problem of a trans-Uranian planet. But the best of methods are no better than the data to which they are applied, and Lowell, well-trained in the intellectual discipline of mathematics at Harvard, recognized that a consistent, coherent body of data for the trans-Neptunian problem was lacking, and that much time and labor would be necessary to acquire such data.

Consequently his first move early in March 1905 was to write Walter S. Harshman, director of the Naval Observatory's Nautical Almanac Office in Washington, D.C., asking him to recommend "a capable computer" to systematically reduce the residuals in longitude of both Uranus and Neptune. Harshman replied that he had given Lowell's letter to William T. Carrigan, "one of our most efficient men" who, he added, "is willing to undertake this work outside of office hours."[62] Carrigan immediately wrote Lowell that he would "do the work you require with such promptness as the time at my disposal outside of office hours will permit, the compensation to be arranged for when I have been made acquainted with the exact character of the work. If these conditions meet with your approval, I am ready to begin as soon as I shall have received the necessary instructions."[63]

Carrigan certainly is a forgotten man in Lowell's Planet X search. His name rarely appears in the roster of early searchers in writings about Lowell's X work published after the discovery of Pluto, a notable exception occurring in A. Lawrence Lowell's *Biography of Percival Lowell* (1935) in which, when he is mentioned in passing, his name is misspelled as "Garrigan" and his entry into the search is put in 1908.[64] Yet he labored for Lowell, reducing the residuals of the outer planets and mathematically attempting to reconcile conflicting planetary theories, for more than four years, only to be told that Lowell had arrived at a theoretical solution without using any of his calculations, and that thus his services were no longer required. Carrigan was, in fact, capable enough as a computer to have published two articles on problems in celestial mechanics in the *Astronomical Journal* just prior to becoming Lowell's part-time employee.[65]

Lowell set Carrigan to work on the exhaustive task of computing the residuals of Uranus and Neptune for a forty-year period from 1780 to 1820. Carrigan soon reported that he had found 151 observations of Uranus made at Greenwich and 171 made at the Paris Observatory during the period. "Of Neptune, however," he noted, "there are but two known observations made before its discovery in 1846. These two observations were made in 1794 [sic] by Lalande who did not suspect the connection of the body observed with the solar system. It was not observed again until Aug. 4, 1846. Please advise further regarding this planet." As to Uranus, Carrigan also inquired "with what degree of rigor you desire to have the residuals computed. I ask this," he added, "in view of the necessity of re-reducing the observations if the utmost accuracy is desired."[66]

Lowell's reply is not contained in the Lowell Observatory archives,* but presumably he directed Carrigan to proceed rigorously with the Uranian observations, for this is what Carrigan did. The work was both laborious and time-consuming and went very slowly indeed. But Lowell, who recognized the problems involved, was not at first impatient for results. He had many other things to do and for the next two years, at least, he allowed Carrigan to carry out his tedious computations at an almost leisurely pace.

Carrigan sent Lowell his first computations in August 1905, but he accomplished nothing more during the following eight months, as Lowell complained mildly early in March 1906. "In your note of Aug. 23 last," he wrote, "you spoke of bending your energies toward completing the computations you were doing for me before the first of the year. It is now well into March and as I have heard nothing from you I beg to call your attention to the matter, and might I ask you to reply by return of post, if you kindly will, letting me know if the work is going on and when I may expect results. I was very well pleased as you know with the first part."[67]

Carrigan apparently explained to Lowell that he had been having trouble with his eyes, for a few days later Lowell advised him: "I beg to say I am

---

*The Lowell-Carrigan correspondence in the archives is quite one-sided on Lowell's side. Only a few of Carrigan's letters survive, while there are scores of Lowell's letters to Carrigan in the files and letter-books. But at least some of the problems Carrigan encountered emerge from Lowell's responses to Carrigan's missing letters.

extremely sorry about your eyes and in consequence I do not want to hurry you in the least with my work. I will therefore ask you to send it from time to time as you suggest, drawing attention if you kindly will by writing in red the discrepancies in the longitudes of both Uranus and Neptune.''[68]

Over the next six months, Carrigan's computations, along with invariable plane plates, arrived sporadically at Lowell's Boston office. But Lowell, in Europe for nearly three months in the summer and busy with other matters on his return, put them aside for later perusal. Late in October 1906, he apologized to Carrigan for not acknowledging his "last batch of computations," explaining: "my only excuse is the press of work."[69] A week later, he sent Carrigan a check for $228.75 for his work, assuring him: "I shall take pleasure in looking over your computations when I get a moment's time—up to now I have been extremely busy.''[70]

As with Slipher's invariable plane plates, however, Lowell did not find time to check over Carrigan's work for another two months and then he was disappointed at the slow accumulation of results. "I have been looking over your computations," he wrote Carrigan on January 23, 1907, "and find that I have the residuals for the years 1801 to 1810 inclusive and 1814 to 1818 inclusive. Is this all up to date?"[71] A week later he asked: "Would it not be possible to get a general preliminary view of the residuals through the centuries by taking up the data in a less exhaustive manner—the work of the last twenty years especially? I am anxious to have such a preliminary survey as soon as possible.''[72]

Carrigan seems to have construed these queries as criticisms of his work, protesting further that Lowell had asked only for residuals for the period from 1780 to 1820. On February 14, Lowell wrote: "You are quite right that I asked only for forty years. But I was now desirous of finding if the later years showed residuals of moment. As for your method of doing it I have nothing to criticize and for the present these latter would take too long. I am curious to see how your residuals for the end of the eighteenth century will come out after your revision. I have, of course, my own theory in the matter.''[73]

Carrigan continued his plodding labors, but Lowell, as already noted, now put aside his Planet X work in favor of observations of Mars during its favorable opposition in 1907, spending the next eight months at his observatory in Flagstaff. It would be another year after his return to Boston in November before he would directly concern himself with the search again, and then only after he learned that another planet-hunter had entered the field. Carrigan and his computations appear on only two occasions in the archival record during this long hiatus. In March 1908, Carrigan finally completed his work on the Uranian residuals from 1780 to 1820 and Lowell, acknowledging his computations, now instructed him to extend them from 1820 up through to the end of the nineteenth century, suggesting also that he pursue a less rigorous course in the work. "I should not take every observation but a few of the best ones only for each [year]," he advised, repeating the suggestion for emphasis in a second letter a week later.[74] And in May, 1908, noting that

Neptune's residuals derived from various theories of the planet differed in their inclusion of long period perturbations, Lowell wrote Carrigan: "What I would like are the longitudes or rather the differences between the longitudes from the sets of elements when all the perturbations are included in each."[75]

Through the winter and spring of 1908, Lowell busied himself largely with publishing the results of his 1907 observations of Mars and Saturn; with publicizing V. M. Slipher's disputed spectrographic determination in January 1908 of the presence of water vapor and oxygen in the martian atmosphere; and with preparing his 1906 Lowell Institute lectures for publication as a book, *Mars as the Abode of Life*. In April he completed negotiations with Alvan Clark and Sons for the construction of a 40-inch reflector for his observatory.[76] And in June, temporarily eschewing astronomy altogether, he abruptly ended his long bachelorhood by marrying, at age 53, his longtime Boston neighbor, Constance Savage Keith, sailing for Europe late that month for a summer-long honeymoon. On his return early in October, he made no move to reopen his theoretical trans-Neptunian exploration. Clearly there was nothing urgent about the X search.

But suddenly, in mid-November, he plunged into the investigation again with furious energy. For he now discovered to his dismay that another explorer had been searching mathematically for an unknown planet beyond Neptune and, indeed, was already well ahead of him in the search. Moreover, this new rival was an old rival—Harvard astronomer William Henry Pickering who had helped Lowell establish his Flagstaff observatory in 1894, had joined in his Mars observations there that year, but had since become a persistent critic of many of Lowell's martian ideas, and most notably his interpretation of the so-called martian "canals."

On November 11, Lowell attended a meeting of the American Academy of Arts and Sciences in Boston at which Pickering postulated the existence and probable location of a trans-Neptunian body which he called "planet O," the first of at least seven hypothetical planets he would propose over the next twenty-four years.[77] He based his prediction on a graphical analysis of the residuals of Uranus, using a method devised by Sir John Herschel in 1849 in his *Outlines of Astronomy* to refine the post-discovery orbital elements of Neptune—a method first applied to the problem of a trans-Neptunian planet by Todd some thirty years later.

Basic to this method is the assumption that the only perturbations of consequence in Uranus' motion resulting from the influence of an unknown exterior planet would occur just prior to their conjunction. Essentially Pickering had simply plotted the residuals of Uranus, taken from Leverrier's 1873 theory of the planet, on graphs, and from the peaks of his curves had estimated the apparent interval between its conjunctions with the unknown disturbing body. From this, he calculated the mean annual motion of the disturber and thus derived some of its orbital elements. Assuming a circular orbit, he concluded that Planet O's mean annual motion was 0°964 and that its mean distance was 51.9 astronomical units from the sun, its period 373.5

years, and its mass twice that of the earth. He also estimated that its angular diameter, assuming Neptune's density, would be $0''.8$ and that its magnitude would be either 11.5 or 13.4 based on the albedos of Neptune and Mars respectively.[78]

Pickering described his method as "a simple graphical process," and pointed out that in its application "no attempt has been made to present an elegant mathematical demonstration in the technical sense of the term."[79] Years later and after many more planetary predictions, he would also concede that his approach was "of more or less 'rule of thumb' character" but would argue that it avoided "many of the complications, and much of the extended labor of the analytical method," which it certainly did.[80]

Pickering apparently had arrived at his Planet O prediction early in 1908 but had delayed announcing it until after unsuccessful photographic searches for O had been made first at Harvard's Boyden Station at Arequipa, Peru, with the 24-inch Bruce doublet there, and then by the Rev. Joel Metcalf at Taunton, Massachusetts, with a 12-inch doublet.[81] After making his announcement, Pickering dropped Lowell a note inviting him to make a search, but Lowell brusquely turned him down. "I am looking up the whole subject myself and have not yet got far enough along to undertake any visual search," Lowell replied tersely. "When I get a position I will let you know."[82]

Pickering's prediction had, in fact, galvanized Lowell into renewed activity, and he now began bombarding a somewhat bewildered Carrigan in Washington with almost daily queries and requests for new data and more residuals, explaining: "I was led to take up the search for X again because of a very striking paper ... by W. H. Pickering the other day before the Amer. Academy. The method of course is as old as Sir John Herschel but I had not supposed the residuals so salient as Pickering found."[83]

Lowell had now resolved to apply the graphical method himself. He went to work immediately on the residuals already at hand, but quickly ran into difficulties. On November 13, he wrote Carrigan asking about inconsistencies in sets of residuals from various theories of Uranus, notably those of Bouvard in 1821, Leverrier in 1873, and the American astronomer Simon Newcomb in 1900. "[U]p to and including 1876 the geocentric residuals in the Greenwich publications are evidently with Bouvard's tables; from 1876 to 1903 inc. with Leverrier's; and from 1904 on with Newcomb's," he noted. "Can these be reduced approximately to either Newcomb's final or Leverrier's theory by any simple calculation or from data you possess? The geocentric places are all I need, but I would like them on one or all of these theories, say Leverrier's and Newcomb's final—from 1781 to 1904."[84]

A few days later, he asked Carrigan if there were "any tables of residuals geocentric or heliocentric of the planets except those in the Greenwich publications from 1836 to the present time? In these," he complained, "it is unfortunate for my purpose that the theories never continue long in one stay. ... Now what I want are the sets of residuals from one and the same theory for

Uranus.... Is it not possible to let me have these with sufficient accuracy and without a great deal of calculation? The things such sets bring out with regard to perturbation, as Jupiter's and Saturn's show, is [sic] very remarkable. I shall greatly appreciate what you can do for me in these matters," he added, "and I know you yourself will be interested when you learn the results."[85]

Lowell was now very impatient indeed for consistent residuals and sought to devise a shortcut for obtaining them as quickly as possible. On November 18, he acknowledged Carrigan's comparisons with Bouvard's tables but added: "What I especially want however are Newcomb's and Leverrier's so that Neptune's perturbations are taken into account.... By calculating Leverrier's for a given year and comparing with Newcomb's and then doing this at intervals of say ten years a sufficiently accurate correction from one to the other can I think be made. Send me first if you will transcriptions of all Newcomb's and Leverrier's residuals with observations on Uranus with Neptune's influence taken into account, and then by the above calculation bridge what gaps there are in both series for the century, especially first 1830–1890."[86]

But the very next day he advised Carrigan that "after sending my letter I realized that the computations for intercomparison between Leverrier's observations and Newcomb's observations would have to be made nearer together, say every three years or less possibly to make the interpolation accurate enough. Now the quickest and good enough way would be I think to take the mean dates of Newcomb's residuals for a given year, calculate what Leverrier's would be for that date and apply the correction, making such direct computations as far apart as the interpolation will bridge. It ought to be possible in the same manner to compare either with Bouvard's observations. ... Will you kindly reduce these to one," he asked, adding: "I would like all this just as soon as possible."[87]

In this same letter, incidentally, Lowell described one of the "very remarkable" results that his graphs had revealed. By plotting the 1836–1877 residuals for Jupiter from Bouvard's tables, he explained, he had derived a curve with three distinct peaks, each corresponding to a conjunction of Jupiter and Neptune. As Neptune was "as yet undiscovered as far as the tables were concerned," he pointed out, "Neptune could have been found apparently from Jupiter's residuals."[88] Pickering had, in fact, already covered much the same ground and had drawn similar conclusions.[89]

On November 21 and 24, Lowell again sent requests for more data, and on November 28, he asked Carrigan for residuals for Uranus and Saturn from Leverrier's theory, beginning at 1876 and going back "as fast and as far as possible. I am greatly in need of these," he explained, "because I am prevented from making any definite conclusions for my curves do not extend far enough. I do not want so elaborate a series as you made for me up to 1820," he cautioned his computer. "One computed place about opposition time a year compared with the mean of the residuals for that year will be quite enough."[90]

Through December and into January 1909, Lowell kept a steady stream of correspondence flowing to Washington, asking questions, giving direction and, monotonously, requesting more residuals. These letters also reveal that he was still learning some fundamentals. On December 1, for example, he complained to Carrigan about discrepancies "which I cannot account for" in residuals obtained from data in the Greenwich and Paris publications.[91] But on December 11, he found that the difficulty "turns out to be [the use of] the mean equinox of date in the British and the true equinox in the French." And he now belatedly also discovered that "all of the Uranus residuals" were contained in the *Annales* of the Paris Observatory, and thus advised Carrigan that he "need not go on with that computation."[92]

At this same time, parenthetically, still another planet-hunter was heard from. For in December, Forbes published a refined set of elements for the inner planet of the pair he had first proposed in 1880, suggesting also that this planet had broken up the comet of 1556 into parts that were observed in 1843, 1880 and 1882. Forbes now assigned this planet a distance of 105.1 astronomical units from the sun, a period of 1,076 years and a high orbital inclination of 52°.[93]

By late December, Lowell was optimistic about his search, informing Carrigan, who was now working on the residuals of Saturn, that "the results so far are both interesting and promising."[94] Early in January 1909 he wrote: "I am delighted with your interest in the investigation and only wish you were nearer so that I might show you the work I have already done."[95] Two weeks later, the wish became an open invitation: "I wish you might get a vacation and come on here for a week," he urged Carrigan. "There are points in the investigation which will interest you. See if you can manage it and I will see to the expense."[96]

By this time, in fact, Lowell had already reached a preliminary conclusion from his graphs and had drafted a brief paper concerning it. In his draft, he deduced from his curves a "true" synodic period for Uranus and X of 113 years, corresponding to a distance of 47.5 a.u., and noted that this distance "agreed strikingly with the evidence from comet aphelia."[97] This paper was not published, however.

Carrigan accepted Lowell's invitation, visiting Boston early in February and getting a guided tour of Harvard College Observatory from its director, E. C. Pickering, William's elder brother, during his stay.[98] But what else he did there, or what specifically he and Lowell discussed is not in the available record, although it would appear that Lowell urged Carrigan to adopt a quicker, less rigorous approach to his computations. Their meeting, certainly, was friendly, and when Carrigan returned to Washington he resumed his calculation of residuals. But his role in the Planet X search was now coming to an end. Ironically, his services were terminated, in part at least, because his work was simply too meticulous.

Late in March, Carrigan forwarded computations of Uranus' residuals from 1876 to 1880 to Boston, Lowell acknowledged them with thanks.[99] But on April 27, he sent another batch of residuals which were, Lowell informed him coldly, "not what I asked for."

> I said, you will remember, that half a dozen observations for any year were enough and I expected to have these grouped in such a manner that but a single reduction to hel. coors. [heliocentric coordinates] would be necessary, taking the mean date and comparing the result with theory. I wished to avoid the time, the trouble and the cost of the method you had before employed. I dwelt on this point to you.
> The R.A. and dec. [declination] for a few (not more than six) observations near opposition can easily be reduced to a mean in very little time and then reduced to hel. coor. Please do this, making one reduction for each year—charging me accordingly and hurrying up the work.[100]

Carrigan was "very much surprised" by the chill tone of this letter and concerned over Lowell's failure to enclose a check in payment for his work:

> On your last communication ... you acknowledged receipt of the sheet containing comparisons for the years 1876–1880 inclusive, but said nothing about this being unsatisfactory. Hence I had no doubt that I was doing right. Had you spoken then it would have saved me time and labor. I did not think you laid such stress on the cost. The cost to me in hours spent at night working is considerable, but it is very much a matter of necessity for me to do it. I am extremely sorry that such a misunderstanding should have existed. The failure to receive your check is a source of much worry as I counted on it to help out my insufficient salary. I wish to do the right thing, however, and will carry out your wishes.[101]

Lowell replied promptly, but in the same cold tone of his earlier letter. "I never supposed you were not doing as I had explained to you I desired done until you stated the amount of work you had put in. As I do not want you to suffer in the matter I enclose you a check for the amount of $170—the cost is *always* and to *everybody* a matter to be reckoned, contrary to your supposition. My personal regard for you deepens my regret at the misunderstanding." But then he added:

> I have now completed my investigation (I regret without being able to use any of your computations) and I find evidence of an exterior planet at 47.5 astr. units from the sun, magnitude 13+, hel. longitude Jan. 1, 1909 287°±. My paper will soon be published. In view of my completion I shall not need any more computations at present, though I shall hope to call on you in the future.[102]

Lowell did not, in fact, ever call on Carrigan again. Nor was the publication of a paper on Planet X forthcoming, although in June he wrote several drafts of such a paper in which he detailed the conclusions he had communicated to his part-time computer. In these drafts, he outlined his method which was a modification of that used by Pickering, at somewhat greater length than in his preliminary draft in January. Assuming a circular orbit for X in the same plane as that of the orbit of Uranus, he plotted the outstanding residuals of Uranus, finding from his curves a synodic period with an unknown planet of 112.5 years, and noting that the corresponding distance of 47.5 a.u. "supports a semi-major axis derived from comet aphelia." And now, in estimating X's mass at two-fifths that of Neptune, he remarked that it was "surprising that such a body should not already have been detected. But from its smaller size it has probably condensed more than Neptune and may be of low albedo [reflectivity]. Lastly, the failure to come upon Neptune by chance shows how easily a fairly conspicuous body may be missed." In summing up his results, Lowell assigned X a period of 327 years, a magnitude of 13, and a heliocentric longitude in 1909.1 of 287°. But this paper, too, was not published.[103]

With Carrigan's dismissal, and the "completion" of his investigation, Lowell's first search for Planet X came to an end. He now turned his attention to a host of other matters. In the spring of 1909, he busied himself preparing his final popular book on astronomy, *The Evolution of Worlds*, for publication.[104] As Mars was coming to another favorable opposition in September, he spent the summer in Flagstaff making plans for observations and urging the Clark firm, futilely as it turned out, to complete his new 40-inch reflector in time for the occasion.[105] Through the fall, he observed Mars and Saturn, finding two "new" canals which he claimed had been constructed by martians within the past few months,[106] and "wisps" crisscrossing the equatorial belt of Saturn, along with two "new" divisions in Saturn's outer rings.[107] Some of his time during the summer and fall of 1909 he spent defending his theories on Mars and planetary evolution, and even his own scientific integrity, against new and increasingly virulent attacks.[108] Early in 1910, he embarked on a tour of mid-western college campuses, lecturing on Mars to chapters of the Society of the Sigma Xi, and in mid-March, he sailed for Europe, returning in mid-May to supervise observations of Halley's comet, which made its perihelion passage around the sun on May 10.[109] Most of the summer he spent in Flagstaff, preparing *Bulletins* and writing articles about the comet, journeying back to Boston in October.[110]

For nearly a year, the subject of a trans-Neptunian planet disappeared from his correspondence. In July 1909, he wrote English astronomer A. C. D. Crommelin to acknowledge the receipt of residuals of Uranus and Neptune for 1908-09 that he had requested back in January,[111] and, he explained: "I have been investigating the matter of a trans-Neptunian planet and I have a solution by variation of the elements of Uranus arising from such perturbations—I assume the radius vector from other indications—which when compared with the outstanding residuals seems to imply such a planet and perhaps its place."[112] But not until July 1910 is there another indication in the archives that Lowell still had Planet X on his mind. Then, a series of letters and postcards from Flagstaff to his Boston office requesting "curve charts" of the perturbations of Uranus and the other outer planets suggests that he had not only resumed the search, but was now approaching the problem in a more rigorous and difficult manner.[113]

The reason for Lowell's virtual suspension of his Planet X work in the spring of 1909 after months of frenzied activity is, perhaps, not too difficult to find. For the urgency that Pickering's first brief Planet O prediction had imparted to Lowell's investigation was dissipated that spring with the publication of Pickering's detailed Planet O paper in the Harvard College Observatory *Annals*.

Lowell read this paper carefully and with a critical eye noting errors of fact and logical inconsistencies in the margins of his copy. Most certainly, Lowell recognized that while Pickering, unlike Todd, had described his method at considerable length, he had arrived at essentially Todd's conclusions. The major variation in the two sets of elements, as Pickering himself pointed out, was in the assigned location of the unknown planet, Todd giving its longitude for 1877.84 as 170° ± 10, and Pickering placing it in 1900.0 at 105°.8, a difference of some 84°. In one of his marginal notes, incidentally, Lowell remarked that Pickering had picked the "worst possible" epoch for his Planet O.[114]

Pickering had also pointed out that his distance for Planet O of 51.9 a.u. agreed roughly with that suggested by comet aphelia. In addition, having assigned O a mass of twice the earth's, he argued that the "fact that ... O has been able to capture only two comets, while Neptune possesses six, and Jupiter thirty-three, tends to confirm us in our belief that the mass of O is small." To this Lowell appended the marginal note: "Error: the distance alone would account for the paucity of prey."[115]

All in all, Pickering's full statement of his work, along with Lowell's own experience with Pickering's graphical method, seems to have convinced Lowell that the problem of a trans-Neptunian planet admitted no easy solution. At the end of Pickering's paper, he wrote in his bold, slanting hand: "This planet is very properly designated O [and] is nothing at all."[116]

CHAPTER 5

THE SECOND SEARCH FOR PLANET X

Carl Otto Lampland, gathering material on Percival Lowell's Planet X search after the discovery of Pluto, wrote in 1935 that he had "a very uneasy feeling that additional data should be in existence somewhere."[1] Lampland's uneasiness is understandable, for the documentary evidence in the Lowell Observatory archives concerning Lowell's X work is in fact both sporadic in time and fragmentary in content. A reason for this, surely, is that Lowell restricted knowledge of his search largely to only those of his assistants in Boston and Flagstaff who were directly involved and with whom he was in frequent and prolonged personal contact. Thus there was seldom any need to conduct broad discussions, give lengthy explanations, or issue detailed directions in his correspondence. Moreover, the fact that Lampland himself was not cognizant of all aspects of the search indicates that perhaps Lowell told his assistants individually only as much as he felt they needed to know. Indeed, Lowell once cautioned Lampland not to concern himself with the mathematical phase of the search, advising him politely but firmly that the "X work wanted are plates and their examination. Your ability is too great to be wasted on computation."[2]

Actually, the archival record is not as deficient as Lampland's remark might suggest, but it must be read in the context of other pertinent material. Thus perhaps the best way to reconstruct Lowell's second search for Planet X is to reverse the usual order of things and start at the end—with his 1915

"Memoir on a Trans-Neptunian Planet." For only in this does Lowell explain why, sometime between July 1909 and July 1910, he decided to exhaust the possibilities of the trans-Neptunian planet problem by undertaking the onerous task of exploring it analytically through the rigorous methods of celestial mechanics. And only in this memoir does he outline his own concept of the problem, and summarize his plan of attack.[3]

This memoir, on the other hand, is not in itself a complete record of his trans-Neptunian exploration or even of his second Planet X search. It contains nothing whatever of Carrigan's long labors, or of the early photographic survey of the invariable plane, or of Lowell's own initial attempts to apply mathematically superficial methods to the problem. Nor does it mention the frustrating second photographic search, carried on from time to time between 1911 and 1916 largely by Lampland. And while it contains page after page of tables listing the results of voluminous computations, it does not record the names of any of the computers who produced these results. Although Lowell certainly intended to credit some of the work to his chief computer, Elizabeth Langdon Williams, she seems to have preferred anonymity.[4]

While Lowell's trans-Neptunian exploration vanishes from the available record from July 1909 to July 1910, it is clear in retrospect that he did not banish it entirely from his mind. During this period of apparent inactivity, he obviously gave the matter some thought and reached some important conclusions. First, from his own and other's experience with empirical methods, he decided that the problem was "so complicated," as he later wrote, that "all elementary means of dealing with it lead only to error," and that the "sole road to any hope of capture lies through the methodical approach of laborious analysis." Secondly, after reviewing Gaillot's and Lau's analyses of the problem and their postulations of pairs of trans-Neptunian bodies in circular orbits, he became convinced that "not only are all summary processes worse than useless, engendering false ideas in their conclusions, but even a proximate analysis proves not to be approximate in its results."[5]

> Thus the simplifying of the problem by the assumption of a circular orbit for the unknown, originally suggested by Tisserand and worked out in the present case with all due reserve by M. Gaillot and more or less likewise by M. Lau, betrays intrinsic evidence of inadequacy. For M. Gaillot reduces the mean of the outstanding squares of the residuals only from 1.04 to 0.88* which, considering that the admission of a disturbing body must necessarily reduce them, is self-confessedly hardly a satisfactory solution. Still more questionable is the attempt to explain planetary irregularities on the supposition of two unknowns, both travelling in circles; as we have no guarantee that the habit of either body is of this obliging simplicity.... To any real solution, the problem must be attacked analytically with all the rigor possible.[6]

---

*Unfortunately, Lowell does not specify to what these figures refer (seconds or minutes of arc?), but it is obvious that the reduction is only slight.

Lowell Observatory

Percival Lowell relaxing on the porch of the "Baronial Mansion," his residence at the Lowell Observatory in Flagstaff.

Lowell's decision to launch such an analytical attack cannot have been an easy one. He was certainly aware that the task he now set for himself was far more formidable than the one which had confronted Adams and Leverrier in 1845 as they set out analytically to find the presumed planet that was disturbing Uranus in its orbit and discovered Neptune. For Neptune, far from contributing to a solution, was now a part of the problem. "We cannot use Neptune as a finger-post to a trans-Neptunian [planet] as Uranus was used for Neptune because we do not possess observations of Neptune far enough back," Lowell pointed out in his memoir. "A disturbed body must have pursued a fairly long path before the effects of perturbation detach themselves from what may well be represented by altering the elements of the disturbed. Neptune has not been known long enough to do this." Lowell, however, did attempt an analysis of such Neptunian residuals as were available, finding that they "yielded no rational result."[7]

Moreover, if an analytical solution must thus be sought in the residuals of Uranus, the inclusion of Neptune's gravitational influence had now reduced these residuals to values so small as to be of problematical significance. "In 1845," as Lowell subsequently noted, "the outstanding irregularities of Uranus had reached the relatively huge sum of 133" [arc seconds]. Today its residuals do not exceed 4.″5 at any point on its path."[8]

There were other factors, too, to compound the trans-Neptunian problem. The presumed unknown's greater distance and Lowell's consequent belief, based on an analogy with the satellite systems of Saturn and Jupiter, that the eccentricity and inclination of its orbit would be large, introduced serious computational complications which both Adams and Leverrier had been able to avoid. This, as Lowell later noted, was due to the fact that Neptune's relative nearness to Uranus and the near circularity of its orbit presented "the simplest possible case of the general problem."[9]

Further, as Lowell clearly understood, even the elements of Uranus given in the various extant theories of its motion were imprecise and were, in effect, unknowns, and thus the residuals derived from these theories were also suspect. "The theory of a planet cannot in the nature of things be exact," he noted in his memoir, because "the observations on which it is founded are more or less in error ... the theory itself may be more or less imperfect ... and an unknown body may be acting of which perforce no account has been taken."

This latter is a particularly insidious pitfall inasmuch as the greater part of the disturbance it causes can be concealed by suitable changes of the disturbed's elements.... Consequently the residuals left by the theory of a planet are not at all the outstanding perturbations but only such small part of them as cannot be got rid of by suitable shuffling of the cards. We have then no guarantee that our proposed elements are the real ones, but only the best attainable under the assumption *that no unknown exists*. Every theory of a planet is thus open to doubt, seeming more perfect than it is. It has been legitimately juggled to come out correct, its seeming correctness concealing its questionable character.[10]

French astronomer Camille Flammarion, *left*, with Percival Lowell, at Flammarion's l'Observatoire de Juvisy, near Paris, in 1908.

Once he had decided to attempt an analytical solution of the transNeptunian planet problem, Lowell set to work with typical energy and determination. He opted for Leverrier's method because, as he later explained, while Adams's approach was "direct and masterful," Leverrier's was "simpler and more complete." Initially, he used residuals from Leverrier's 1873 theory of Uranus brought up to date to 1906 by the Greenwich observations. Later he substituted residuals from Gaillot's 1910 theory which, "in addition to its general excellence, has two qualities which specifically commend it." The first of these concerned the masses of the known planets Gaillot had adopted and which Lowell considered the best available. "The full merit of this can be appreciated only by those familiar with most works on celestial mechanics," he said, "in which, after excellent analytical solutions, values of the quantities involved are introduced on the basis, apparently, of the respect due to age. Nautical Almanacs abet the practice by never publishing, consciously, contemporary values of astronomic constants; thus avoiding committal to doubtful results by the simple expedient of not printing anything not known to be wrong." Gaillot's other "mark of merit" for Lowell was in giving the residuals between his theory and the observations.[11]

Basically, Lowell sought to establish two things in his analytical investigation of the trans-Neptunian planet problem. First he had to find that the perturbations in the motion of Uranus, after the uncertainties of the available theories and the probable errors of observation were taken into account, were sufficient to indicate the presence of some unknown exterior body, that there was indeed "a planet out there as yet unseen by man," as he had proclaimed in his 1902 lectures. "The test of the existence of an unknown body thus left out of account," he noted in his memoir, "lies in the substantial reduction of the outstanding squares of the theory by its subsequent admission. The reduction must be substantial to be indicative, because a slight betterment necessarily follows the introduction of another factor among the adjustable quantities." And in fact he found such an indication from his solutions, the admission of an unknown body into his calculations reducing the residuals of Uranus "from 90% to 100% nearly."[12]

Secondly, he had to find the most probable position of the perturbing body on some particular date, for this was the datum which would indicate where best to direct the telescope in his photographic search. But to do this he must first determine as accurately as possible—and the problem permits only approximate solutions—the elements of the unknown's elliptical orbit. Of the elements to be determined, Lowell assumed the value of a', the mean distance, and consequently that of μ, the mean annual motion which is derived from a' by Kepler's third law. Eventually, he sought to find the most probable value for a' by applying the method of least squares to various solutions for various assumed distances, thus treating X's distance as a variable. Similarly, he derived the most probable longitude at epoch ϵ', by least squares solutions for values chosen at intervals around the 360° of the circle. In addition, as he assumed that X's orbit lay in the same plane as that of Uranus, he was not initially concerned with the inclination of its orbit, i', or the longitude of its ascending node, Ω', although these elements entered into his later, unsuccessful attempts to estimate X's latitude. With these assumptions then, the unknown elements of X to be determined were e', its eccentricity; $\bar{\omega}'$, its longitude of perihelion; and m', its mass. And e' and $\bar{\omega}'$, could be stated in terms of m' in his equations.[13]

The details of Lowell's analytical investigation are quite complex and are, indeed, beyond the scope of this study. The mathematically knowledgeable reader who is interested in following his procedures in detail is referred to Lowell's trans-Neptunian planet memoir. But it is, nonetheless, necessary for any general understanding of his work to describe his approach here at least in its broad outlines.

Lowell started, as did Leverrier, by forming a series of observation equations for the residuals of Uranus for various dates, or the means of groups of dates, which were of the general form:

$$\frac{\delta v}{\delta n} \Delta n + \frac{\delta v}{\delta \epsilon} \Delta \epsilon + \frac{\delta v}{\delta e} \Delta e + \frac{\delta v}{\delta \bar{\omega}} \Delta \bar{\omega} + P = r \quad\quad (I)$$

where v = the true heliocentric longitude of Uranus, P = the perturbation in Uranus' motion caused by the unknown planet, and r = the difference between the observed and theoretical positions of Uranus, that is, the residual. The innocent-looking symbol P in this equation, it may be noted, conceals a multitude of computational problems.

Lowell's basic solutions involved 24 such equations covering observations of Uranus from 1752.81 to 1903.47. In other solutions, he added 1709.79, the mean date of the five earliest prediscovery observations of Uranus in 1690, 1712 and 1715 by Flamsteed, and the date 1907 and 1910.5, representing residuals from 1904 to 1910 supplied him by Gaillot. Also, in some solutions, he inserted ten additional dates, interpolated from graphs of the Uranian residuals, between 1840 and 1900. In all, solutions were made variously for 24, 25, 26, 27, 34 and 37 observation equations.[14]

To these groups of equations, Lowell then proceeded to apply Leverrier's method, which he outlined briefly in his X memoir as follows:

What he [Leverrier] did to begin with was to form equations like (I) for the several residuals taken in groups from 1690 to 1845; then selecting four of them suitably spaced, to eliminate Δn, $\Delta \epsilon$, Δe and $\Delta \bar{\omega}$, expressing these in terms of the perturbations. Then combining all the remaining equations, except certain of the earlier ones, in two groups he solved for e', $\bar{\omega}'$, m' and constants. Then finally substituting his solutions in the omitted equations, he found for what value of ϵ' their residuals became a minimum.[15]

With this as a basis for procedure, Lowell initially worked out several solutions which "differed in the number of perturbation terms used, in the manner in which they were considered and in the exactness in which the elements of Uranus were incorporated...." Eventually, he chose "one of the simplest" of these, which he designated [H_{14}], to be worked out for thirty specific longitudes of epoch completely around the zodiac. Other solutions were made only for a few values of ϵ' around 0° and 180°, the problem having a dual solution which yielded two possible positions for an unknown planet, each about 180° apart.[16] Generally, as Lowell noted in his memoir, the "work was done in duplicate, and as a further check, the functions involved were plotted so that any important deviations from their proper curves could be detected by sight."[17]

From its beginnings in the summer of 1910, Lowell's analytical investigation became progressively more rigorous through its off-again-on-again, five-year course. As the full complexities of the problem became clearer from successive solutions, and as Lowell's knowledge of the finer points of celestial mechanics increased, he realized that he would have to adopt far more difficult and complicated procedures than had been required of either Adams or Leverrier.

For example, initially he followed Leverrier in using a semi-least-squares method that had proven to be adequate in the case of Neptune. But he soon found that this would not do for his Planet X, and in taking up the problem again, as he later noted, he deemed it "advisable to pursue the subject in a different way, longer and more laborious, but also more certain and exact: that by a true least squares method throughout.... In consequence," he added, "the whole work was done *de novo* in this more rigorous way, with results which proved its value."[18]

Also in his earlier solutions, Lowell, like Leverrier, had carried his computations only through the first powers of the eccentricities of Uranus and X, and again this had been sufficient for Neptune. But in the final stages of his investigation, he undertook the arduous task of exploring second-order forms in the solutions as well. "After the problem had been solved for the first powers," Lowell explained in his X memoir, "the large coefficients of the higher powers of e' in the perturbative function remained a source of much mental perturbation as to the validity of the result. I therefore decided to take them into account.... The only sure way was to consider the problem as directly involving e'^2, unpleasant as the complication was bound to be, a difficulty which Leverrier had envisaged and fortunately in his case found he could evade."[19] Lowell even briefly probed the third power terms in e', but gave up the attempt because to solve for them "would have introduced still more uncertainty than for the squares."[20]

For the extensive computations required by an analytical attack on the problem, Lowell enlisted the assistance of the aforementioned Miss Williams, a mathematically talented young woman who had been employed in his Boston office since at least January 1905, performing general office chores and handling some of the details involved in the publication of his books and the observatory's *Bulletins*.[21] In addition, she undertook calculations in connection with his various astronomical projects, and she had apparently reduced and plotted some of Carrigan's ill-starred residuals during the final months of Lowell's first X search.[22]

Lowell began his second search where Gaillot left off—probing the analytical problem initially by assuming a circular orbit for X. This work was underway in July 1910 when he left for a summer's visit to Flagstaff,[23] and was continued briefly in October after his return to Boston.[24] But the result was apparently unsatisfactory for he soon abandoned the assumption of circularity in favor of an elliptical orbit for X.

The computations went slowly at first, and through the fall and winter of 1910–11 Lowell found little to report about the work to his assistants in Flagstaff. On Dec. 10, 1910, however, he confided to Lampland that "Miss Williams and I have been pegging away at it, have constructed the curve of perturbation due X if at 47.5 astr. units including all terms of the first power of the eccentricities and examining the most important terms in their squares. We find some interesting things," he added, "... but we do not find the planet for except in periods the theoretic curve swears as the residual one. It is of course possible that Leverrier's theory is not sufficiently exact."[25]

Oddly enough, Lowell never really fully explained the basis for this initial assumption of 47.5 a.u. as X's mean distance from the sun, even in his memoir. Yet this is a key assumption in any solution of the problem and, in fact, he preferred this figure throughout most of his analytical investigation, abandoning it only at the end after his least squares solutions for other distances consistently indicated a lesser value for a'. But it is nonetheless clear that he relied on no less than three separate lines of evidence to support his preference. The first of these, of course, was the presence of cometary aphelia at about this distance, as Flammarion first pointed out in 1879 and as he himself noted in his 1902 lectures at M.I.T. Secondly, his graphical solutions, as recorded in his unpublished papers of January and June 1909, had also indicated a similar distance. The third and final confirmation for Lowell, however, came in the concept of commensurability of periods, and this will require some further elucidation here.

Commensurability, in general, relates to two bodies revolving around the same primary, whose periods of revolution can be represented by simple ratios such as 2:1, 3:2 and 5:3. In such cases, the maximum mutual gravitational attraction of the two bodies is exerted periodically at the same points in their orbits, resulting in resonances which tend to shift the bodies out of such orbits. Peirce, of course, used the concept in the Neptune controversy, and it had been used to account for Cassini's division in Saturn's rings, as well as by Daniel Kirkwood in the 1860s and 1870s to explain the gaps in the belt of asteroids between Mars and Jupiter, which are still called "Kirkwood gaps."

Lowell himself worked with the concept from at least 1907 in his continuing studies of the Saturnian ring system and Saturn's satellites.[26] Moreover, in June 1909, in his unpublished summary of the results of his first X search, he had noted in passing that the period of a planet at the distance indicated by his curves on his graphs "would have near commensurability with that of Uranus."[27] He did not develop this idea more fully, however, until four years later. Then, in a memoir on "The Origin of the Planets" published in 1913 by the American Academy of Arts and Sciences, he used commensurability to formulate what he considered to be a "general planetary law" by which he sought to explain the observed structure of the solar system. This declared that:

> Each planet has formed the next in the series at one of the adjacent commensurable-period points, corresponding to 1:2, 2:5, 1:3, and in one instance 3:5, of its mean motion, each of them displacing the other slightly sunward, thus making of the solar system an articulate whole, or organic organism which not only evolved but evolved in a definite order, the steps of which celestial mechanics enables us to retrace.[28]

Lowell likened this law to Mendeliev's Periodic Law of the Elements in chemistry, noting that "it, too, admits of prediction."

> Thus in conclusion I venture to forecast that when the nearest trans-Neptunian planet is detected it will be found to have a major axis of very approximately 47.5 astronomical units, and from its position a mass

Lowell Observatory

Workmen erecting the Lowell Observatory 40-inch reflector and its dome during the summer of 1909. Standing on the pier, in white shirt and coveralls, is C. A. Robert Lundin of the Alvan Clark and Sons optical firm which built the instrument.

comparable with that of Neptune, though possibly less; while, if it follows a feature of the satellite systems which I have pointed out elsewhere, its eccentricity should be considerable with an inclination to match.²⁹

The only part of this prediction that can be derived from Lowell's "law" is, of course, his postulated planet's distance. And the 47.5 astronomical units that he here gives as the approximate distance for a trans-Neptunian planet is, in fact, "slightly sunward" of a major commensurable orbit at 48.4 a.u. which yields simple ratios with the periods of both Neptune and Uranus of 2:1 and 4:1 respectively. These ratios, expressed as the terms 4n'-2n and 4n'-n appear frequently both in the correspondence between Lowell and Miss Williams regarding the analytical search for X, and in his final X memoir itself.

By early March 1911, Lowell's calculations apparently had progressed to the point where he felt that a satisfactory solution was at hand. Thus on March

114   *The Second Search*

13—his 56th birthday—he telegraphed Lampland at Flagstaff: "Please begin to photograph ecliptic where south with forty-inch. Hope to wire position in a few days. Calculations tremendously long."[30]

Lampland replied promptly, offering congratulations and bringing up an old problem that had plagued the earlier invariable plane work. "I was greatly pleased to learn that your calculations on the position of planet X are coming along so well," he wrote. "I have a faint idea of the prodigious amount of work you have done. Let me hope that your efforts will be crowned with the proper reward—the discovery of the planet." But, he added: "The examination of the plates will mean much work. I hope we will be able to devise some other method than the superposition of plates. It seems to me that this way is very laborious and somewhat uncertain. I have thought of an instrument for the purpose, but it will be impossible to undertake work on it now...."[31] Five years before, it will be recalled, Lampland had thought of an instrument to facilitate the examination of the search plates but Lowell had not looked on his proposal then with favor. Now, however, he not only appreciated the suggestion, as Miss Leonard informed Lampland a few days later,[32] but he acted on it. Within months, a Zeiss "blink" comparator had been added to the observatory's equipment, and henceforth it was used both in Boston and in Flagstaff to compare X search plates. It would indeed, although with some modifications, eventually reveal a ninth planet—Pluto.[33]

Miss Leonard's letter to Lampland also communicated some other information regarding the new search. "Dr. Lowell has no objection to your going in for the hunt there as long as he tells you where to look from this end," she wrote. "It has been a long computation and Dr. Lowell and Miss Williams have been faithful to it all winter. We feel rather heavy with it but *when* it is found it will lighten our hearts and make us glad! I am sure you appreciate what the work has meant and now your part in it will soon follow."[34]

Lowell, however, in his telegram to Lampland, had apparently been somewhat oversanguine in his hope that a position for X would be forthcoming in a few days. Or perhaps, with his fine sense of history, he simply wanted to launch his new search on a significant date—March 13 being not only his birthday but the 130th anniversary of Herschel's 1781 discovery of Uranus.

In any case, the computations continued and on April 3, Miss Leonard advised the elder Slipher in Flagstaff: "Dr. Lowell is working like a slave on Planet X and just now he has a very severe cold—but he won't stop working!"[35] Not until late April is there an indication in the record that he and Miss Williams had arrived at a position for X. "I telegraphed you last week to look in heliocentric longitude 235°," he wrote Lampland on April 27. "Our residuals, from necessities in the case very uncertain, seem to indicate for Planet X a heliocentric longitude of 239°.... Please take plates in neighborhood for 2° or 3° on either side of the fundamental plane and devise best method of comparing them for the stranger!"[36]

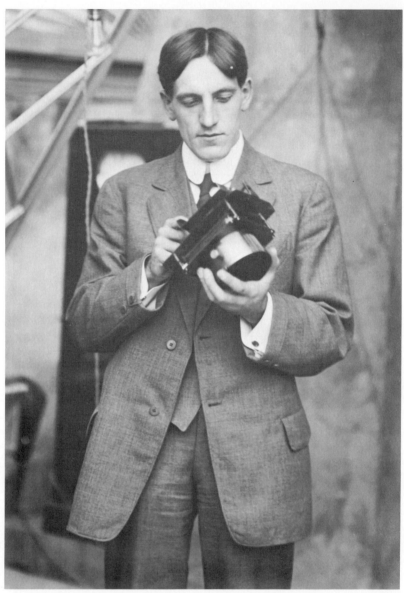

Lowell Observatory

C. A. Robert Lundin of the Alvan Clark and Sons firm of telescope-makers, *above,* and again during the installation of the Lowell Observatory's 40-inch reflector in 1909, *top right.* The instrument was used in the second search for Percival Lowell's Planet X, and twenty years later, Lundin produced a superb 13-inch objective lens for the refractor that was used to discover Pluto.

Lowell Observatory

At this point, Lowell decided to go to Flagstaff himself to oversee the observational phase of the new search, arriving on May 8 and remaining at his observatory through June. From time to time during this period, he dispatched instructions to Miss Williams concerning the computations, and in mid-June he set her to work on solutions using the residuals from Gaillot's theory of Uranus in place of those from Leverrier's.[37] Presumably during this sojourn, the new Zeiss comparator arrived in Flagstaff. For on July 3, with his return to Boston, he wired Slipher to "Hold comparator and star plates. Hope to be back soon."[38] And on July 8 he telegraphed Lampland: "Results from Gaillot's residuals indicate search should be carried farther east.... My telegram to hold the comparator arrived too late I see. I shall ship it back to you. Don't hesitate to wire anything by day letter."[39] Only five days later, incidentally, he reversed himself on the direction of the search, advising Slipher that "Lampland should photograph west not east with probable longitude two hundred ten."[40]

Late in July, Lampland's plates provided a brief excitement in the search. "Suspicious object retrograding on plates," Lowell wired him on July 27, giving its coordinates on June 22–23. "Calculate present position and photograph in vicinity."[41] The record does not reveal what this object was; it was not, of course, planet X.

Lowell's X search disappears again from the archival record and for nearly a year. Lowell returned to Flagstaff in mid-August, staying at the observatory until February 1912. And then, after only a few weeks in Boston, he sailed for Europe early in March, to return only at the end of May. During his prolonged visit to Flagstaff, the X work was virtually suspended as Lowell turned his attention to other astronomical projects which now became more pressing.

The first of these, as always, was Mars, for its November 25, 1911, opposition provided a first opportunity to observe the planet at a relatively favorable aspect with his new 40-inch reflector. In addition to visual observations, Lowell and his assistants obtained some 3,000 photographs of Mars during this opposition.[42] A second was Uranus. In the summer of 1911, Slipher made a spectrographic study of this planet which resulted in the detection of a retrograde rotation in a period of 10¾ hours, and Lowell promptly undertook the publicizing of this considerable observational feat and the preparing of an observatory *Bulletin* describing it.[43] At Flagstaff, too, he worked on the problem of precession both in relation to Mars and to the Egyptian pyramids,[44] and continued his investigation into commensurability.[45]

Parenthetically, during Lowell's temporary suspension of his Planet X search, W. H. Pickering published predictions for three more trans-Neptunian planets in addition to his 1908 Planet O. Pickering designated his presumed Planets P, Q and R, and provided them with more or less complete sets of orbital elements on the basis of his lengthy statistical analysis of cometary orbits. To his Planet P, he assigned a distance of 123 astronomical units and a period of 1,400 years, making it roughly analogous to the inner planet of the pair Forbes had first suggested in 1880, also on the basis of cometary orbits. But his other two new planets were far more intriguing. He declared that Planet R was 6,250 a.u. from the sun with a period of half-a-million years and a mass of 10,000 earths. And Planet Q, at a distance of 875 a.u. and a period of 26,000 years, had a mass of 20,000 earths and thus, as Pickering himself pointed out, was "practically a dark companion" to the sun. Moreover, Q was a most erratic planet as the eccentricity of its orbit was 0.54 and its orbital inclination was 86°.[46] If Lowell had any thoughts or comments to make about these latest Pickering prognostications however—and one suspects that he did—he left no record of them.

With his return from Europe late in May 1912, Lowell took up the X work again, directing a resumption of the photographic search and then, over the next month, firing a barrage of questions and instructions at Lampland in Flagstaff regarding it. On June 5, he requested Lampland to send him "as near as you can the values of the limiting magnitudes on the Planet X plates. Also send any of the new astrographic charts that bear upon the regions you have taken."[47] The next day, he wired Lampland to "try to make the plates of the same density no matter what the exposure. Also send them to me the day each pair is completed."[48] On June 8, in a letter, he asked his assistant to "let me know with a minimum of trouble to yourself what the magnification of the plate is. How many millimeters to a degree and any other matters that may help me in my search. The plates are certainly an improvement over what they were before. I notice some big stars marked. What do these mean? They certainly are not strangers. Are they reference stars?"[49]

Lowell now wanted the pace of the photographic search stepped up. On June 12, he complained to both Lampland and Slipher: "I have not received any X plates in a long time."[50] And in a letter to Lampland that day he explained that it "is important that they should be taken at once and received at once so that if anything is found it can be followed up. The plates of last year that I have examined do not need to be duplicated. I think I should go more south than north of the fundamental plane. I have found nothing so far which cannot be interpreted as variable and what strikes me as very odd: I have run across but one asteroid. In a short time we shall not be able to make these plates at all as the motion of X would not be apparent. The pair of plates should be only a day apart—if possible."[51] Two days later, he advised Lampland to take his plates "two or three degrees south of the ecliptic to begin with rather than north along the invariable plane because of greater chance of X there having been overlooked."[52]

On June 20, Lowell found a variable star on a pair of Lampland's plates. Wiring its position to Flagstaff, he directed: "If not already found notify me."[53] Subsequently he wrote Lampland: "With regard to the variable, you should by this time have received the photographs of the region I sent you which will enable you to send me its exact position as also that of the variable found last year. When you have time, and only when," he added, "you may photograph this region and so discover the period."[54]

The demands of this accelerated search now became heavy on Lampland. Duncan, who had begun the invariable plane survey back in 1905, had since earned his Ph.D. degree and was now teaching astronomy at Harvard. In mid-July, Lowell dispatched him to Flagstaff to help with the photographic work during the summer.[55] "Throw as much as you can on him," Lowell advised Lampland.[56]

Lowell and Miss Williams, meanwhile, began a new round of computations, extending their earlier longitude solutions and expanding the work now by attempting the ominous task of determining X's position in latitude as well. "That you have not heard from me about Planet X," he wrote Lampland August 9, "is because we are hard at work on the latitude—for the longitude $L = 155°$ in 1850 we get least squares residuals smaller than Gaillot's from the best amelioration he could make of the Uranus elements without recourse to any unknown body. The $e' = .20 \pm \bar{\omega}' = 347°$. So you can place X approximately. We are also at work on a more precise longitude determination. Do the best you can with these data till further notice."[57]

Two weeks later, Lowell wrote Lampland that "it was the mean longitude I gave you.... I am very sorry—first that the planet should now lie so near the sun due to being near its aphelion—this sounds contradictory but it is not—and secondly that the latitude determination is not yet finished. In consequence I have delayed sending you any instructions as to where to look. For I feel convinced that the inclination is considerable, both because of the size of the residuals in latitude and because of the great eccentricity, the two going together and diminishing mass also."[58]

The long, laborious calculations continued. In September Miss Leonard, returning from a vacation in Maine, reported to Lampland that she had found Miss Williams "a mere shadow from her perple [sic] Xing computations."[59] On September 12, Lowell wired: "Solution Uranus residuals in latitude including X gives considerably smaller residual squares than Gaillot's solution without it. Resulting longitude ascending node one hundred eighty-five degrees. Inclination uncertain owing to necessary uncertainty of mass,"[60] A few days later he advised Lampland: "We are revising and extending our work on Planet X; results of which we are, unfortunately, not yet in a position to communicate to you. Every new move takes weeks in the doing."[61]

The exhausting computations of the X work did, in fact, have two direct consequences about this time. First, Lowell began to expand his staff of computers in his Boston office. By early November, Miss Leonard could report to Flagstaff: "We have four assistants here now plus the old standbys. It is a very busy place!"[62] The names of these assistants, however, do not appear in the archival record for nearly two years when the mathematical phase of the X search came to an end. Then, Lowell sent two of them, Thomas B. Gill and Earl A. Edwards, to Flagstaff to assist in the continuing photographic search,[63] subsequently recommending the other two, Johnson O'Conner and Herbert H. Tucker, for positions with other employers.[64]

Second, toward the end of October, Lowell himself became "quite ill" and was forced not only to postpone a planned visit to Flagstaff in November, but to suspend his active participation in the X work temporarily. "We have not seen him since," Miss Leonard wrote to Flagstaff on October 30, "only a word now and then on the phone.... Dr. Lowell is not seriously ill but he is weak and run down and must needs be careful and quiet. I fear he has overworked this summer!"[65] By Nov. 8, Lowell's wife advised Slipher that Lowell was improving.[66] But more than a month later Miss Leonard was still concerned about his condition. "I hope it may not be so prolonged as his breakdown ten years ago!" she confided to Lampland. "He worries about the work—he wants to be *in it*!"[67] And to Slipher she noted that Lowell had been absent from his office for seven weeks. "It is nervous exhaustion," she explained, "and he is *up* and *down*! Some days he cannot even telephone. He gets nervous about the work and impatient for things to come from Flagstaff."[68]

Meanwhile the computations by his augmented staff went on, and, as Lowell's health slowly improved through December and January, he gradually resumed a more active direction over the work. The revised and extended analysis he had begun back in September was now nearing completion. On January 22, 1913, he wired Lampland two new possible positions for X as of January 1, 1913, advising him to search "well south of the ecliptic....

Examine plates yourself telegraphing discoveries," he added. "Rush as planet getting in bad position."[69] Another telegram that same day informed Lampland that X's magnitude was "probably ten or eleven" and that "probably five minutes exposure enough."[70]

Over the next few weeks, telegrams giving new data and instructions went to Flagstaff every few days. On January 29, Lowell wired: "Latitude uncertain. Seven degrees north of ecliptic most probable. Increase longitude by fifty three minutes for precision."[71] On February 3, he advised: "Probable descending node of X eighty degrees twenty minutes. Inclination two degrees thirty. Epoch nineteen thirteen. But these elements uncertain. Think five minutes exposure enough."[72] Two days later, he repeated the data on the descending node and inclination, adding: "Refer to X orbit on Uranus," and "Major axis X forty seven point five."[73]

Finally on February 8, he began a letter to Lampland by announcing: "Here are our new elements of a new planet," and listing them for the epoch January 1, 1850, as: mean longitude at epoch, 11°40′; longitude of perihelion, 186°; eccentricity, .228; mean distance 47.5 a.u.; longitude of ascending node, 110°59′; inclination of X orbit on Uranus orbit, 7°18′; and mass, 2.40 (unit = 1/50,000 of the sun). "You will be interested to know," he added, "that by solving for other distances as well, my original supposition of = 47.5 turns out almost exactly right...." Then he calculated X's heliocentric longitude for February 15, 1913, at 58°4′ and instructed Lampland: "Please note everything interesting as you may catch other fish and notify *us* of the haul. Good luck!"[74]

Almost immediately, however, Lampland himself fell ill and on February 21, Lowell wrote him: "Now you must do as I did: go off and get well. I cannot afford to loose [sic] you. The X work can be carried on by Dr. Slipher with Mr. Slipher till your return."[75] Four days later, Lowell decided that a brief visit to Flagstaff might be in order, telegraphing his closest friend and confidant there, attorney Edward M. Doe, to "expect the worst blizzard of the season: Me—by Limited March 8."[76]

While Lowell's trip undoubtedly was motivated in part by Lampland's indisposition and a desire to keep the X search going, there were in fact some other important developments at his observatory that called for his personal attention. During the preceding two months, the elder Slipher had completed two spectrographic investigations that proved to be of fundamental significance in astronomy.

The first of these was his discovery, in December 1912, that the nebula in the Pleiades near the star Merope was not intrinsically luminous but shone by the reflected light from nearby stars, thus demonstrating the existence of dust, or "pulverulent matter," as Lowell called it, in interstellar space.[77]

The second was Slipher's discovery in January 1913, with observations made from September through December 1912, that the great spiral nebula in Andromeda was approaching the solar system at the then astonishing speed of 300 kilometers per second.[78] "It looks as if you had made a great discovery,"

Lowell wrote him on February 8, the same day he had forwarded his new elements for Planet X to Lampland. "Try some more spiral nebulae for confirmation."[79] This Slipher did, soon finding that the edge-on spiral in Virgo, NGC 4594, was receding from the solar system at the then incredible speed of 1,000 km/sec.[80] Over the next four years, he measured the radial velocities of some twenty other spirals, finding not only similarly high velocities but a preponderant movement away from the sun, thus providing the first observational evidence for the subsequent development of the modern theory of an expanding universe.[81]

Both of Slipher's discoveries were widely hailed by leading astronomers all over the world. Lowell himself certainly realized their importance. From Slipher's results on the Andromeda and Virgo nebulae, he quickly concluded that such nebulae were not insipient solar systems, as some had speculated, but "something larger and quite different, other galaxies of stars."[82]

While Lowell was at his observatory, his staff of computers in Boston labored on, and indeed telegraphic queries from Miss Williams regarding their calculations were waiting for him on his arrival.[33] And although he stayed in Flagstaff barely a month, he also managed to get in some observations of Saturn with the younger Slipher, and to complete work on his paper on commensurability and the origin of planets which he presented to the American Academy on his return to Boston in mid-April.[84]

Sometime in the spring of 1913, Lowell apparently decided to make one final all-out attack on the trans-Neptunian problem, now extending the time span of his solutions by including observations up through the year 1910 and investigating the higher powers of the eccentricities of Uranus and X.

Late in May, he wrote Gaillot asking for the residuals in latitude and longitude from observations of Uranus since 1903,[85] and after receiving these in June,[86] he requested similar data for Neptune, which Gaillot sent in August.[87] Miss Williams and her colleagues were already hard at work on the squares solutions.

Through July and most of August, Lowell vacationed at the summer home of his sister, Mrs. Katharine Bowlker, on Mount Desert Island, Maine, keeping in touch with the X work through telegrams and letters from Miss Williams in Boston and forwarding brief progress reports and probable positions to Lampland in Flagstaff. On July 4, he acknowledged a squares solution from Miss Williams, suggesting to her that "perhaps a cubed solution had better be tried."[88] A few days later he asked Lampland: "Generally speaking what fields have you taken? Is there nothing suspicious?" adding that he could not yet send a better latitude figure for X "as we have been solving including the second power of the eccentricities."[89] Late in August Lowell informed Lampland: "So far best determination for first power e prime for present position is two hundred thirty nine degrees and for second power ditto two hundred forty one degrees. Use these," he added. "Suspect inclination large probably south. Am personally still on the retired list. Await another excitement proving true. Any news grateful."[90]

There was no news, however, and soon an old priority once more interrupted the photographic, if not the mathematical, search. For Mars was again brightening in the sky and, as the elder Slipher was continuing his spectrographic researches on spiral nebulae with the 24-inch refractor, Lowell telegraphed instructions to Flagstaff to "rig up the 40-inch for the best possible visual observations at coming Mars opposition."[91]

Actually the 40-inch was far from being an ideal telescope for the X search, as Lampland had pointed out to Lowell nearly a year before. "Now in regard to the Planet X problem," he had written then, "you shall have every assurance that the ground will be covered as rapidly as possible with the reflector. But if it becomes necessary to photograph a large area, it will take a long time with that instrument; and also, in the great number of plates that must be examined there is considerable danger in overlooking an object near the edges of the field on account of the elongated images and the small overlap of fields. If then," he suggested, "a powerful doublet with a large field could be had it should be excellent for this problem."[92]

Lowell had not acted on this suggestion at the time, but after his decision to switch the 40-inch from the X work to visual observations of Mars, Lampland apparently broached the subject again. Consequently, in late November 1913, just prior to leaving for Flagstaff, Lowell wrote John A. Miller, director of Swarthmore College's Sproul Observatory, to ask if such an instrument might be obtained temporarily. "We are very anxious to get hold of a rapid doublet for a few months on work we have in hand," he explained. "Mr. Lampland suggests that you have a good one nicely mounted and wonders if you would let it go for awhile if we could get good negatives of certain parts of the sky that you might want. Do you feel like making such an arrangement and on what terms?"[93]

Miller was obliging, although there is no indication in the Lowell archives of his terms, if any. The Sproul 9-inch Brashear photographic doublet subsequently arrived in Flagstaff and by April 1914 was being used as the primary instrument in the continuing X search.

Mars did not slow the pace of the mathematical search. Lowell took up residence at his observatory December 2, and thereafter through the winter a steady stream of telegrams and letters, the latter usually containing charts and tables in Miss Williams' meticulous hand, flowed to Flagstaff. His computers were now working variously on first power least squares solutions for different distances of X, on the latitude problem, and on solutions in the squares of the eccentricities. Lowell himself, preoccupied with his observations of Mars and later of Saturn, reviewed their work and occasionally—very occasionally, it would appear from the record—issued instructions or gave suggestions.

By January, Miss Williams was working on second power solutions which involved biquadratic equations with four roots and required a preliminary determination of approximate values of the unknowns for their final solution, and had run into difficulties with what had been designated the "b" root. On

January 3 and again on January 7, she wired Lowell asking advice and instructions, explaining that the elements of X were "indeterminate" in the b root solution because the value of X's mass turned out to be "almost zero."[94] Subsequently, she forwarded a solution, along with an explanatory table, to Flagstaff and appealed: "We are really at a loss as to what you wish done next—might we not have a telegram from you?"[95] Some two weeks later Lowell acknowledged this solution, remarking that "the result is certainly extraordinary—how things do point to about the same place," and directing that new first and second power solutions now be made including the mean dates 1907 and 1910.5 from Gaillot's residuals. "Whatever else occurs to you," he added, "please wire me being as explicit as possible since I forget what has been done."[96]

The work continued through February and into March, with Gaillot's residuals and the b root giving minor problems,[97] and with the first power least squares solutions now indicating that his assumption of 47.5 a.u. as X's mean distance was not "exactly right," as he had concluded earlier, but that the unknown should be somewhat closer to the sun.[98]

At this point, unfortunately, the archival record regarding the X search becomes very fragmentary indeed, and, in fact, from March 1914 until Lowell's death in November, 1916, there are hardly a dozen letters in the observatory's files relating to it. Thus a reconstruction of the final events in Lowell's long trans-Neptunian exploration must, of necessity, rely heavily on inferences made from his other activities and from his subsequent statements in his Planet X memoir.

It is clear, however, that the intensive computational phase of the search came to a virtual end by April 1914, and that the various solutions worked out by his computers in the previous months had yielded results which in a general way were quite similar, as far as X's probable position was concerned, to those obtained from his earlier solutions. Consequently, Lowell now decided that further computations, at least on such an extensive basis, were no longer required, and that the best hope for finding X lay in the systematic photography of those regions of the sky generally indicated by his many solutions. Thus, after his return to Boston early in April, he sent two of his computers—Gill and Edwards—to Flagstaff to help in the photographic search there, and they were soon exposing X search plates under Lampland's supervision with the newly installed 9-inch Sproul doublet.[99]

At this time, Lowell seems to have been briefly optimistic for the success of the photographic search. Before sailing on what would be his final European trip in May, a trip delayed for a month by his wife's surgery for an ulcer, he wired Lampland that the "most probable place for X now seems to be: hel. long. 261° and perhaps a little north of the ecliptic," adding: "Don't hesitate to startle me with a telegram—FOUND!"[100] But little was accomplished in the search during Lowell's three-month sojourn abroad because of an unusually long siege of Flagstaff's annual "July rains" and because Mrs. Lampland, in turn, now required surgery for an ulcer.[101] Back in Boston in August,

Lowell confided ruefully to the elder Slipher: "I feel sadly, of course, that nothing has been reported about X, but I suppose the bad weather and Mrs. Lampland's condition may somewhat explain it."[102]

Sometime during the summer of 1914 Lowell also decided, reluctantly it would seem, to publish the theoretical aspects of his second Planet X search despite their inconclusive nature. By now, his analytical investigation was completed, or at least carried as far as he felt it was practical to go, and his corps of computers had been dispersed, with only Miss Williams remaining to handle such calculations as might be required in updating the data for the photographic search and in summarizing the work for publication.[103] The photographic search was to continue, and on September 15, Lowell wrote Lampland asking for "some rough statement of the fields covered for X; merely to make sure that they connect with the old ones and also that they extend eastward well beyond the most probable $L=262°$," and suggesting that he photograph "more to the north of the ecliptic."[104]

By the end of September, Lowell was back in Flagstaff, remaining at his observatory through mid-November. While the record does not indicate specifically what occupied him there, he undoubtedly spent some of his time organizing and writing up his theoretical X work. Just as surely, he approached the task with little of the enthusiasm he usually brought to such projects. Years later, Lampland recalled that during the preparation of his X paper, Lowell "remarked more than once that he did not 'feel up' to the task of summarizing and bringing to a wholly satisfactory conclusion all the results of his research."[105] But even more indicative of his mood at the time, perhaps, is the fact that early in December he announced his X paper to John Trowbridge, the president of the American Academy, only as a casual postscript to a long letter dealing with unrelated matters:

P.S. Now that I am writing you I have a paper that I should like to present at the January [1915] meeting on:

A TRANS-NEPTUNIAN PLANET

As I am going away later in the month I should like to get this off my mind.[106]

Some two weeks later, Lowell sent Lampland longitudes for X of 84° and 88°.4, advising him that "with this as finger-post go ahead. Latitude ? is all I can say." He added dryly: "I am giving my work before the Academy on January 13. It would be thoughtful of you to announce the actual discovery at the same time."[107]

CHAPTER 6

A TRANS-NEPTUNIAN PLANET

The meeting of the American Academy of Arts and Sciences on January 13, 1915, was a frustrating one for Percival Lowell in a number of ways. For one thing, of course, Lampland had no announcement to make regarding the discovery of a new planet. For another, Lowell's brief summary of his analytical search for Planet X generated little interest or comment among the Academy's members or elsewhere. For still another, although Lowell had submitted his paper for publication by the Academy two days after the meeting,[1] it was not published until nine months later and then it appeared as the Lowell Observatory's first *Memoir* rather than under the Academy's imprimatur.[2]

Finally, Lowell had embroiled himself in a picayunish dispute with the Academy's president, John Trowbridge, over a gold medal which he proposed to sponsor to be awarded by the Academy to an "astronomer" working in what Lowell considered to be the more traditional areas of the discipline. The medal, in fact, had been the subject of his letter to Trowbridge in December in which he had announced his Planet X paper as a postscripted afterthought.

"In this connection," he had written then, "I beg to call your attention to the fact that astronomer is used in its technical sense, and not including astrophysics which, of late years, owing to the effect that pictures have on people, has usurped to itself the lime-light to the exclusion of the deeper and more profound parts of astronomy proper. Astrophysics,'' he added, ''I thoroughly welcome but as it is now getting the lion's share of endowment and help I wish to bring back the recognition due its elder sister whom I might add is still young.''[3]

Lowell had also proposed to have the medal awarded by a three-man committee composed of himself, the amiable S. W. Burnham of the Yerkes Observatory, and mathematician W. E. Storey of Clark University, one of Lowell's oldest and closest friends.[4] But when Trowbridge suggested a somewhat different makeup for the committee, Lowell threatened to withdraw the medal unless he was appointed chairman and allowed to name one of the three members.[5] The Academy apparently was not interested in a committee thus stacked to do Lowell's bidding, and the matter was quietly dropped.

Lowell's reaction to these various frustrations seems to have been quite bitter. ''The stagnation and old-fogyism in consequence of which the Academy is run by a set of men certainly not broad [of] view or judgement is greatly to be deplored and ought to be changed,'' he fumed to an acquaintance some weeks later.[6]

The meeting, however, was not a total loss from Lowell's standpoint. Lampland was elected to membership, culminating several years of campaigning by Lowell on his assistant's behalf.[7] Moreover, Lowell himself largely succeeded in getting the problem of a trans-Neptunian planet off his mind—his expressed intention in announcing his X paper to Trowbridge in December. From this point on he would give the subject only occasional, cursory attention.

Early in February 1915, Lowell returned to Flagstaff, but Saturn and commensurability, rather than his elusive planet X, were now uppermost in his mind. Indeed, Miss Williams was now computing Saturnian orbits commensurate with those of the planet's two inner satellites, Mimas and Enceladus, instead of solar orbits for a trans-Neptunian planet.[8]

Lowell's particular interest in Saturn dates back to at least 1907 when the planet's ring system was presented edge on to the earth. Late in October of that year W. W. Campbell reported that observers at the Lick Observatory in California had seen pairs of ''knots'' or bright condensations in the rings symmetrically placed on either side of the planet.[9]

Lowell, then at Flagstaff, immediately confirmed the observation, but in so doing plunged himself into a brief imbroglio in the public press with no less a figure than the U.S. Naval Observatory's Simon Newcomb, then the dean of American astronomy. In reporting his confirmation, *The New York Times* quoted Lowell as saying that the condensations indicated that collisions were occurring between particles of matter making up the rings and were thus ''proof that ... the ring system is in the process of falling in upon the planet.'' The *Times* made this statement the basis for a provocative headline which implied, to some readers at least, that an unprecedented calamity was in progress on the ringed planet.[10] Lowell's statement was quickly disputed by Newcomb and by John Brashear of the Allegheny Observatory, both of whom seem to have assumed that Lowell's "falling in" process was a precipitous one. In another *Times* article two days later, the two astronomers sharply criticized Lowell's conclusion and cited a long list of authorities from Laplace through Clerk Maxwell and Otto Struve to J. E. Keeler in defense of the stability of Saturn's rings.[11]

Lowell, in turn, retorted in still a third *Times* article that these criticisms were "founded on a misapprehension of what has been observed at Flagstaff," arguing that the observed condensations were evidence of collisions of particles in the rings and that such collisions would necessarily reduce the speed of some of the colliding particles in their orbits around Saturn. "This must reduce the orbit of the particle bringing it at last down upon the planet," he declared. "No demonstration of celestial mechanics is more certain than this. Indeed Clerk Maxwell himself foresaw it. There is no catastrophe involved as Prof. Newcomb supposes. The falling in upon the planet is a very slow process, but has been going on ever since the ring system formed and will go on for a long time to come."[12]

Privately, Lowell chortled incredulously to Lampland: "Newcomb has been talking celestial mechanics rot to the papers to the effect that Saturn's rings are stable! Think of it in the presence of collisions. Clerk Maxwell knew better. I have settled him, however, with the *Times*."[13]

In a subsequent observatory *Bulletin*, and in articles in the *Scientific American*, the *Philosophical Magazine*, and *Popular Astronomy*, Lowell described the "agglomerations" in the rings, which he now called "tores" because rather than flat rings they now appeared to be "rings after the manner of anchor rings encircling the planet." Continued observation, he found, showed these "thickenings of the rings" to be permanent in position and from micrometric measurements, he concluded that they resulted from the commensurable action of the satellites Mimas and Enceladus.[14] His observations and conclusions were not subsequently confirmed by any other astronomers, however, and it may be noted here that when the rings were again presented edge on to the earth a few years after Lowell's death, no "knots" or bright condensations were seen. Interestingly, Saturn's rings again became a matter of intense concern to astronomers in the 1970s.

Lowell's next major observational campaign on Saturn came in the fall of 1909 when its rings were tilted obliquely at a narrow angle to the earth. Now he discerned "wisps" angling across the planet's equatorial belt that were similar to those he had reported on Jupiter early in 1907, and found two "new" divisions in ring A, the outer ring. The wisps he described in an observatory *Bulletin*, the divisions in a paper in *Astronomische Nachrichten*.[15]

Not for another four years did Lowell again turn his attention seriously to Saturn. But then, in 1913 with the younger Slipher, Earl, as his assistant, he began a new series of studies of the planet that were continued off and on through the spring of 1915. Indeed, the completion of this investigation was a major reason for his return to Flagstaff in Feburary 1915, for on March 21 of that year, Saturn's rings would be presented to the earth at their maximum obliquity—their angle of inclination then being 26°56′.8.

His observations of Saturn in 1913 and 1914 produced several "new" divisions in the rings, and these Lowell reported in a short *Bulletin* dated February 4, 1915,[16] four days before his arrival in Flagstaff. Now, in observations of the planet made through May, Lowell and Slipher discerned and measured micrometrically or from drawings no less than ten "new" divisions in the outer rings A and B in addition to the well-known Cassini division.[17]

Back in Boston for the summer, Lowell analyzed these measurements and decided that the positions of the divisions corresponded to orbits commensurable with those of Mimas and Enceladus, except that the observed divisions seemed shifted systematically away from the commensurable orbits predicted by computation. His mathematical analysis of this latter circumstance, in turn, convinced him that some kind of gravitational anomaly was involved. Thus, he concluded that Saturn was composed of "concentric confocal spheroids of differing densities increasing inward," and that the planet was "actually rotating in layers with different velocities, the inside ones turning faster."[18] Or, as he later explained it somewhat more colorfully, Saturn was rotating "like an onion in partitive motion."[19]

One incidental result of this investigation of Saturn was to turn Lowell's attention again to Jupiter for apparently he felt that this largest member of the sun's family might also be composed of layered spheroids rotating differentially and that this might be determined from observations of the motions of its satellites. In August 1915 at any rate, he sent instructions to Flagstaff to have the younger Slipher measure the positions of the four largest Jovian moons during the planet's opposition in mid-September. "I want these," he explained to the elder Slipher, "for comparison in a problem I am working on Saturn from which it appears that the attraction of Saturn is greater than has been supposed; not on account of its mass but on account of its distribution inside Saturn."[20] Lowell and the younger Slipher were still studying Jupiter and measuring its satellites when Lowell died some fourteen months later.[21]

Lowell's activities during the last two years of his life, both astronomical and otherwise, are poorly documented in the Lowell archives, probably because he spent so much of his time in Flagstaff. In 1915, he was at his observatory from early February through May completing his investigation of Saturn, among other things. He returned to Flagstaff in mid-September primarily to observe Jupiter, remaining through October. And from late December 1915 through April 1916, he again sojourned at his observatory, now devoting most of his time to Mars, which came to a least favorable opposition on February 9, 1916. Although the telescopic martian disc then was barely 13 arc seconds in diameter, half its maximum apparent size at most favorable oppositions, Lowell and his assistants took 4,000 photographs of the red planet.[22] At this time, too, Lowell inaugurated the device of the "observation circular" to announce the results of the work at Flagstaff, issuing four during March 1916 regarding Mars, including one in which he described secondary features which suggested to him "a new order of Martian markings," and a fifth announcing V. M. Slipher's spectrographic discovery of "a permanent Aurora."[23] The discovery of Pluto in 1930 was announced in such a circular.

Through this period, too, the photographic search for Planet X continued with the Sproul 9-inch doublet, but without apparent success. The word "apparent" is necessary here, it must be noted, for on March 19 and April 7, 1915, images of a trans-Neptunian body which fifteen years later proved to be the ninth planet, Pluto, were recorded in one-hour exposures made by Lowell's former computer, Thomas B. Gill, but were not found on the plates at the time.[24] Observing conditions on these nights were only fair, and in his observation log for April 7, Gill noted: "Haze! Finally obscured by clouds."[25] Only after the discovery of Pluto in 1930 were the images recognized for what they were. Then, Lampland reported, they were "very faint," about a half a magnitude below Pluto's discovery magnitude of 15+, and were found only because preliminary orbits for Pluto indicated where to look on the 1915 plates.[26] As the elder Slipher also later noted, the observatory was "then not so well equipped for examining the plates," and in 1915 the observers there were "expecting to find it [Planet X] somewhat brighter than it turns out to be."[27]

It is at least possible that Lowell examined these plates himself, but it is not really probable for he had long since turned this tedious chore over to his assistants. But in any case, there is more than enough irony in the circumstance. That the images of Pluto were missed in 1915, the English astronomer A. C. D. Crommelin pointed out after Pluto's discovery was "rather a tragedy ... as it would have cheered him to know that his prediction was verified."[28]

During the spring of 1915, in fact, the photographic search was carried on under something less than optimum observing conditions, as Lampland complained to Miller late in June in requesting an extension of the observatory's use of the 9-inch doublet. "Even with our utmost efforts we have been unable to cover the fields we planned to photograph," he wrote. "With observing weather as we usually have it we should have had our program cleaned up in good shape before now. We have had a bad year, winter almost without end, I was about to say. The snow kept coming, even through April and the first days of May brought us a full twenty-four inches for good measure.... The discouraging part of it is that with us it is almost time for the rainy season."[29]

But it was Lowell, rather than Lampland, who was now discouraged. From Boston in mid-July, he inquired almost plaintively of his assistant: "No news from X?" giving a heliocentric longitude of 264° on July 1, 1915, and adding: "Most probable position I think 10°± from the ecliptic. What news have you?"[30]

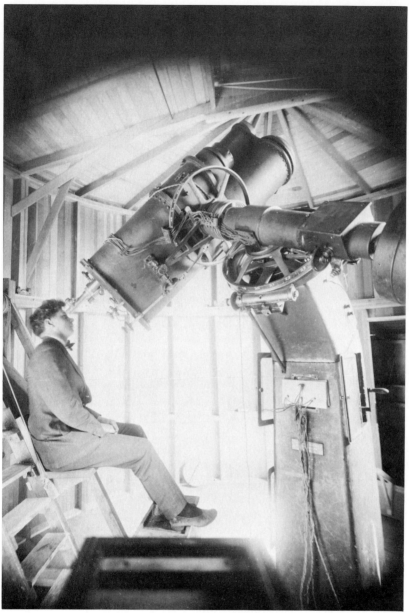

Lowell Observatory

Thomas Gill at the eyepiece of the Sproul Observatory's 9-inch photographic doublet, *ca.* 1915, in its makeshift dome at the Lowell Observatory in Flagstaff.

Lampland, in turn, sought to restore some of Lowell's waning confidence. "X is not yet in sight," he wrote in mid-August, "though you may well believe that I am in hopes that he is not far away.... This is no time to be discouraged."[31]

Lampland's hopes, if not Lowell's, were still high a month later. "The distant planet has not yet been located but for all of that we are not discouraged," he confided to Miller in a long, chatty letter on September 15. "It is, of course, a laborious undertaking. The by-products of the work, such as variables and asteroids, should be of some value. The investigation should manage to partially earn its living, as it were, aside from the main object in view. I suppose you are getting tired of extending the time of stay of the 9-inch," he added. "But you see we are a hopeful lot—in some things at least. Each day brings the hope that a little more work may turn the trick.... After so much work on a problem how one hates to give in!"[32]

By this time, however, Lowell apparently had given in. He was not a man to admit defeat in so many words; his usual reaction was to ignore it and turn his attention to other matters. It is notable here in this regard that after September 1915 and the publication of his trans-Neptunian planet memoir, references to Planet X or the search no longer appear in his available correspondence. Nor does he again refer to it in his subsequent writings, lectures or statements to the press. The photographic search did, in fact, continue for another ten months, but there is little in the record to indicate that Lowell actively concerned himself with its progress. A note in Gill's observation log for October 8, 1915, records the fact that "Dr. Lowell paid a visit to the 9-inch" that night, but after this there is nothing except routine notations concerning search plate exposures.[33]

It may well be that Lowell had now decided that even the Sproul 9-inch doublet was inadequate to the demands of such a photographic search. Years later, in noting Lowell's expectation that X's magnitude would be 13, Lampland recalled that "as the search proceeded, he more than once remarked that the planet might be considerably fainter ... and that very likely it would be necessary to search for a fainter object as soon as more suitable instrumental equipment should be available. At this time (1914–1915–1916), there was little prospect of carrying to a successful conclusion the erection of a new photographic telescope of the size and type best suited to the problem." Lampland also remembered that at his suggestion, Lowell had written Max Wolf in Germany to inquire if he could rent one of Wolf's 16-inch Bruce photographic objectives. But, he noted, under the "existing circumstances this did not appear to be feasible. Lowell's unexpected and untimely death in 1916 and the disturbed conditions of the Great War much delayed further progress on providing instrumental equipment."[34]

The photographic search with the Sproul 9-inch began on April 14, 1914, and continued through July 2, 1916, with Gill bearing the brunt of the observations, although Lampland and Edwards also took their turns at the telescope. During its course, nearly a thousand plates were exposed, with the fields being duplicated at approximately two-week intervals.[35] Although the subsequent examination of these plates, primarily by Lampland, failed to turn up a trans-Neptunian planet, the search did record 515 asteroids and some 700 variable stars.[36] The final entry in the 9-inch observation logs, incidentally, is "Lunch," indicating that the search was not over, but only suspended temporarily for more mundane concerns. It proved to be a long lunch break for Flagstaff observers, however, as the search for Lowell's Planet X was not resumed for another thirteen years.

Lowell spent the late spring and summer of 1916 in the east, and the few documents that relate to this period in the Lowell archives show that he had indeed turned his attention to matters other than Planet X. In June, he telegraphed the elder Slipher: "Please see that wild flowers on [Mars] hill are never picked. Put up notices."[37] He also suggested that his chief assistant search spectrographically for the presence of carbon dioxide in the atmosphere of Mars, Uranus and Neptune.[38] During this period, too, he set Lampland to making photographs of some spiral nebulae to determine if comparisons of these with earlier photographs made by others might reveal "rotary motion."[39] And late in July, he informed the elder Slipher that he intended to return to Flagstaff about October 15 "to observe Jupiter's V^{th} satellite."[40]

In August, Lowell spent some of his time preparing a series of four lectures in which he defended his embattled theory of the probable existence of intelligent life on Mars and philosophized somewhat bitterly over the difficulties encountered historically by scientists in getting major discoveries accepted by the scientific establishment. These he delivered variously to student audiences at universities and colleges in Idaho, Washington, Oregon and California during late September and early October.[41] By mid-October, he was back at his observatory and was still observing Jovian satellites there a month later when, on November 12, 1916, he died of a massive stroke at the age of 61.[42]

Lowell's final words on Planet X are found in his "Memoir on a Trans-Neptunian Planet" which was published in September 1915 in 105 handsomely printed quarto pages, with nine plates graphically displaying some of his findings. They are, moreover, carefully qualified words, a fact sometimes forgotten or ignored in subsequent discussions and debates over the relation of his analytical investigation and its outcome to the actual discovery of Pluto. Clearly, he recognized the uncertainties of his data, the limitations inherent in the problem itself, and the consequent indefiniteness of even his "best solutions"—he does not call them "predictions" in his memoir.

If Lowell presented these final conclusions with little of the confidence that usually marked his astronomical writings, he nevertheless considered that his analytical investigation had settled several questions raised in the aftermath of the discovery of Neptune. One of these stemmed from attempts to explain the fact that while Adams and Leverrier had been incredibly close in predicting Neptune's location in the sky, the former placing it $1°5$ and the latter only 55 minutes of arc from its discovery position, their postulated elements for the planet bore little resemblance to Neptune's orbit.

"A point in passing which these results serve to establish," Lowell wrote, "is the fallacy that though the elements of a disturbing unknown may not be found, the direction of that unknown can be discovered. The excuse for the statement seems to have been a mistake of Sir John Herschel's.... Sir John asserted that Neptune was found at the best time for its detection because its residuals from having reached a maximum had begun rapidly to decrease and that this showed it, the then unknown, to have been in conjunction with Uranus on or about 1822, such being the date of the maximum disturbance. Because at conjunction the tangential component of the perturbing force changes sign, he argued that the excess of longitudes must there be greatest."[43]

But not so according to Lowell. "The maximum deviation is not reached until the perturbing force or component after its reversal has had time to annul its previous effect on the velocity, a time long after its own reversal." Nor was there ground for the assertion for any other reason, he contended, citing his table of computed heliocentric longitudes for X to show that "the direction of the unknown varies according to the a' [mean distance] selected and unless all possible a's were taken as we have done in order to find the most probable distance, the deduced position would certainly be erroneous, as much wrong as the elements themselves."[44] That both Leverrier and Adams obtained fairly large eccentricities for their orbits," he noted, "was due to the erroneous value of a' they adopted, the various elements adjusting themselves to do their best to counteract its influence."[45]

More importantly, Lowell considered that his analytical investigation refuted Peirce's contention in the controversy following Neptune's discovery that commensurability affected the solution of the problem. Initially, Lowell seems to have accepted Peirce's conclusion that, because there were alternate solutions, and because Leverrier, in particular, had not taken commensurability into account in setting limits on his presumed planet's distance, the discovery of Neptune was a "happy accident." In 1909, for example, in summarizing Peirce's arguments in his book, *The Evolution of Worlds*, he wrote:

He [Peirce] proved this first by showing that Leverrier's two fundamental propositions —
1. That the disturber's mean distance must be between 35 and 37.9 astronomical units;
2. That its mean longitude for January 1, 1800, must have been between 243° and 252° —
were incompatible with Neptune. Either alone might be reconciled with the observations, but not both.
In justification of his assertion that the discovery was a happy accident, he showed that three solutions of the problem Leverrier had set himself were possible, all equally complete and decidedly different from each other, the positions of the supposed planet being 120° apart. Had Leverrier and Adams

fallen upon either of the other two, Neptune would not have been discovered.

He next showed that at 35.3 astronomical units an important change takes place in the character of the perturbations because of the commensurability of period of a planet revolving there with that of Uranus. In consequence of which, a planet inside this limit might equally account for the observed perturbations with the one outside of it supposed by Leverrier. This Neptune actually did. From not considering wide enough limits, Leverrier had found one solution, Neptune fulfilled the other.[46]

During Lowell's subsequent investigation, however, he specifically probed the problem of commensurability and convinced himself that his old mathematics professor at Harvard had erred, at least on this point. In introducing the subject in the early pages of his X memoir, Lowell declared:

In the matter of limits, neither of the old geometers was successful. Adams did not originally touch upon it at all, and Leverrier thought he had assigned limits but in this he was mistaken. Leverrier's work on the limits is so difficult to follow as to make its incorrectness a mitigated misfortune. In pointing out the fallacy, Peirce took for granted that the solution must fail for points of commensurable period, because the divisor of the coefficient here becomes infinite. In consequence, Sir John Herschel suggested that this was a question somebody should look into; which, of course, nobody did. Long afterward, in 1876, Adams wrote a note on the subject basing his proof on the supposed smallness of the residual terms, but, as we shall see later, these terms are large.[47]

Later, in discussing the perturbative action of X on Uranus, Lowell expanded on this point, explaining that the largest terms in the perturbative function were "always those whose period is either very long or nearly equal to that of the disturbed body. Owing to the great distance of X from Uranus, these terms in the present case are peculiarly embarrassing, consisting of those in $4n'-n$ and $4n'-2n$ which besides being very large, introduced the eccentricities to the third power. For as $a' = 47.5$, $4n'-n = 0$ nearly," he pointed out, "while $4n'-2n = -n$ approximately. Since in the integration the most important of these terms are divided by $4n'-n$, their coefficients become very large."[48]

And still later in his memoir, after detailing his first-power solutions for distances of X ranging from 51.25 a.u. sunward to 40.5 a.u., he came to "an important conclusion" on the point. "The various values of a' were chosen," he wrote, "first to secure a sufficient range in distance for the unknown, and secondly to include within that range an important commensurability point: namely, that for the period $4n'-n$ which occurs at $a' = 48.4$. By including this point within the field of exploration," he explained, "the solution would show whether the commensurability points affected the continuity of the perturbative function or not."

The outcome is conclusive. From the charts constructed from the several values calculated, it is evident that there is no discontinuity in the results at the points of commensurable period, the function passing through them without a break. Thus the *a priori* opinion at the time of the Neptune controversy that such points barred transit from one region of space to another proves to be unfounded and Sir John Herschel's question to be answered—and in the negative.[49]

Lowell here also enlarged on his conclusion that Adams' "supposed proof that commensurability introduces no discontinuity is not such in fact."

In his note on the subject ... he bases his argument on the statement that the remainder terms, after all that can be has been thrown on the elements of Uranus, are negligibly small, and in consequence continuity exists. But contrary to his supposition these remainder terms are not small; some of them are among the largest in the perturbation series when the time element is considered.... It is only near the epoch that the coefficient is small. But for solution it is imperative to go as far from the epoch as possible. Thus the residual terms in (4n'-2n) dwarf most of the other terms of the perturbation.[50]

This statement he followed with an equation demonstrating an analytical proof of the continuity of solution.

Before proceeding to his solutions, Lowell noted that the sum of the squares of the residuals of Uranus left by the theories of Gaillot and Leverrier ranged from 50.4 to 207.9 seconds of arc, depending on the number of observations and the time span considered.[51]

"These are the only theories of the planet worked out with sufficient completeness in their residuals to be at once available for further research," he pointed out. "The two differ chiefly in the values adopted by Leverrier and Gaillot respectively for the masses of the other planets used in the computations, though Gaillot has otherwise perfected Leverrier's theory besides bringing it up to date."[52]

But, he added, it was "too much to expect of any unknown to account for all these residuals. For in addition to inevitable errors in the theory of Uranus there are bound to be errors of observation. It is important to evaluate their probable amount as this will permit us to assign a minimum beyond which we cannot hope to go in rectification of the residuals recorded."[53]

This he did, deriving values for the probable observational errors by applying the method of least squares to sets of observations of Uranus grouped in time. He relied particularly on the prediscovery observations made by Flamsteed in 1715 and by Lemonnier in 1750 and 1768–69 at the beginning of the series, and on those made at Greenwich at the end. Parenthetically, he noted here that "during Flamsteed's three observations the planet moved nearly two degrees ... and during Lemonnier's eight, one degree. That it was not discovered then shows how vital scientific imagination is to science."[54]

The values he found for the sums of the squares of the probable errors of observation, by computation and from the curves of plotted residuals, ranged from 10.63 to 14.86 seconds of arc. And from these, and the least squares residuals from Gaillot's theory, Lowell drew a major conclusion:

When we compare the probable errors from observation with the actual residuals from Gaillot's theory we see that the latter are four times as great as the former and five times what they should be were they attributable to errors of observation alone. But his theory has been constructed with the greatest care from the most accurate data. So, was Leverrier's in his day. Yet the difference in the two sets of residuals is great. We cannot, therefore, assert that the present theory is not responsible; but we can say from the doctrine of probabilities that the present outstanding irregularities in the motion of Uranus are probably *not* due to errors of observation. This is a primary point of some importance.[55]

The primary purpose of Lowell's long analytical search was, of course, to find some indication of the most probable position of an unknown trans-Neptunian planet in terms of its heliocentric longitude at a particular point in time. Here he found that both his first and second power solutions yielded sets of orbital elements and positions for X that were, at least, in general agreement.

To show this, he tabulated thirty-three heliocentric longitudes for X for the date July 0, 1914,* according to the method of solution and the number of observations of Uranus used, for longitudes of epoch around 0° and 180° and for distances of 51.25, 47.5, 45.0, 42.5 and 40.5 astronomical units. For both first and second power solutions, the positions fell between $225°\!.1$ and $275°\!.2$ when ϵ' was around 180° and between $41°\!.9$ and $115°\!.9$ when ϵ' was around $0°\!.$[56]

Applying the least squares method to these solutions, Lowell determined the more probable positions and these further narrowed the ranges and, additionally, indicated that X's distance was somewhat less than his preferred 47.5 a.u.[57] The masses, eccentricities and perihelion longitudes corresponding to these positions also fell within quite narrow limits, particularly for the six most probable [H_{14}] solutions.[58] In finally choosing his two "best solutions," it may be noted, Lowell opted for two derived by his first power [H_{14}] method.

"Assembling the various solutions, got by considering such diverse terms and in such different ways," he wrote, "we are first struck by the general concordance the results exhibit. This is significantly displayed in the several sets of elements, in the comparatively close resemblance of the excentricities, masses, perihelion places and mean longitudes of epoch deduced. Even the resulting heliocentric longitudes of the unknown for a given date, July 0, 1914, come out in rather striking accord."[59]

He cautioned, however: "That they are not more so will perhaps be one's second thought.... Even in his problem, Leverrier discovered that his preliminary solution was entirely altered by the omission of two apparently

---

*This was Lowell's method of noting the mid-year point in time. It is the equivalent in modern usage of 1914.5.

trifling terms. Much more then should this be the case in the present investigation where we are perforce obliged to neglect terms, the coefficients of which we know to be large."[60]

If the longitude results were thus less than certain, the results of his investigation into X's probable latitude were so exponentially, as he noted in his memoir in briefly outlining his work on this problem. "To determine the inclination of the orbit of the unknown from the residuals in latitude of Uranus has proved as inconclusive as Leverrier found the like attempt in the case of Neptune," he reported. "The cause of the failure lies in the fact that the elements of X enter into the observational equations for the latitude. Not only e' and $\bar{\omega}'$ are thus initially effected, but ϵ' as well. Hence as these are doubtful from the longitude results, we can get from the latitude ones only doubtfulness to the second power."[61]

Before summarizing his final conclusions in his memoir, too, Lowell made some general remarks regarding the overall problem, his own and other's investigations of it, and the indefiniteness of his results. There were, for example, no shortcuts to a solution:

"That Leverrier's solution gave him limits which were erroneous shows how necessary to a full comprehension of the problem is the rigorous and more complete method of solution," he wrote. But this, he added, "does not detract from the great analytical skill displayed by both Adams and Leverrier in their masterly attack on the problem. That alone deserved success. Why it attained it is nevertheless a cause for some surprise, for Leverrier left out terms bigger than two he retained. The explanation would seem to lie in the nearness of Neptune and the near circularity of its orbit. Neptune turns out to have been most complaisant and to have assisted materially to its own detection."[62]

Simplifying the problem by assuming circular orbits and more than one perturbing body was also inadequate, he noted. "How important the circularity or non-circularity of the orbit is to the finding the unknown may be seen from Gaillot's circular solution which, although it gives a major axis for X (45 astr. units) not inconsistent with that from the elliptical one, is entirely discordant (295°) with the latter when it comes to assigning the present position of the perturbing body."[63] And the assumption that there was only one such body was "the only practical and workable supposition.... For though other bodies doubtless exist beyond, their effect is not likely, owing to distance, to be such as seriously to modify the primary solution. If the matter should be distributed in an asteroidal ring, although the theory would remain correct, the matter itself would escape detection," he added.[64]

Lowell also touched again on the quality of the data available for the trans-Neptunian planet problem vis-à-vis that used in the discovery of Neptune. "It becomes pertinent here to note what both Leverrier and Adams considered a satisfactory betterment of the Uranian residuals, and which, in view of the then relatively imperfect theories of the other planets it certainly was. The sum of the squares of the residuals left by Adams' theory was 348, and by Leverrier's 615. This was an enormous improvement on Bouvard's, or even Bouvard's corrected by Leverrier. Yet both are much larger than the sum with which we at present start as our initial data."[65]

Moreover, he declared, "It is vital to notice how it is the far residuals [in time] that give us information of the presence of an unknown not the near ones. These [near ones] have been made small by proper shift of the elements of the perturbed body and can always be so minimized. It is only the distant ones which cannot be thus reduced and which therefore constitute our guideposts in the case."[66]

Lowell then presented for consideration what he called "the credentials" of his solutions insofar as they indicated the presence of a trans-Neptunian planet and its probable position. Here he tabulated the sums of the squares of residuals from Gaillot's theory of Uranus, which did not take an unknown body into account, with those from his six most probable first power [H_{14}] solutions which did, along with the probable errors of observation. From this he concluded that "the outstanding squares of the residuals on the theory of no unknown body have by the admission of such an outsider been reduced 90%." And, he added: "Of the same claim to confidence are the credentials ... of all other solutions; some indeed attaining practically 100% elimination."[67]

Nevertheless, Lowell himself was not at all confident about his results, and before stating his final conclusions he entered some caveats:

> But that the investigation opens our eyes to the pitfalls of the past does not on that account render us blind to those of the present. To begin with, the curves of the solutions show that a proper change in the errors of observation would quite alter the minimum point for either the different mean distances or the mean longitudes. A slight increase of the actual errors over the most probable ones, such as it by no means strains human capacity for error to suppose, would suffice entirely to change the most probable distance of the disturber and its longitude at the epoch. Indeed the imposing "probable error" of a set of observations imposes on no one familiar with observation, the actual errors committed due to systematic causes, always far exceeding it.[68]

The duality of the solution, which yielded two probable positions each about 180° apart, provided another uncertainty, he noted. "If we go by the residuals alone, we should choose solutions which have their mean longitudes at the epoch in the neighborhood of 0°, since the residuals are there the smallest. But on the other hand this would place the unknown now and for many decades back in a part of the sky which has been most assiduously scanned, while the solutions with ϵ' around 180° lead us to one nearly inaccessible to most observatories, and, therefore, preferable for planetary hiding. Between the elements of the two, there is not much to choose, all agreeing pretty well with one another."[69]

Owing to the inexactitude of our data, then, we cannot regard our results with the complacency of completeness we should like. Just as Lagrange and Laplace believed that they had proved the eternal stability of our system, and just as further study has shown this confidence to have been misplaced; so the fine definiteness of positioning of an unknown by the bold analysis of Leverrier or Adams appears in the light of subsequent research to be only possible under certain circumstances. Analytics thought to promise the precision of a rifle and finds it must rely upon the promiscuity of a shot gun after all, although the fault lies not more in the weapon than in the uncertain bases on which it rests. But to learn the general solution and the limitations of a problem is really instructive and important as if it permitted specifically of exact prediction.

For that, too, means advance.[70]

Lowell summarized the findings of his analytical investigation in fourteen briefly stated, carefully worded points. These will be given here entirely in his own words, for his conclusions have been the subject of a lengthy debate within astronomy since the discovery of Pluto and the subsequent realization that its orbit agrees quite closely with that of one of Lowell's two "best solutions" for his Planet X.

His investigation, he wrote, established that:

1. By the most rigorous method, that of least squares throughout, taking the perturbative action through the first powers of the excentricities, the outstanding squares of the residuals from 1750 to 1903 have been reduced 71% by the admission of an outside perturbing body.

2. The inclusion of further terms yielded solutions in accordance with the first.

3. Solutions taking the years 1690–1715 also into account agreed substantially with those from the years 1750–1903.

4. So did those in which the additional years to 1910 were considered.

5. The second part of the investigation, in which solutions were made for the second powers of the excentricities as well, gave comfortable results.

6. When the probable errors of observation were reckoned, the outstanding squares of the residuals of theory excluding an outside planet proved to have been reduced by its admission from 90% to 100% nearly...."

7. Though this would indicate an absolute solution of the problem, it must be remembered that the actual as against the probable errors of observation might decidedly alter the result; and so might the terms above the squares in e and e' necessarily left out of account.

8. The investigation disclosed two possible solutions in each case, one with ϵ' around $0°$, one with it around $180°$; and that this duality of possible place would necessarily always be the case.

9. On the whole, the best solutions for the two gave:

ϵ' around $0°$	ϵ' around $180°$
$\epsilon' = 22°.1$	$\epsilon' = 205°.0$
$a' = 43.0$	$a' = 44.7$
$m' = 1.00$	$m' = 1.14$
$e' = .202$	$e' = .195$
$\bar{\omega}' = 203°.8$	$\bar{\omega}' = 19°.6$
hel. long. July 0, 1914	
$84°$	$262°.8$

the unit of mass = 1/50000 of the Sun.

10. It indicates for the unknown a mass between Neptune's and the Earth's; a visibility of the 12–13 magnitude according to albedo; and a disk of more than 1″ [arc second] in diameter.

11. From the analogy of the other members of the solar family, in which excentricity and inclination are usually correlated, the inclination of its orbit to the plane of the ecliptic should be about 10°. This renders it more difficult to find.

12. Investigations on the perturbation in latitude yielded no trustworthy results. This is probably because the excentricity e' as well as the planet's other elements enter as data into the latitude observation equations.

13. The perturbative function is not discontinuous at the commensurability of period points, a fact hitherto in doubt.

14. That when an unknown is so far removed relatively from the planet it perturbs, precise prediction of its place does not seem to be possible. A general direction alone is predicable.[71]

CHAPTER 7

LOWELL'S LEGACIES

At his death Percival Lowell left many legacies, not the least of which was an endowment for his observatory to assure that the work begun there in his lifetime would continue. There is no better testimony to the loyalty and respect his assistants held for him, or the dedication he inspired in them, than the fact that this work was continued and, indeed, produced significant results over the next fifteen years.

The problem of a trans-Neptunian planet, however, was not at first a part of this work, although clearly it was recognized as one of Lowell's legacies. "Some of the questions we shall have to answer," C. O. Lampland confided to his former astronomy professor, John A. Miller, after Lowell's death, "will be as to what efforts we have made to find a photographic doublet for continuing the trans-Neptunian work...."[1]

Lowell himself, as has been noted, apparently had concluded by late 1915 that the telescopes used for his earlier Planet X searches had been inadequate for the purpose. His assistants, too, from their own experience, were now convinced that a larger instrument, specially designed for sky survey work, would be necessary if such a search was to have any chance of success. But their initial attempts to obtain such a telescope, or at least the lens for one, were frustrated by several circumstances beyond their control.

The first of these involved the disruptions attendant on World War I and its aftermath, as Lampland recognized in October 1918 in a letter to Elis

Mrs. Constance S. Lowell, widow of Percival Lowell, at the 24-inch refractor at the Lowell Observatory shortly after her husband's death.

Strömgren in Copenhagen. In this, he asked the Danish astronomer to contact Max Wolf at Heidelberg, Germany, about the possibility of leasing one of Wolf's 16-inch Bruce photographic objectives after the war. "We are very much in need of a large photographic doublet for a problem we have had in hand for some years past," he explained. "It is at present impossible to procure suitable glass discs for so large an objective and it will in all probability be some considerable time after the termination of the war before such an instrument could be completed."[2] In June 1919, Lampland also inquired of two English glassmaking firms about the availability of astrographic discs 12 to 18 inches in diameter "to be used in a photographic telescope."[3] And about this same time Guy Lowell, the observatory's first trustee after Lowell's death, took up the matter of photographic doublets with the Reverend Joel Metcalf of Taunton, Massachusetts, who was considered to be one of the finest optical craftsmen in America, writing him "to ask you seriously whether you will help me obtain one for the Lowell Observatory."[4] These initial inquiries, however, came to naught.

A second and more important factor delaying the resumption of the "trans-Neptunian work" was the fact that despite Lowell's bequest, the observatory's staff was now required to operate under financial constraints resulting primarily from extended litigation over Lowell's estate, initiated by Lowell's widow, Mrs. Constance S. Lowell. This complex and somewhat bitter legal battle, the details of which are not germane here, dragged on for more than ten years and not only cast a shadow over the provisions Lowell had made for his observatory in his will, but progressively sapped the resources of his estate through court costs and what the elder Slipher later

Sketch of Guy Lowell, a cousin of Percival, made in 1917.

Lowell Observatory

declared to be "very excessive" attorneys' fees.[5] Through this period, the Lowell staff was under continuous pressure to keep expenses to a minimum and to husband the observatory's limited funds. Consequently, as a practical matter, the considerable expenditure required to obtain a large photographic telescope was then, and in fact later, simply beyond the observatory's financial means.

Lowell died childless and in his will, after providing quite handsomely for his wife, he left his residuary estate of more than $1 million in trust for the observatory, with the proviso that ten percent of the annual income be added each year to the principal. He also set up machinery for selecting a sole trustee to preside over the observatory's affairs, expressing the hope that "preference will be given to a male descendant of my immediate family if a suitable one exists." Initially, he named his brother-in-law, William L. Putnam, as trustee, but in subsequent codicils, he designated Harcourt Amory and George Putnam. At Lowell's death, none of these men was available to serve, and consequently Guy Lowell, one of Percival's cousins, accepted the trusteeship. Part of the observatory's problems after Lowell's death, incidentally,

Lowell Observatory

C. O. Lampland at the Lowell Observatory 40-inch reflector with equipment used in his radiometric studies of the planets.

stemmed from still another codicil to his will in which he declared that his widow "shall be consulted about the conduct of the observatory." Mrs. Lowell on occasion seems to have construed her role as something more than a consultant, however.[6]

Guy Lowell was a prominent New York architect, known for his design of the New York State Supreme Court Building on Manhattan's Foley Square, and a socialite whose interests leaned more toward the arts than the sciences. Despite this, however, he proved to be conscientious in the discharge of his responsibilities to the observatory, although he confined himself almost entirely to administrative matters, leaving the scientific decisions up to the elder Slipher, who had become acting director at Percival's death in accordance with an expressed wish in Percival's will. Guy Lowell visited Flagstaff a number of times during his ten-year tenure as trustee, but he much preferred Italy, where he had served during World War I as a major in the U.S. Army Intelligence Corps. He was on board ship at the island of Madeira en route to an Italian holiday when, in February 1927, a massive stroke ended his life. In accordance with Percival's will, he had designated one of Percival Lowell's nephews, Roger Lowell Putnam, son of William L. Putnam and Lowell's sister, Elizabeth, as his successor as the observatory's sole trustee.[7]

One of the more important effects of the lengthy litigation over Percival's will was its impact on the morale of the observatory's staff. This is perhaps best reflected in the comments of Roger Lowell Putnam when he assumed his new trusteeship duties, and while some of the legal issues in the will content were still in doubt. "It will be my ambition, as it is yours," he vowed to Slipher, "to make the observatory of the greatest benefit to the advancement of science, and I hope there will be no more legal battles to take thoughts away from this main issue."[8] And a few weeks later, as the matter of attorneys' fees was still being argued in court, he pledged: "It is going to be my ideal to get all lawyers out of the picture just as promptly as possible."[9]

It was not until April 1927, however, that the lawyers finally took their leave, and only then could the observatory staff, with the active and enthusiastic support of the new trustee, give serious attention once again to the problem of a trans-Neptunian planet. Still the financial constraints remained, as Putnam noted in August 1927. "The Lowell Observatory is not rolling in wealth," he confided then to Samuel Boothroyd, a former Lowell assistant in 1897–98, "and it has to conserve its resources as much as possible."[10]

Yet during the troubled decade following Percival Lowell's death, much solid work was accomplished by his assistants on a variety of problems to which, indeed, he had given high priority and which, in the last few years of his life, he had clearly felt held greater promise than his long and fruitless search for Planet X.

While the trans-Neptunian problem was, of necessity, now placed in limbo for ten years, his assistants kept busy with a number of other unfinished projects that could be continued with the existing instrumentation and within the established observational programs of the observatory. Some of their results proved to be of considerable significance in astronomy. But unfortunately, although almost nothing has been published concerning this work, it can be described here only briefly and to the extent that it illustrates what occupied the Lowell staff during the long hiatus in the Planet X search, and gives insights into the characters of the men who carried it to its eventual conclusion.

One major point should be made about this work at the outset. It was almost entirely observational and there were few attempts to synthesize or theorize from the new data, as Lowell had done so provocatively during his controversial career. Such speculation as his assistants now permitted themselves was confined largely to their private correspondence. They published their observational results or reported them at astronomical meetings only infrequently and sporadically, and in carefully qualified terms usually unencumbered by any attempts at interpretation. Lampland in particular was reluctant to go beyond his observations, and his circumspection in discussing his work prompted Miller at one point to caution Slipher: "Lampland is so conscientious that he is apt to put an element of doubt in his listener's mind that his results do no warrant."[11]

One effect of this conservatism was that the Lowell Observatory now began to acquire a reputation for accuracy and quality in its work that it had not always enjoyed when Lowell had guided its affairs. And soon other astronomers, more imaginative perhaps and less loath to speculate, found that the data being accumulated at Flagstaff were well worth their attention.

In March 1921, for example, Lampland reported that the Crab Nebula (M1) in Taurus had undergone changes in configuration over the preceding eight years.[12] His discovery, based on comparisons of fifteen photographs of the Crab with the 40-inch reflector between 1913 and 1921, generated considerable interest among astronomers and was, as Slipher informed Harvard's Solon I. Bailey, the "fruit of a carefully conducted piece of work."[13]

Lampland, however, carried this work no further and thus it remained for his friend John C. Duncan, the first Lawrence Fellow at Lowell Observatory in 1905–06, to follow up his discovery. One month later, using the 60-inch reflector at Mount Wilson, Duncan obtained a plate of the Crab which, when compared with one made in 1909 by G. W. Ritchey with the same instrument, not only confirmed the variability of the nebula, but permitted the measurement of the approximate rate of its expansion. This, in turn, led to the eventual identification of the Crab with the brilliant supernova reported by Chinese and Japanese observers some nine hundred years before, in A.D. 1054. As Duncan himself noted at the time, his work was a direct result of Lampland's discovery, but subsequent literature on the subject often ignores Lampland's contribution.[14]

The Crab Nebula, incidentally, also provides a typical example of V. M. Slipher's scientific conservatism. The spectrum of the Crab, he informed spectroscopist E. A. Fath shortly after Lampland's discovery, "is the most extraordinary one known."

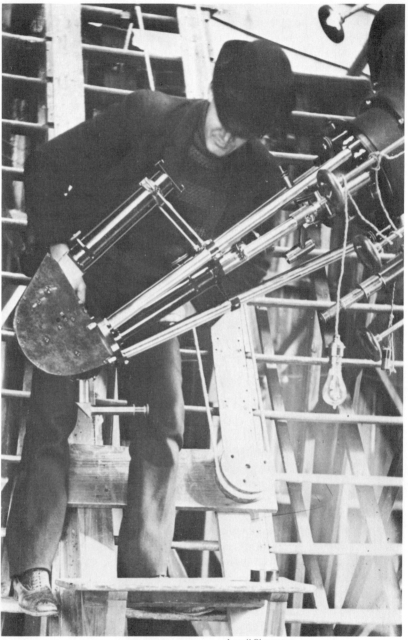

Lowell Observatory

Derby-hatted V. M. Slipher adjusts spectrograph attached to Lowell Observatory's 24-inch refractor during his work on spiral nebulae, *ca*. 1912.

V. M. Slipher as he looked about 1909.
Lowell Observatory

I have since 1915 made several spectrograms, which verify one another, but none is quite as good as can be made and I have postponed publishing a discussion of the spectrum until I might improve the observations. Thus only a brief announcement of the peculiar spectrum has yet been published from here.[15]

Lowell's three assistants differed from Lowell in many other respects besides their scientific conservatism. In high contrast to Lowell's ebullience and impulsiveness, they were reserved and cautious. Where Lowell had gloried in public attention and thrived on controversy, they studiously shunned the public eye and avoided even scholarly debate whenever possible. Nor did they possess Lowell's easy articulateness, wry humor, or charismatic personal attraction. Only the youngest of them—Earl C. Slipher—displayed anything of Lowell's exuberance and extrovertism, and then only outside of astronomy. After Lowell's death, he blossomed into a considerable politician, on the local level at least, and served as a Flagstaff city councilman, as mayor and later as an Arizona State Senator. The younger Slipher also gained a reputation as a popular lecturer and as a genial and entertaining host to astronomers and other scientists who visited Flagstaff and the observatory.[16]

All three men came from bucolic backgrounds, a far cry from Lowell's cosmopolitan upbringing amid the affluence of New England's late nineteenth-century industrial aristocracy. V. M. Slipher, born in 1875, and Earl, born in 1883, spent their childhood and youth on a farm in Mulberry,

Indiana. Lampland was born in 1873 in a log cabin near the village of Hayfield, Minnesota. There he also worked on a farm, clerked in a store, and played clarinet in the village band before matriculating at Valparaiso Normal School in Valparaiso, Indiana, where he graduated in 1899.[17]

All three men, too, learned their astronomy at Indiana University under John A. Miller and Wilbur A. Cogshall, with the elder Slipher graduating in 1901, Lampland in 1902, and the younger Slipher in 1906. Each, on completing his degree, came to the Lowell Observatory where, collectively, they spent a total of 160 years in its service. Lampland died in 1951 at age 78, and E. C. Slipher died in 1964 at 81, both still active in the observatory's work. V. M. Slipher retired in 1954, and died on November 8, 1969, three days before his ninety-fourth birthday.[18]

Lowell's assistants, it can also be said, were all gentlemen, physically robust, hardworking, tenacious of mind, modest, and intensely loyal to Lowell, his observatory, and to each other. Lowell was, as it turned out, extremely fortunate in his choice of men to administer his astronomical legacies.

Mars, of course, had held top priority for Lowell, and the systematic study of the planet he had initiated during his lifetime was not only continued after his death, but pursued in innovative ways with results that sometimes seemed to support some of his earlier conclusions. Astronomers at other observatories, too, began to find indications that Mars might not be as hostile to life as Lowell's critics had long contended. Indeed, in the years immediately following the 1924 martian opposition, when the planet made its closest approach to earth in the twentieth century, Lowell's much-disputed postulate of a habitable Mars came as close to general acceptance as it ever would.

Although the martian oppositions of 1918 and 1920 were not particularly favorable, both visual and photographic observations were made at Flagstaff. The young Slipher obtained some 6,000 images of the planet. The visual work was the responsibility of George Hall Hamilton, who joined the staff in 1917 and remained until June 1922, when he married Lowell's chief ally in the Planet X search, Elizabeth L. Williams, who was then working at the observatory in Flagstaff.[19] Results at the two oppositions, however, were quite routine.[20]

The martian oppositions of the early and middle 1920s, and particularly the highly favorable 1924 event, proved to be the most productive in terms of Lowellian theory. The planet was now observed not only visually and with the camera and the spectrograph, but first at Lowell and then at Mount Wilson with the radiometer in attempts to measure quantitatively surface temperatures on the planet. The results of these radiometric observations lent temporary credence to Lowell's earlier contention, based on deductions from insolation and light-scattering data, that nearly earthlike temperatures were reached on Mars.

Beginning radiometric observations of stars at the Lick Observatory in 1914, W. W. Coblentz of the U.S. Bureau of Standards was the pioneer in this work. Early in 1920, he joined forces with Lowell Observatory and, after Lampland had adapted his radiometers to the Lowell 40-inch reflector, Coblentz visited Flagstaff in October 1921 and carried out a brief program of stellar radiometry.[21]

The possibilities of radiometry for planetary studies were not, of course, lost on either Slipher or Lampland, and Slipher managed to persuade Coblentz to return to Flagstaff during the 1922, 1924 and 1926 martian oppositions to make his radiation measurements not only of Mars but of other planets.[22] To Slipher's great satisfaction, Lampland worked with Coblentz in making these observations, and carried on the work in the long periods between Coblentz' opposition-time visits. "I am very glad that this could be so easily agreed upon," Slipher wrote Mrs. Lowell shortly after Coblentz' 1922 sojourn at Flagstaff, "for as I remarked to Mr. Lampland his joining with Dr. Coblentz stamps it more as an undertaking of the Observatory and also that it will join him in the work and he will be credited in any future work he does of the kind with having had experience in the field; it introduces him as a worker in the field and that is I am sure important."[23]

Slipher's optimism notwithstanding, however, the Coblentz-Lampland collaboration proved to be one of the most turbulent in the history of astronomy, and a particularly painful experience for Lampland who carried it through, nevertheless, to its abrupt and bitter ending soon after the 1926 opposition.

It is clear from the archival record that the problem here arose from the personalities of the collaborators themselves. Coblentz appears to have been an arrogant, domineering, abusive man who ranted, raved and swore at the sensitive, mild-mannered Lampland during their brief but strenuous observing sessions at Lowell's unwieldy 40-inch reflector. In addition, he consistently criticized Lampland's careful preparation for these sessions, belittled his efforts to improve the radiometric apparatus, and ridiculed or ignored his suggestions and reservations concerning the work. A further complication in their partnership was the presence on its periphery of Donald H. Menzel, then a young astronomer at Princeton. With Coblentz' approval but to Lampland's deep dismay, Menzel undertook to mathematically reduce radiation data obtained at Flagstaff to thermometric degrees, publishing his results with a specificity that Lampland, at least, felt was far from justified by the data.[24]

This unhappy but not entirely unproductive collaboration dragged on, with Lampland largely suffering in silence, until November 1926 when Coblentz sent Lampland a letter concerning their work in which, among other things, he characterized Lampland's contributions as "drool," and even accused the Lowell astronomer of deliberately smashing delicate equipment. At this point, Slipher took up the institutional cudgels on behalf of his friend and colleague, writing to George K. Burgess, director of the Bureau of Standards and Coblentz' superior, and to Coblentz himself to complain about the latter's behavior and unilateral actions. Slipher succeeded in establishing an uneasy truce on the institutional level, and drew apologies all around, but the Coblentz-Lampland collaboration was over.[25]

From the standpoint of Lowellian theory, the significance of the planetary radiation studies in the mid-1920s lay in their indication that surface temperatures on Mars, at times and in places, at least, reached well above the freezing point of water. After the 1924 observations, Coblentz and Lampland, in a joint paper, reported that their results appeared to be "in harmony with other observations of Mars," presumably a reference to Percival Lowell's early insolation studies which had yielded a maximum temperature of 22°C for the planet and a mean of 9°C. Reductions of their 1924 data, they declared, "indicate that the temperature of the brightly illuminated surface of Mars is not unlike that of a cool bright day on this Earth with temperatures ranging from 7° to 18°C or 45° to 65°F."[26]

Moreover, there was independent reinforcement on this point in the results of radiometric observations made at the Mount Wilson Observatory during the 1924 opposition. There, astronomers Edison Pettit and Seth B. Nicholson reported an average absolute temperature at the center of the planet's full disk of 280° [7°C] or, as they put it, "a little above freezing in the tropics at noon."[27]

Pettit and Nicholson's findings, however, differed in some other details from those obtained at Flagstaff, and this was cause for additional concern for Lampland. Henry Norris Russell of Princeton, then a regular summer sojourner at Flagstaff with his family, sought to reassure him: "What ever discrepancies there may be between your results and those of Pettit and Nicholson will probably be cleared up at the next opposition," he wrote the Lowell astronomer. "Compared to our ignorance of the matter before 1924, I think that it is the *accordance* of the results of the two observatories and not the points of difference between them, which are [sic] impressive."[28]

The radiometric observations of Mars during the favorable oppositions of the 1920s created considerable interest both in and out of astronomy, indicating as they did that maximum temperatures on the red planet approached those encountered in terrestrial temperate zones. But other, better-known techniques also produced some results that tended to support some of the deductions Percival Lowell had made concerning the physical conditions of Mars.

Visual and photographic observations were pursued at many observatories including, of course, Lowell Observatory. Most notable, or at least the most publicized of these were observations at the Lick Observatory where astronomers drew and attempted to photograph linear markings on the planet that were somewhat reminiscent of the Schiaparellian and Lowellian "canals." R. J. Trumpler's work was important in this regard, and the elder Slipher subsequently cited it on occasion to defend Lowell's observations of these much-disputed martian features.[29]

The spectrograph was also turned toward Mars during these close approaches, and in 1924–25 produced two significant results, one favorable to the Lowell view of the planet and the other—obtained at Flagstaff by the elder Slipher—unfavorable. The latter involved a new attempt to determine the presence of chlorophyll in the dark areas and, like Slipher's earlier studies in 1905–07, yielded negative results.[30] "I have been disappointed," he wrote Guy Lowell of this second failure, "in not finding certain evidence of chlorophyll in the spectrum of the dark regions of the planet. Had not found any in times past but these earlier observations I did not get under as good conditions as regards plates and seasons on the planet.... If chlorophyll showed itself now along with the radiometric evidence of a terrestrial-like climate on Mars, the body of knowledge of Mars would all be in harmony. But the chlorophyll spectrum in light reflected by vegetation is not too conspicuous and may be it is especially so with certain varieties of plants."[31]

Slipher's disappointment may have been assuaged somewhat a few months later when astronomers at Mount Wilson reported spectrographic evidence that seemed to confirm Slipher's much-disputed 1908 claim to the detection of water vapor and oxygen in the martian atmosphere. In February 1925, Walter S. Adams and Charles St. John used a six-prism spectrograph with the Mount Wilson 60-inch reflector to observe Mars, and found that the amounts of water vapor and oxygen in the martian atmosphere were of the order of 5 and 15 percent respectively, of the amounts in the atmosphere above Mount Wilson at the time of the observations.[32] Comparable values from Slipher's 1908 spectrograms, as measured by Frank W. Very, had been 22 and 15 percent, considerably higher for water vapor but in full agreement with Adams and St. John's result for oxygen.[33]

Overall, the observations of Mars during these favorable oppositions brought more pluses than minuses for Percival Lowell's controversial martian theories. But Slipher, unlike Lowell, published no popular books or articles proclaiming the fact, although he now made the point privately whenever the opportunity presented itself in his correspondence. When for example, an Elmira (New York) College student in March 1926 asked him for a disclaimer of life on Mars for use in a classroom debate, Slipher replied:

> Unfortunately for your side recent investigations of Mars here and elsewhere tend more and more to confirm Lowell's conclusions regarding the planet by adding further evidence of atmosphere and water although both less than on Earth, and temperature such as would sustain organic life. While the canali markings [the "canals"] are best interpreted by assuming Mars possessed of intelligent beings, yet the great distance between us and Mars renders the telescope incapable of showing directly objective presence of living beings.[34]

And again, in another letter summing up the results obtained at recent martian oppositions, he declared that "the heat measures here first and later at Mt. Wilson Observatory have shown for the planet a temperature rising as high as Dr. Lowell had concluded from his work the planet must enjoy. Also the detection here in 1908 of water vapor on Mars with the spectrograph was verified in 1925 at the Mt. Wilson Observatory. And in addition, observations here ... as well as some made at the Lick Observatory imply even more atmosphere for Mars than Dr. Lowell's estimate. In short, Lowell's conclusions as to conditions on the planet have been very satisfactorily confirmed with the new data and additional methods."[35]

By the end of 1926, Slipher could detect a "somewhat different general attitude" on the part of astronomers in regard to Mars and the Lowell Observatory's early martian observations. "Years ago," he wrote to an English correspondent, "most astronomers looked with suspicion upon reports of observations concerning the planet and rather looked on such work as unscientific and injurious to the reputation of those engaged in it. The pendulum now seems to have swung oppositely and there is a tendency to look upon things observed on Mars as 'discoveries,' although they may be only phenomena recorded many times in the past.... No doubt the work done here more recently in other lines ... has in a large degree corrected the erroneous opinions held concerning the planetary work...."[36]

In assessing such statements, it is important to know to what degree Lowell's assistants actually accepted his thinking, particularly in regard to Mars. And it is clear that both the Sliphers and Lampland were Lowellians to a considerable degree, although none of them unqualifiedly embraced his most sensational conclusion of the probable existence of intelligent martians. Rather, during Lowell's lifetime and after his death, they accepted the framework of Lowell's Mars theories as working hypotheses which, through their various investigations, they sought to confirm in some of its points, and especially those relating to the physical conditions of the planet. That it was politic for them to do so goes without saying. But as has been shown, they were genuinely encouraged when their observations, or those of others, proved to be compatible with the Lowellian view of Mars.

The young Slipher, perhaps, traveled farthest along the Lowellian road. After almost sixty years of visual and photographic observations of Mars he, like Lowell, remained convinced of the reality of the linear markings, and that vegetation alone could explain the periodic changes observed in the telescopic appearance of the planet. And although he publicly declared shortly before his death in 1964 that there "is no evidence whatsoever that intelligent animal life exists on Mars," he nevertheless argued the possibility that such life *could* exist there, and his conclusions as to the planet's habitability paralleled almost point by point those detailed by Lowell in 1895 and thereafter.[37]

The querulous Lampland apparently never put himself squarely on record either publicly or privately on the subject of martian life in particular or extra-terrestrial life in general even though he, like Lowell, had observed and photographed the so-called "canals" and the seasonal changes of Mars and, as noted above, viewed the planet as being quite earthlike. The elder Slipher, too, publicly avoided the question of martian or extra-terrestrial life, and indeed only after Lowell's death did he even discuss it in private correspondence. Then, however, he was willing enough and in one letter in particular, written early in 1923 to a botanist, he revealed that he was a staunch believer in the probability of "extra-mundane" life, and expressed his belief with what was for him unusual eloquence:

> If life be thought to result from the substance and forces of nature then we could expect from what we have observed of distribution of life on the earth to find it in the main where favorable conditions repeat themselves; or if life be thought of as due to Divine intelligence it is for me just as difficult to think of its being confined alone to the earth; for if life has importance in the Creator's mind it would seem strange that all the vast universe (with doubtless many fertile and genial fields for life) should be left barren ... except this earth. For me to think of life as limited to the earth would mean either that we minify God and magnify man, or we would need largely to discard the results of astronomical research....
> A thousand million rose bushes and one rose! It is neither according to religious or scientific teaching, but reminds us of the time when the earth was regarded as the center of the astronomical system.[38]

Mars, of course, was not the only planet occupying the attention of the Lowell Observatory staff in the years following Percival Lowell's death. In 1921, Slipher reopened his 1903 spectrographic investigation of Venus, confirming to his own satisfaction at least, his earlier conclusion that its rotation must be considerably longer than 23 or 24 hours, as had been long believed, but failing to find any evidence of oxygen or water vapor in its murky upper atmosphere.[39] In this same year, he used the 100-inch Hooker reflector atop Mount Wilson in an unsuccessful attempt to determine spectrographically the rotation period of Neptune.[40] In 1924, Slipher observed a transit of Mercury, explaining to an inquiring newspaper reporter that transits of Mercury "are not extremely rare, as are those of Venus, but the individual's opportunity to observe one is rare enough that he should not let it go unimproved."[41]

Other solar system objects were also of concern in the years after Lowell's death. Comets appearing from time to time were routinely recorded photographically and spectrographically and in 1927, Lampland attempted radiometric observations of Skjellerup's comet to mark, as Slipher noted, "the first time heat measurements have been applied to a comet."[42] During this period, too, Lowell Observatory mounted two expeditions to observe solar eclipses—the first to Syracuse, Kansas, in June 1918, and the second to Ensenada, Mexico, in September 1923. The major result of the first of these, in which Mrs. Lowell participated, was the elder Slipher's spectrographic observations of the solar corona, but the second apparently yielded no results worth publishing.[43]

While studies of Mars and other solar system objects occupied much of their time, Lowell's assistants nonetheless managed to pursue a number of other investigations, and some of these, particularly the elder Slipher's continuing work on spiral nebulae, proved to be of fundamental importance in astronomy. Its importance was such, indeed, as to provide perhaps an additional explanation for the temporary suspension of the trans-Neptunian planet search by the Lowell staff.

In the years following Lowell's death, the focus of astronomical interest fell far beyond the solar system and its problems. Astronomers then were largely caught up in what has been called the "great debate" over the question of the size and structure of the universe itself, a debate usually epitomized by the inconclusive confrontation between Harlow Shapley and Heber D. Curtis in 1920 before the U.S. National Academy of Sciences.[44] The enigma of the spiral nebulae was basic to this long controversy—whether they were a part of the Milky Way and perhaps incipient solar systems in the early stages of evolution, or whether they were vast, distant aggregations of stars, "island universes" as philosopher Immanuel Kant suggested in the mid-eighteenth century, and analogous to the Milky Way itself.

Slipher's studies of spiral nebulae held considerable pertinence for this debate, although they were strangely neglected by the participants for many years. Nor did Slipher actively attempt to bring their significance to the attention of the debaters, preferring as always to eschew controversy and let his results speak for themselves. This they eventually did, and his discovery of the enormous radial velocities of these faint, difficult-to-observe objects in 1913 and thereafter proved to be a prerequisite to the final resolution of the great debate.

In August 1914, Slipher had published radial velocities for fifteen spirals, noting that all but two of them were receding from the earth at velocities of up to 1,100 kilometers per second, a startling figure at that time, and suggesting that this indicated "a general fleeing from us or the Milky Way."[45] In 1917, the year Holland's Willem de Sitter first postulated on theoretical grounds that the universe might be expanding, Slipher's list contained twenty-five objects, their velocities continuing to be "preponderantly positive," that is, away from the earth, and he declared that the "island universe" theory "gains favor in the present observations."[46] By this time, enough of Slipher's velocity measurements had been confirmed at other observatories so that there was little doubt either of the accuracy of his work, or the reality of the phenomenon.[47]

By 1918, however, the spirals were still an enigma to most astronomers. During visits to Mount Wilson and Lick Observatories, Slipher found "nebulae as the subject most discussed, especially at Lick. All seem eager for any bit of information concerning them and are interested to exchange views concerning them. All seem open-minded and willing to accept new suggestions and ideas ... for it is puzzling the minds of the best of them to develop a satisfactory concept of just what nebulae really are."[48]

Slipher continued his tedious, time-consuming work of measuring radial velocities, and in early 1921 announced that the spiral N.G.C. 584 in Cetus "has the highest velocity known, almost two thousand kilometers per second, receding."[49] After 1921, however, Slipher's observations of nebulae dropped off sharply, in part because he had exhausted most of the objects bright enough to give a readable spectrum with the 24-inch refractor, and in part because he was now into other things. Yet he remained almost alone in the field, and while he did not publish the few radial velocities he subsequently obtained, he freely gave permission to others to use his unpublished data in their publications.[50]

Thus in 1929 when Mount Wilson's Edwin P. Hubble derived his famous constant relating the radial velocities of spiral nebulae to their distances that has allowed modern astronomers to approximate the dimensions of the known universe, he used, as he later reminded Slipher, "your velocities and my distances."[51] And in fact, of the forty-six radial velocities available to Hubble at the time, forty-one of them had been obtained by Slipher. The work of Mount Wilson's Milton Humason along this line, so often linked with that of Hubble, had then only just begun.[52]

A second aspect of Slipher's nebular studies had a somewhat more immediate, decisive impact on the "great debate." In 1913, in determining that the spiral N.G.C. 4594 in Virgo was receding from the earth at about 1,000 kilometers per second, Slipher also noted that the faint, red-shifted absorption lines in its spectrum seemed to be slightly inclined. Typically, he waited a full year to obtain what he considered to be a satisfactory confirmatory spectrogram before announcing his discovery that the Virgo Nebula was rotating.[53]

Over the next three years, Slipher found rotation in a number of other spirals, including the Great Nebula in Andromeda, and late in 1917, in publishing the results of his work on the spiral N.G.C. 1068, he concluded that they were all rotating in the same relative direction, that they were all "turning into the spiral arms like a winding spring," as he put it.[54] Unfortunately however, this conclusion was in direct conflict with results obtained by Adriaan van Maanen at Mount Wilson who, since 1916, had reported evidence for rotation in the opposite direction, that is, with the arms unwinding, on the basis of comparative photography of spirals over time. Van Maanen's work was of deep interest because of its implications to the "great debate." For if internal motion in spirals could be detected in photographs taken at relatively short intervals then the spirals could be at no great distance from the earth. Many astronomers for many years cited van Maanen's work to argue against the "island universe" theory and the idea that spiral nebulae were distant galaxies of stars like the Milky Way.[55]

Slipher was aware of this conflict from the beginning. A month before van Maanen's first publication, Duncan had confided to him that van Maanen, "a very enthusiastic Dutchman," had measured photographs of the spiral M101 made some years apart and had gotten what seemed "to be certain evidence of motion *along* the arms of the spiral. He has measured a few points on M81 also, with a somewhat similar result."[56]

Slipher, in publishing his spectrographic findings a year later, had no comment to make on the conflict with van Maanen's results, nor did he subsequently mention it publicly. Even in his private correspondence, he remained neutral and circumspect. Thus, for example, he answered an inquiry by New Zealand astronomer A. C. Gifford:

> It is unfortunately a fact that the results from the spectrographic observations of nebulae show the central parts of the spirals to be rotating in a direction opposite to that indicated by van Maanen's observed motions.... No results have been got here spectrographically since to modify the conclusions drawn from the earlier spectrographic rotations.... I have heard some expressions of doubt as to whether van Maanen's results might not somehow be in error. Then perhaps there are some astronomers who might think some other interpretations might be applied to the spectrographic observations. It is perhaps natural that I would not hear the spectrographic results questioned.[57]

Van Maanen's results, as it turned out, were in error, as subsequent work by other astronomers made clear and as van Maanen himself conceded some years later.[58] The situation in regard to the direction of nebular rotation, moreover, was complicated by the question of which edge of the spiral was nearest the observer, Slipher opting for the edge showing dark, absorbing matter silhouetted against the bright nucleus of the spiral. Again, it was many years before he was shown to be right.[59]

Lampland, too, had concerned himself with comparative photography of nebulae in an attempt to detect proper motion and/or internal motion in these objects. He had begun these studies in 1913 largely as a result of Slipher's radial velocity discoveries. A spinoff of this work, as noted earlier, was Lampland's discovery of the variability of the Crab Nebula. But in announcing this, Lampland also had reported changes he had observed in the spiral nebula M99. He had published preliminary results indicating internal motion in this nebula in 1916, at the same time as van Maanen's initial publication of internal motion in M101, but his 1921 report stirred renewed interest because of the heightened "great debate." Lick Observatory's J. H. Moore, for example, wrote Lampland: "Your discovery of variations in the nucleus of a spiral nebula is one of the most interesting and most important of recent observations. It looks bad for the 'island universe' theory doesn't it? I must say it is one of the strongest points yet brought out against the hypothesis."[60]

Van Maanen, whose observations of internal motion in spirals were still unconfirmed elsewhere, was particularly interested in Lampland's 1921 observations of M99, and requested further details. The cautious Lampland replied: "I should not perhaps at this time make any more definite statements than that the nucleus undergoes changes in its form and internal structure. There may be some variability in the light of the nucleus but that is uncertain.... The suspected changes outside the nucleus are not of motion but variation in brightness." And, he added to van Maanen's very probable chagrin: "It may not be possible to establish from the present material whether or not the apparent differences are real."[61]

Lampland's few published conclusions regarding internal motion in nebulae were similarly hedged about with qualifications. Van Maanen, on the other hand, publishing his findings far more frequently and far more confidently, with the result that they attracted far more attention. The Dutch astronomer's work, however, became subject to progressively sharp criticism through the years and after 1921, perhaps aware of the growing skepticism over van Maanen's claims and certainly preoccupied with preparations for his stormy collaboration with Coblentz in planetary radiometry, Lampland largely abandoned the comparative photography of nebulae.

One other aspect of Slipher's spectrographic studies took on increasing importance in the years following Lowell's death and indeed kept him busy even after the trans-Neptunian planet problem once again came to the fore. This was his investigation of the spectra of the aurorae and the light of the dark night sky.

This work had begun serendipitously in the late spring of 1915 when Slipher launched a long-delayed program to obtain spectrograms of the night sky designed "to see how well the composite features come out in it as a suggestion of what we might expect from a galaxy observed from a great distance."[62]

What turned up on his first long-exposure plate, however, was the distinctive yellow-green emission line of polar aurorae, although no auroral display was in progress during the observation. Subsequent plates of the dark night sky also showed this prominent line and thus in March 1916, as noted earlier, Percival Lowell proclaimed Slipher's discovery of a "permanent aurora."[63] The following November, in a Lowell *Bulletin,* Slipher reported that the chief auroral line appeared on more than fifty plates exposed to different regions of the night sky, even though no auroral activity was evident.[64]

By early 1919, Slipher obtained a series of high dispersion spectrograms of the night sky with a 15-inch camera, including one exposure of 115 hours, and determined from them that the wavelength of the chief auroral line was 5578.05Å, some seven Ångstrom units above the previously accepted value. From this he concluded that the line could not be due to nitrogen, as others had proposed, and some years later it was independently found to be due to atomic oxygen.[65] Slipher continued his night sky and auroral observations through the 1920s, discovering many new features in their spectra that were later shown to be due to oxygen, nitrogen, sodium and other components of the earth's upper atmosphere in various states of ionization. In the early 1930s, despite the responsibilities he carried as the result of the discovery of Pluto, he extended this work to the zodiacal light, discovering auroral features in its spectrum as well.[66]

Thus, during the decade following Percival Lowell's death, not only Lowell's assistants but astronomers in general were primarily concerned with other, more pressing matters that held far greater significance for astronomy than any hypothetical trans-Neptunian planet. This latter problem was not wholly forgotten, however. In 1919 Harvard's William H. Pickering aroused passing interest in it by announcing a new orbit for his 1908 Planet O, this one based on Neptune's residuals, and by calling on all observatories to join in a search for his planet at its predicted opposition during the last week in December.[67]

This call, as will be seen, was heeded at Mount Wilson Observatory and was at least noted at the Lowell Observatory, although no action seems to have been taken at Flagstaff. On December 30, in response to a query from the Los Angeles *Times*, Hamilton and Orley Hosmer Truman, who had begun a brief service as a Lowell assistant earlier in the year, declared that they had "been expecting for some days to make a search for the suspected extra-Neptunian planet, but have been delayed until now by unfavorable conditions." They added that they intended to make a search visually with the Lowell 24-inch refractor, but if they carried out this intention, there is no indication of it in the 24-inch observation logs in the Lowell Observatory Archives. In any case, a visual search must have been futile in the context of the observational problem. It is notable in this connection that the elder Slipher, asked by the press to comment on Pickering's prediction, said nothing about any search being undertaken at Lowell, and confined his remarks to his belief that such a trans-Neptunian planet existed, and to Percival Lowell's long work on the problem.[68]

Pickering's announcement, incidentally, may have inspired another trans-Neptunian planet prediction of a more frivolous nature. In March 1920, one P. W. Gifford of Chicago, Illinois, who described himself as a "Co-ologist" and as a "Humanitarian Harmonist," proclaimed the existence of two trans-Neptunian planets. These he identified as Styxia, with a period of 228 years, and Janus, with a period of 426 years, adding that they were "presently 23° and 27° in the astrological sign of Taurus" respectively. "The above information," he explained, "was ascertained by means of Co-ologic Astrology...."[69]

One other event during this period proved to have significance for the trans-Neptunian planet problem. This was the death, on February 21, 1925, of the Reverend Joel Metcalf who, his obituarist Solon I. Bailey of Harvard declared, "had no superior and probably no equal in this country" in the field of applied optics. Bailey revealed that Metcalf, at the time of his death, was working on a 13-inch triplet, "the largest telescope of this type ever attempted. It is doubtful," he added, "whether at the present time in this country there is anyone who can complete the work on this lens and bring it to the degree of perfection which Mr. Metcalf would have achieved."[70]

The partially worked glass discs for this lens were subsequently purchased from Metcalf's estate by Guy Lowell, with personal funds, and became the property of the Lowell Observatory.[71] A few years hence, they would be figured into a superb objective lens for an unusually fine photographic telescope to carry out a third and, as it turned out, final search for Percival Lowell's Planet X.

CHAPTER 8

THE THIRD SEARCH FOR PLANET X

While the search for Percival Lowell's Planet X was suspended for more than ten years following Lowell's death, another erstwhile planet-hunter, William Henry Pickering, continued his sporadic efforts to solve the trans-Neptunian problem.

During these years, Pickering issued two new sets of orbital elements both for his 1908 Planet O and his 1911 Planet P, postulated still another trans-Neptunian planet which he designated S, and suggested a "rather massive unknown satellite of Saturn" which he later elevated to full planetary status as Planet U. All in all, from 1908 to 1932, Pickering proposed no less than seven hypothetical planets for the solar system beyond the asteroids—O, P, Q, R, S, T and U—and as his final elements for O and P define completely different bodies than did his original ones, the total can be set at nine, certainly the record for planetary prognostication.[1]

In all these predictions, early and late, Pickering relied on his graphical method supplemented by his analyses of cometary motions—an empirical approach which, as noted earlier, he conceded to be "of a more or less 'rule of thumb' character."[2] But by 1919 at least, he could add Neptune's residuals to those of Uranus, Saturn and Jupiter on his graphs.

Except for his 1919 and 1928 orbits for Planet O, which figured briefly in the controversies that followed the 1930 discovery of Pluto, Pickering's planetary predictions are only of passing interest as astronomical curiosities.

His 1911 elements for Planets P, Q and R have been described in Chapter 5, and of these three proposed planets, only P seriously occupied his attention in later years. He subsequently declared that Planet Q would never be found because, despite its great mass and high inclination, it was "not well situated to produce perturbations" in the motions of other planets. And Planet R he also later dismissed as being "doubtful."[3]

In 1928, however, he published a circular orbit for a Planet P which reduced its presumed distance from 123 to 67.7 a.u. and its period from 1400 to 556.6 years. Now he gave P a mass of twenty earths and a magnitude of 11.[4] And three years later, after the discovery of Pluto, he issued another, elliptical orbit for P, putting its distance at 75.5 a.u. with a period of 656 years and a mass of fifty earths. Its magnitude was still 11, and its eccentricity and inclination of orbit were 0.265 and 37° respectively, close to the values in his 1911 orbit.[5]

His Planet S, first proposed in 1928 and furnished with a partial set of elements in 1931, is more interesting perhaps, if only because its distance of 48.3 a.u. is close to the 47.5 a.u. long preferred but eventually rejected by Percival Lowell for his Planet X. Its period Pickering put at 336 years, its mass at more than five times the earth's and its magnitude at 15. He also pointed out that the aphelion points of at least four periodic comets were between 47.6 and 59.1 a.u. and thus their orbits "might connect with S."[6] Some people later thought S was similar to Pluto.[7]

Pickering's 1929 postulation of a Saturnian satellite became in 1932 a prediction for a barely trans-Jovian planet U which circled the sun at a mean distance of 5.79 a.u. in 13.93 years. Its mass was 0.045 earths, and he noted that its eccentricity of 0.26 and perihelion distance of 4.28 a.u. swung it inside Jupiter's orbit briefly during its journey around the sun. He also speculated that U might have passed within the Roche limit of Jupiter and been "torn to pieces." It might "exist therefore merely as a swarm of gigantic meteors, or perhaps we might say small asteroids, like Saturn's rings."[8]

The least of Pickering's planets is surely planet T, a suggestion advanced in 1931 in the course of publishing his final predictions for P and S and abandoned even as it was made. On his graphs he noted that he had found "small perturbations of Uranus following the large ones caused by planet *P*" and he thought these might be due to a second body with a period of less than 188 years and a corresponding distance of 32.8 a.u., just beyond Neptune's orbit. But he quickly added: "We can hardly as yet accept *T* as a real planet until it presents further evidence of its own existence."[9]

Pickering's 1919 and 1928 elements for his Planet O however, in light of Pluto's subsequent appearance on Lowell Observatory's Planet X search plates, have somewhat greater importance in the history of planetary discovery. For while these clearly described two entirely different bodies, both later proved to resemble Pluto in one or another respect and thus were used, by Pickering and others, to argue short-lived claims that Pluto was actually Pickering's Planet O.

In 1908, it will be recalled, Pickering placed Planet O at a distance of 51.9 a.u. with a period of 373.5 years, a mass of twice the earth's and a magnitude of 11.5 or 13.4, depending on albedo.[10] His 1919 orbit for O, with Neptune's residuals now included on his graphs, extended its distance to 55.1 a.u., its period to 409 years, and its magnitude to 15, while retaining the same mass. And in 1919, he suggested that while he had "little direct information" on the longitude of O's ascending node or the inclination of its orbit, these elements might "perhaps be estimated at $100°\pm$ and $15°\pm$ respectively." In addition, he predicted O's position at opposition December 30, 1919, and urged all observatories to make a search for the planet at that time.[11]

These 1908 and 1919 predictions are not so far apart, relatively speaking, and can be considered as referring to the same hypothetical planet. But Pickering's 1928 Planet O, in contrast, was as much intra-Neptunian as trans-Neptunian, spending half its course around the sun inside of Neptune's orbit. On the basis of a circular orbit, Pickering first placed this O at a distance of 35.23 a.u., with a period of 209.2 years. Its mass, he now found, was only half that of the earth, its magnitude was 12, and its angular diameter was $0.''5$ [seconds of arc] or the equivalent of 10,160 kilometers [6300 miles]. He also worked out some elements of an elliptical orbit with an eccentricity of 0.195, noting that such an orbit called for a shorter period. Neptune's period of 164.8 years "was found to give a satisfactory result," he wrote, "and at once implied that O was really not an exterior, but a companion planet to Neptune, the case being analogous to the sixth and seventh, and to the eighth and ninth satellites of Jupiter." This object, he added, "is in no sense a satellite of Neptune, revolving around it in 165 years, but a true primary planet."[12] Harvard College Observatory actually made a brief photographic search for this Planet O in January 1928, prior to the publication of Pickering's prediction, but no planet was found.[13]

During the long hiatus in the Planet X search, the astronomers at Flagstaff undoubtedly followed Pickering's planetary prognostications with some interest, although there is nothing in the available record to indicate this. Even Pickering's various 1928–29 predictions failed to bring a reaction on Mars Hill, although the observatory's staff then was already deep in preparation for a third search for Percival Lowell's Planet X. After 1919, there are apparently no references to Pickering's predictions in the archival correspondence until this new search turned up Pluto. Then, when the point was raised, the elder Slipher minimized the effect of Pickering's work on the X search. To H. H. Turner, the veteran editor of *The Observatory,* for example, he declared that "we ... have been quite uninfluenced by Pickering's publications, except possibly to press forward the search a little more vigorously."[14]

This third search for Planet X began in April 1927, two months after Guy Lowell's death. He had named Percival Lowell's nephew, Roger Lowell Putnam, to succeed him as sole trustee, a choice that came as a surprise to Putnam. "I had no idea until Mr. Guy Lowell's death," he wrote Slipher shortly after the funeral in Boston, "that he had named me as his successor, and feel very much unprepared for the duties of Trustee, particularly as I am very busy here. On the other hand, I do look forward to the opportunity with a great deal of pleasure.... I presume that the reason Mr. Guy Lowell appointed me ... was due to the fact that I specialized in mathematics in college, and have always been very fond of it. I am afraid, though," he added, "that I have forgotten most of what I knew, but hope to find that I know enough to at least listen intelligently when we talk."[15] A few days later, he wrote: "I am looking foreward to coming out and meeting you and your associates and working together to carry out Uncle Percy's wishes and to carry on his work."[16] And on March 15, he notified Slipher that he would arrive in Flagstaff on April 2 aboard the Santa Fe Railway's famous train, "The Chief."[17]

Putnam's selection as trustee proved to be an excellent one and he would serve the observatory well for four decades. In many ways, he was much like his "Uncle Percy." A graduate of Harvard University, he was already a successful businessman in Springfield, Mass., where he lived and ran a large package machinery corporation, at the time of Guy Lowell's death. His many interests included an active concern with politics, although his persuasion here was liberal and Democratic in contrast to Percival's ardent conservative Republicanism. Eventually he would be elected mayor of Springfield and serve as an official in the national administration of President Harry S Truman in the late 1940s. Putnam also had Percival's receptive mind and quick intellectual grasp, his extrovert personality and, importantly, his enthusiasm for the project at hand. Indeed it can be said that Putnam's enthusiasm for the trans-Neptunian planet problem was primarily responsible for the resumption of the long-suspended Planet X search.

There is no record of the conversations that Putnam had with Slipher and other members of the staff during his early April visit to the observatory. But there is no doubt that Percival Lowell's planet X was discussed at length, and that a decision was made to launch a new search as soon as a suitable photographic telescope could be built. The cost here was still a concern, but it is clear that this was not to delay the start of the project. Expenses would be trimmed by building the mounting and the dome in Flagstaff under the supervision of Slipher and the observatory's fine instrument-maker, Stanley Sykes. And the observatory already had the set of partially-finished 13-inch glass discs, purchased two years before by Guy Lowell from the Reverend Metcalf's estate, that could be turned into an objective lens.

On his return to Springfield, Putnam immediately set to work to find an optical specialist to figure the 13-inch discs into a photographic objective. On April 26, he wrote Slipher that he had talked the matter over with Harvard College Observatory director Harlow Shapley who felt that J. W. Fecker of Pittsburgh "was really the only man in the country who could figure the lens. Fecker is doing a lot of work for the Harvard Observatory," Putnam added, "and Professor Shapley thought that, by writing to him, he might get more

Lowell Observatory

Stanley Sykes, Lowell Observatory's instrument-maker for nearly fifty years, who was responsible for building the mounting and the dome for the 13-inch Lawrence Lowell telescope used in the final Planet X search.

Lowell Observatory

The dome for the 13-inch Lawrence Lowell refractor under construction at Lowell Observatory in 1928.

results and he has, therefore, written to see if Fecker could take it on, how long he thinks it would be, and how much it would cost."[18]

Three days later, Putnam reported to Slipher that Fecker "feels confident that he can figure and grind the 13″ lens all right, but he wants $4,000 for doing it. This seems to me quite a lot of money, especially in our present financial condition. Professor Shapley suggests getting in touch with Sir

Howard Grubb of Parsons & Company in Newcastle [England], and I think I shall write him, at least to get his figures. Meanwhile, I have a bright idea of trying to get somebody to give a lens to the Observatory, but I have got no further than having the bright idea, and may not succeed. Somehow or other, though, I feel sure we will get this done, and done fairly soon"[19]

The time for Putnam's "bright idea" came two months later. He had, in fact, written Grubb,[20] but on June 27 he advised Lampland that progress was slow. "It takes a long time to get letters back and forth from England. I think that Fecker of Pittsburgh is still the best bet," he declared, adding as an afterthought that "one of the things I am going to Boston for this afternoon is to see if we cannot get someone else to finance it."[21] This Boston visit, as it turned out, proved to be very productive indeed. On June 30, he informed Slipher:

> Both you and Lampland will be particularly glad to hear that President [Abbott Lawrence] Lowell of Harvard is going to give the thirteen-inch telescope. I am still waiting to hear from Sir Howard Grubb, but as soon as I do hear from him, will take steps to start getting the lens figured. I have done nothing about the rest of the instrument—the mounting, etc.—because I don't know what to do. The estimate for figuring the lens was $4,000 from Fecker, and I told Uncle Lawrence that, by rule of thumb, from that the total cost of the instrument would not be over ten thousand dollars. I hope I am right, because I would hate to go back and ask him for any more.[22]

Putnam's letter was good news on Mars Hill. "This is a splendid gift and a fine spirit of helpfulness," Slipher replied. "We hope this instrument will have a useful career and that President Lowell will find satisfaction in the results attained through its use." Slipher also reassured the trustee on the cost. "Your estimate ... ought to cover the job very nicely.... The housing of the instrument need not be very costly as such an instrument is short of focus and so does not require a large shelter. A good driving clock with a good big worm wheel so as to give accurate running of the instrument are important features, and then a good sized guide telescope...."[22]

Slipher by this time had already been at work on plans for "the rest of the instrument" and had, in fact, ordered a set of worm gears for the new refractor from the Philadelphia Gear Works, writing the firm "to suggest that you take plenty of time to finish the job. We are, as you know, particularly anxious that the gear and worm be very carefully cut so as to insure their giving very accurate driving for our telescope."[24] And in responding to Putnam's letter about President Lowell's gift, he dwelt at length on the mounting for the refractor. "One feature of the mounting ... is of very great importance, namely it must be mounted so as to allow passing through the meridian for all declinations without having to interrupt to shift the pointing of the instrument to the other side of the pier as would have to be done with the usual equatorial mounting for stars north of the zenith. In this respect, the design we have used in the mounting of the small reflectors here is quite satisfactory. In this case the polar axis is rather long and supported at both ends with the declination axis mounted about midway of the polar axis. In this way the whole sky can be reached and exposure can run right through the upper culmination of the object without any interruption. Such a mounting would not be costly to build."[25]

Sir Howard Grubb's estimate did not come in until early in August and Putnam then advised Slipher that it was "slightly more than half the price quoted by Fecker, but I have yet to hear how much the duty will be. It seems to me that, if the two prices come out pretty nearly equal, it would be better to stick close to home. I should, however, very much like your thoughts and Mr. Lampland's on the whole question of the new telescope. Do you think it would be better to wait until I can get out and talk the matter over, which may not be for a couple of months yet—or had we better try and get started quickly?"[26]

The import duty for the lens proved to be 50 percent and thus the total cost under Grubb's estimate, to Slipher's mind at least, seemed high.[27] By the end of August, Putnam concluded that a more thorough discussion was required, writing Slipher that he was "holding off doing any work on the 13-inch telescope until we can get together."[28]

Putnam did not get to Flagstaff until the second week in October, and again, there is no record of the conversations he held there. Yet obviously it was decided to seek still another estimate for figuring the 13-inch lens, this from C. A. Robert Lundin of the Alvan Clark and Sons telescope-making firm of Cambridgeport, Massachusetts. Indeed it is curious that the Clark firm and Lundin were not considered in the first instance, for the firm had made Lowell's fine 24-inch refractor, Lundin himself had supervised the construction of the Lowell 40-inch reflector and its conversion in 1925 to a 42-inch aperture,[29] and both Slipher and Lampland had worked closely with him on optical matters over the years. Perhaps Putnam at the beginning had President Lowell and the Harvard connection in mind and simply preferred to follow Shapley's thinking first.

At any rate, with his return to the east, Putnam began negotiations with Lundin and quickly reached an agreement. On November 15, in a letter confirming their conversations "in regard to the making of the Cooke type triplet lens for the Lowell Observatory," he declared that the observatory "wants to have as perfect a lens as possible, with as wide an aperture as you can get with the discs that you have. I understand that you are planning to make the focal length approximately sixty-four inches." And as to price, Putnam noted: "We do not want the Alvan Clark & Sons Company to lose money doing this work for us, but we expect you to do it as cheaply as possible. We understand that your estimate for giving us a first-class instrument is $3,500, and that you can give us definite assurance that it will not be more than $4,000. This includes mounting the lens in a duralumin cell."[30] Lundin fulfilled this agreement in every detail except the price, as will be seen.

Slipher was glad to know that a satisfactory bargain had been made with Lundin. "We have confidence in his doing it well," he wrote.[31] And in passing on word of the agreement to onetime Lowell assistant W. A. Cogshall at Indiana University's Kirkwood Observatory, he expanded somewhat on plans for the new instrument. "Of course, our first big piece of work with the 13-inch camera will be to survey the ecliptic for planet X," he wrote. "We expect to use a second smaller camera as a check on the larger plate."[32] When the Planet X survey did get underway again, it may be noted, the 13-inch plates were checked with a 5-inch camera loaned to the observatory by Cogshall, and one of these smaller plates helped to confirm the discovery of Pluto.

Once a decision on figuring the 13-inch discs had been made, work on the new telescope began. By early February 1928, Lundin could report some startling, but welcome news.[33] "A few days ago," Slipher wrote to Putnam on February 18, "we had a letter and a blue-print of the 13-inch from Mr. Lundin. He stated that the lens had been ground and that it was then being polished. We are rather surprised at the size of the plate, it being a considerable [sic] larger than we had imagined. He states that he does not expect the definition to be perfectly sharp over the full size but he thinks the field will be useable (At least for some purposes) for 20° in diameter. And he suggests we may want a 22 × 22 inch plate!! Please do not let him know we feel the least doubt on this point, but it sounds too good to be true. Of course, if it gives good useable definition over a 14 × 17 (this is a commercial standard size of plate) or a 17 × 20 it will be a marvelous instrument for making the search for Planet X. We shall be very glad to get it in action on that survey."[34]

There are only about a dozen letters in the Lowell archives dated in 1928 that relate directly to the 13-inch project. Lundin in Cambridgeport and Sykes in Flagstaff worked steadily at their respective tasks, but the former apparently did not keep the astronomers at Flagstaff advised of progress on the lens, and Slipher reported only sporadically to Putnam on the work on the instrument itself, the mounting and the dome.

The location of the 13-inch on Mars Hill was decided upon in late March 1928 after Slipher surveyed the observatory's grounds and concluded that "the best compromise" was to put the new telescope on a small rise several hundred feet northwest of the observatory's main building on land leased from the U.S. Forest Service. "If we make use of that land there is no doubt about our being able to hold it," he pointed out to Putnam, "but if we should not put some instrument on it we might not be able to prove its being necessary to the work the Observatory is doing."[35] The trustee quickly agreed, replying that "I feel very sure that we want to locate the new instrument on the land leased from our Forest Service."[36]

Slipher confidently expected that the new telescope would be ready by the fall of 1928,[37] and Sykes, at least, kept to this schedule. By late August, Slipher could report that "we hope soon to be able to begin assembling the parts" of the instrument,[38] and at the end of September he advised Putnam that "the building is now completed and the mounting well along."[39] But Lundin did not finish the 13-inch lens until late in January 1929.[40]

During this period, of course, the Lowell astronomers had other projects in hand, and other matters in mind. So much so, in fact, that early in May 1928, when Slipher proposed four of the observatory's programs to Henry Norris Russell for grants from the National Research Fund, he actually forgot to include the new Planet X search in the list.[41] He quickly realized the omission, however, advising Russell on May 12: "Since writing you a few days ago in regard to the National Research Fund another important piece of research has been thought of that I quite overlooked although it is one that we are very anxious to follow up. As you know Dr. Lowell had done a lot of work on the question of a planet beyond Neptune 'in a theoretical way' and some searching had been done here. But the work was laid aside until we could have a more suitable telescope with which to continue the search. Now as you know Lundin is finishing soon for us a 13-inch aperture camera of short focus type with which we expect this fall to again take up the search.... And as things stand it would be very desirable to have another man to take up the work as we other men have rather full programs as it is.... There would be important incidental results that would come from such a search, quite enough to make the work very profitable should it fail to yield the object sought."[42] Coincidentally, Putnam wrote Russell about the Planet X search the same day, and the information, Russell subsequently advised Putnam, "came just in time for me to include mention of the search for a trans-Neptunian planet in the Lowell proposals."[43] But no National Research Fund grant was forthcoming.

Slipher, indeed, was particularly busy during the year, completing his new spectrograph and carrying out extensive observations with it of the light of the night sky and aurorae both from Mars Hill and the observatory's new mountain station at the 11,500-foot level on the San Francisco Peaks north of Flagstaff. A word about this station is perhaps of interest here because for a while it was apparently the highest astronomical facility anywhere in the world. Approval of Slipher's plans for the station late in 1926 was one of the final services Guy Lowell performed for the observatory.

Slipher, in the fall of 1908 and again in 1909, had laboriously packed his spectrograph and other equipment high on the Peaks for observations of Mars which he had hoped would confirm his disputed finding, in January 1908, of water vapor and oxygen in the martian atmosphere. His efforts were unsuccessful, partly because atmospheric conditions were unfavorable on his first ascent, and a winter storm cut short his observations on his second.[44] But he did not abandon his belief that the Peaks might provide a useful site for infrared spectrographic observations, although he did not again act on it until many years later when an event occurred that made such a venture more practical. In 1926, a scenic road, built by pioneer Flagstaff resident J. W. Weatherford, was completed to within fifteen hundred feet of the summit of the Peaks, thus making the heights easily accessible to motorized vehicles. On October 2 of that year Slipher outlined a proposal to Guy Lowell:

> There are a few lines of observing that require as little atmospheric obstruction as possible—those concerned with the very short waves of light and some concerned with the long waves of heat. We can, of course, gain quite an important additional height by going up the Mountains here. Thus we can get observations at a very much more favorable height than can any other observatory and it is an advantage that we should take. This is not a new idea. ...The opportunities now are much more favorable and the need of a high vantage point now more important. We can use the station for both spectrum work...and the heat measures of the planets. A scenic highway has now been completed up to an elevation of nearly 11,000 and it ought not to be very difficult for us to get equipment up for using the 15-inch reflector on a near peak at an elevation of close to 12,000....We think we could take our home-made 12-inch mounting up as it is modified to carry the 15-inch....[45]

Lowell immediately telegraphed his approval of the plan, advising Slipher to "try and keep down the cost and if possible put it where it can be reached by motor."[46] Slipher obliged, economizing by using the stump of a sawed-off Ponderosa pine as one of the piers for the 15-inch reflector and cutting a small cabin into the side of the slope of the mountain so that only a roof and front wall were needed to make it a snug haven for observers using the station.[47] The cost of labor and materials, he reported to Lowell in December, came to $375 for the cabin and a tin-roofed instrumental shelter. And the station, located on a rounded, southeastern summit called Schulz Peak at an elevation of 11,500 feet, was also accessible from the observatory "in a little less than an hour under good conditions of road."[48] The following year the observatory received a license to operate a radio, with the call letters KGCC, to communicate with observers at the mountain station.[49] The Peaks site was used quite frequently through 1931, primarily by Slipher for his night sky, aurorae and planetary spectra observations, but then only sporadically until 1936 when the Forest Service discontinued maintenance of the Weatherford road and the remaining equipment at the station was dismantled and brought down the mountain to the observatory. Actually, the Lowell astronomers found there was no appreciable gain in seeing conditions at the Peaks station, vis à vis the observatory itself on 7,250-foot Mars Hill.[50]

There were some other activities as well. On April 1, for example, Putnam was finally able to close Percival Lowell's State Street office in Boston, more than eleven years after his death.[51] Late in May, Slipher participated in the founding of the Northern Arizona Society of Science and Art and its Museum of Northern Arizona, subsequently serving for many years on its board of directors during the period of its growth into a major research institution in the Southwest.[52] In June, Lampland, fretting about the possibility of seasickness, sailed for Europe aboard the White Star Line's S.S. *Laurentic* to attend the International Astronomical Union's sessions in Leiden that summer.[53] Just before the meeting, incidentally, Slipher had written to I.A.U. president Willem de Sitter and vice president Arthur S. Eddington to propose a separate I.A.U. commission on radiometry, citing the work of Coblentz and Lampland, Pettit and Nicholson, and C. G. Abbot of the Smithsonian Institution as evidence of the increasing importance of the field.[54]

No progress was made in 1928, or later for that matter, on one project that seems to have been of particular interest to Putnam—the writing of a book about Mars that would point up those observations at the 1922, 1924 and 1926 oppositions which tended to support some of Percival Lowell's earlier findings regarding the planet, and especially those relating to the planet's atmosphere and temperature. Slipher had made the point to Putnam in their first exchange of letters shortly after Guy Lowell's death. "If you have been following the trend of developments in regard to Mars," he had written the new trustee early in March 1927, "you no doubt have been pleased as we have with the fact that the results of the more recent oppositions ... have yielded here and elsewhere results that have more or less confirmed Dr. Lowell's conclusions as to the conditions of Mars."[55]

Just when the idea of writing a book about Mars was suggested does not appear in the archival record, but it may well have been discussed during Putnam's first visit to Flagstaff the following month. Certainly the time was ripe, for the 1920s had seen a considerable outpouring of books on astronomy, and public interest in the subject was high.[56] At any rate, the Mars book shows up often enough in Putnam's correspondence with Slipher and Lampland in 1928 and 1929 and from those brief references in letters dealing with other matters, it is apparent that the astronomers were reluctant to become authors. Usually, they answered Putnam's inquiries or requests for copy for the book with promises or apologies pleading the pressures of other work.[57]

Putnam's own enthusiasm for the project, on the other hand, is clear, and even in the excitement that followed the discovery of Pluto, it remained very much on his mind. "Trustee grants temporary respite from all work on Mars book," he telegraphed Slipher two days after the momentous announcement by the Lowell Observatory, "and guarantees not to mention it in letter for some time."[58] The Mars book, however, was never written. The nearest thing to it was the younger Slipher's publication in 1962 of his photographic studies of Mars over fifty-five years in which he included an assessment of post-Lowellian martian results within the framework of Lowell's earlier work.[59] But by 1962, such a comparison was far less favorable to Lowellian theory than it would have been in the late 1920s, given what was then known or believed about the red planet. Only a few years after the younger Slipher's book was published the Mariner spacecraft missions to Mars began demolishing once and for all Lowell's more sensational concepts of the planet.

One other incident occurred in the fall of 1928 that is of some interest. Early in October, Mount Wilson's Edwin P. Hubble wrote the elder Slipher that George Ellery Hale and Mount Wilson director Walter S. Adams had asked him to make a preliminary investigation of observing conditions in Arizona in connection with a "proposed plan for another observing station in the Southwest. The subject," he added, "is in a rather confidential stage at present so I will explain fully when I see you.... We all agreed the first step was to run over to Flagstaff for a talk with you and while there, to request permission to calibrate the small telescope to your scale of seeing.... Mr. Hale is rather anxious for me to start as soon as possible."[60]

The subject of this letter, as it turned out, was the 200-inch reflector, and it is rather odd that Hubble thought the project was in a "confidential stage" for it apparently was common knowledge at the time to some astronomers. At least in mid-September, Putnam reported to Slipher on conversations he had had with Harvard's Shapley and Yale University's Frank Schlesinger, noting that both were "full of praises of the work you have done," and that both were "vehemently damning the new project for a two-hundred inch telescope, sponsored by Hale for the benefit of Cal. Tech. [California Institute of Technology]. It is apparently going through, and [Henry Norris] Russell will be somewhat identified with it. They feel it is a terrible waste of six million dollars and so do I."[61]

Hubble did visit the Flagstaff observatory and made tests of the atmospheric conditions and the seeing in the general area late in October, and he also visited Tucson where he talked with Andrew E. Douglass, who had helped Lowell found his observatory in 1894 and who since 1916 had been director of the University of Arizona's Steward Observatory. On his return to Mount Wilson, Hubble advised Slipher: "Best results are the series of observations to be expected from the Grand Canyon and from Cameron [an Indian trading post fifty-four miles north of Flagstaff] which can be compared with your seeing at Flagstaff.... Douglass had a great deal to say about the effects of local conditions, which will be of value when the general location is decided." And, he added: "You doubtless saw the announcement of the proposition last Monday. The confidential stage is past now and a 200-inch for the California Institute of Technology is definitely assured."[62]

Slipher's reaction to all this was one of bemused skepticism. Early in December, he wrote Cogshall about Hubble's visit, noting that it came

"just before the publication of the announcement of the 200-inch reflector. He was making tests with a 2-inch telescope! It was quite apparent that the new Big Bertha had already been located definitely at no very great distance from Mount Wilson. I amused myself by saying that of course with a proposition of that magnitude and importance that choosing a location is in a way simplified in such a case for with it the best possible observing conditions must vastly outweigh all other factors, as it is made accessible by the auto and the airplane in the most out-of-the-way nook in the world, today and even more so in the future! Hale is the man behind the undertaking and it is sure to be placed within sight of Mount Wilson. To my way of thinking it will be an unfortunate act to place it so near the 100-inch, for both under the same weather conditions reduces very considerably the effectiveness of the world's two largest telescopes."[63]

Early in January 1929, Adams published an article in *Science* describing the plans for the 200-inch reflector, and Slipher wrote Putnam to call his attention to it. "Other observatories are very much out of the picture," he noted. "In that regard Lowell Observatory got more mention than any, excepting of course Mount Wilson. Our observing conditions and those of northern Arizona come in for consideration, as well as some lines of our work.... Doubtless those behind the project have their minds pretty well made up that the best available elevation rather near Pasadena will be found to have 'everything considered as favorable observing conditions as could be found.' In other words," he added, "I do not think tests of conditions in Arizona mean very much. Dr. Hubble was over this fall making some tests evidently for publicity material. He had only a two-inch aperture telescope for the work! And he thought he was getting worth while tests. General opinion would place more weight on tests made with a telescope of more aperture, especially as it is for locating a 200-inch telescope."[64] Putnam agreed. "I have no doubt," he noted, "that California, being of a grasping nature, will not let the Observatory outside her borders, which is a great mistake."[65]

The Mount Wilson astronomers, parenthetically, had a special technique for testing with the 2-inch, and southern California skies, to most astronomers, have proven to be superior for astronomical seeing.

The 200-inch reflector, of course, was subsequently located on Palomar Mountain south of Pasadena—"at no very great distance from Mount Wilson," as Slipher had predicted. But despite his skepticism over Hale's site surveying for the project, he was well aware of the implications of a 200-inch telescope for astronomy, and more particularly for Lowell Observatory. Even as he penned his private jibes at Hale and Hubble, he confided to English astronomer James R. Worthington, an old Lowellian, that "there are some planetary problems Lowell Observatory would like to look into before the 200-inch takes the field."[66]

But Slipher at this time was far more concerned with the 13-inch refractor, for Lundin was taking considerably longer to finish its objective lens than the Lowell astronomers had expected. The impending resumption of the X search was very much on his mind, as his letters in January 1929 clearly show. On January 5, for instance, he wrote to a correspondent in Michigan: "We hope to have completed and in operation in a few weeks a new photographic telescope to continue under much more favorable instrumental conditions the search for the planet beyond Neptune the prediction of which from theoretical considerations Dr. Lowell occupied in no small way the latter years of his life...."[67] A few days later, to Putnam, he noted his impatience to get on with the X search. "Then after the big objective comes from Lundin it should not take long to have the instrument where some preliminary test exposures can be made, and if no obstacles are met it will be possible before long to get the instrument at real work on the Planet X search. We shall be very glad," he added, "to get that work going again with such a good instrument as we can expect the 13-inch to be. Dr. Lowell put so much into that investigation that it will be a great pleasure to carry on the search."[68] And to his old professor and friend, Miller, he explained on January 21 that the new 13-inch would "first be used in charting the ecliptic in a search for the suspected planet beyond Neptune to which Dr. Lowell devoted a lot of theoretical work and computation. This is the same problem we used your 9-inch on when we borrowed it some years ago."[69]

The long wait was now over, however. On January 22, Putnam advised Slipher that Lundin had the lens ready to ship to Flagstaff, and asked him "to give us the real cost of the 13″, so that I can collect it from my uncle."[70] But any enthusiasm this news may have generated was quickly dampened when Lundin sent Putnam his bill three days later. "We want to put our cards on the table," he wrote, adding that the cost of finishing the 13-inch discs, with a 15 percent profit for the Clark firm, totaled $5,667.41, well above his original estimate of $3,500 or his agreed-upon ceiling cost of $4,000.[71]

Putnam, apparently without consulting Slipher, replied to Lundin on January 31, offering to "split the difference between $5,667.41 and $4,000, paying you $4,883.70. I want to do everything fairly," he declared, "and yet it is not my money we are talking about."[72] Lundin accepted this offer the next day, calling it "most generous" and adding that "we appreciate the fairness you have shown us."[73] Putnam, in turn, mailed Lundin a check for $2,383.70 "to complete the payment," having previously advanced $2,500 toward the cost of the lens.[74]

Clearly, Slipher did not receive the news about the amount of Lundin's bill until these negotiations had been completed, for on February 6, two days after Putnam mailed the final payment, he wired the trustee: "Lundin situation prompts desire to test lens quality before settlement."[75] Then, after learning that a settlement had already been made, he penned a long letter to Putnam. "We are disappointed, of course, but we think your offer is fair.... If the lens is good we will all soon forget about this part once we get it going and doing good work in the search for planet X." He then reported that $4,000 would be "a fair figure to use under the circumstances" for the cost of the instrument itself, the mounting and the dome. "Lundin's bill rather knocked the breath out of us," he declared, adding that he had been hoping that A. Lawrence Lowell's $10,000 gift would also cover the $1,200 to $1,500 cost of a Ross type lens for a guide telescope for the 13-inch.[76] A week later, Putnam replied that he was turning his "commission" as trustee of "about $1,500" and "about the cost of a Ross type lens, back to the observatory—let me know what you want."[77]

The 13-inch lens arrived in Flagstaff on February 11 and Slipher was still fretting about it. "The 13-inch objective is here this morning," he wrote his brother, Earl, who was then serving in the Arizona State Senate in Phoenix, "and we are anxious to get it up and [make] some test plates with it soon. Lundin has never said a word as to the field or quality of the objective. He has only complained of the difficulties he had in getting anything out of the material because one of the disks being thin. He got a thousand more for his work than it was expected to cost."[78]

Slipher's fears, however, quickly proved unfounded, and he soon wired Putnam: "New lens arrived safely. Has been mounted and first test exposures tonight very encouraging as to its quality."[79] Putnam replied also by wire: "Your telegram is fine news."[80] Lampland, too, hailed the news. From Princeton University, where he was spending the spring 1929 semester as visiting professor in astronomy, he wrote Slipher on February 28: "I hope all goes well with the new telescope. I am glad that the preliminary tests were so encouraging."[81]

Actually, some minor problems did turn up, but with the drive mechanism of the telescope, not with the lens. On March 3, Slipher wrote Lampland that the weather had prevented them from getting any long exposures with the 13-inch, except one of an hour. "This one showed a string of little images, as if something was shifting by steps. We went over the instrument to make sure things were tight, but have not been able to get another plate to see if we caught the trouble."[82] A few days later, Slipher advised Putnam: "I had hoped to write you earlier about the new telescope. But it has taken longer than was expected to get it into good enough adjustment to feel the results are fairly indicative of what it is capable of doing.... The new lens seems to be able to reach the same magnitude of stars [as Max Wolf's photographic charts] with a little less exposure time, for the center of the field. For the outer portions of the field, the lens is very much in the lead...."[83]

Putnam was "delighted" at Slipher's report on the performance of the 13-inch lens. "I have sent a copy of your letter to Uncle Lawrence, and he has given us the full $10,000, which leaves something like $1200 to apply to a guiding telescope or something of the sort."[84] Putnam was, in fact, so delighted with the lens that two months later, despite the considerable cost overrun for the 13-inch, he ordered an 8-inch Ross type lens from Lundin who, incidentally, left the Clark firm in April.[85] The cost was to be $2,200. "We are giving you this order," Putnam informed him, "because we are so pleased with the thirteen-inch Cook [sic] type lens that you have made for us. . . ."[86] Slipher had already been in correspondence with Frank E. Ross, the Yerkes Observatory astronomer who had devised the lens type, explaining only, however, that such a lens was needed for a "program of observing for the 13-inch instrument [that] is in the main that of charting the ecliptic zone."[87] The 8-inch Ross lens, parenthetically, was eventually delivered, but was never used in the third search for Lowell's Planet X.

On March 13, 1929, the seventy-fourth anniversary of Percival Lowell's birth, Slipher had dutifully descended Mars Hill to the campus of Northern Arizona State Teachers College in Flagstaff (later Northern Arizona University) to present the Lowell Prize to the top-ranking student there, an award established in 1918 by Lowell's widow in his memory and presented annually ever since by the observatory. The following day, he wrote a long letter to

Mrs. Lowell. "Was at the College Assembly yesterday and spoke of Dr. Lowell, emphasizing his research on the trans-Neptunian Planet and the search for it with the new telescope," he reported. "I am happy to say that the new telescope appears to be exceptionally good, and is going to enable us to make a very accurate search for Dr. Lowell's Planet X. With good weather, which ought to come before long, the work ought to move along steadily. I am hoping that we can get both of the promising fields Dr. Lowell designated, more or less well covered before the summer rains set in, even if we are not able to keep up with the examination of the plates. Fortunately the new lens covers a large field with each exposure and that expedites the search. It also reaches very faint stars with moderate exposures, and that is helpful."[88]

But poor observing conditions, along with problems relating to the 13-inch's clock drive mechanism, continued for another three weeks, as Slipher complained to Lampland on April 5. "Observing weather has been rare and we have not been able to get much observing done with the 13-inch this lunation so far," he wrote. And, after describing the mechanical difficulties that had been encountered, he added: "It is to be hoped we will soon reach the last of this kind of obstacle and get a real start with the work."[89]

The real start of the third search for Planet X came the very next night, April 6, 1929. Then, what is recorded in the first entry in the 13-inch telescope observation log as "Negative No. 1" was exposed for one hour, starting at 10:20 p.m., with the telescope guiding on the star δ Cancri. It was a brilliant night, fair with a light wind, although the steadiness of the seeing was rated at only three on a scale of from one to ten. The temperature was 22°F.

The historic entry ends with the notation, "Observer: C. W. Tombaugh."[90]

CHAPTER 9

A YOUNG MAN FROM KANSAS

When V. M. Slipher belatedly proposed the Lowell Observatory's new Planet X search for a National Research Council grant to Henry Norris Russell in May 1928, he was thinking in terms of hiring an additional assistant to undertake the tedious observational chores that he knew the search would entail. "As things stand," he had advised Russell, "it would be very desirable to have another man to take up this work as we other men have rather full programs as it is."[1]

Slipher had definite ideas on some of the prerequisites for a career in astronomy and expressed them on occasion in his correspondence with young, aspiring astronomers seeking a position at the Flagstaff observatory. Physical stamina was one. In 1924, for example, he answered one such inquiry by noting: "Unfortunately, the work of Astronomy involves rather irregular hours and it is therefore more tiring physically than one might at first suppose. ... Therefore one of robust health has a decided advantage."[2]

Nor should material reward be a primary concern, as he once cautioned Robert S. Richardson, who subsequently became a well-known astronomer and writer of books on astronomy. "It is probably true that the pay in the field of Astronomy is lower than in other lines of intellectual endeavor of a more commercial relation," he pointed out. "In other words if the matter of remuneration weighs much now or may do so in the future in your choice

[178]

of work, I doubt whether one could recommend to you the pursuit of Astronomy. However, Astronomy, as well as many other sciences, holds out something in addition to the pay alone...."³

More specifically, for the man who was to carry out the third search for Percival Lowell's Planet X there was an additional requirement: a willingness to follow instructions. By December 1928, Slipher felt that he had found a man meeting these criteria. Early that month he confided to Cogshall:

A young man has been writing us from west Kansas about employment in the Observatory. He is a farmer boy, with high school training, much interested in astronomy, planetary work particularly, and has made two or three reflectors—6 to 9 inches aperture. He sent us some drawings of Jupiter that look fairly good for such a chap working all alone. We are wondering if he might not be able to make exposures with our new 13-inch camera. What would you guess such a fellow would do? ... Our experience with the highly trained fellow is that ... he is apt to look to us for assistance in what he wants to do, and the result is we do not get help in our work that is what we most need.⁴

A few weeks later, Slipher wrote Putnam at length along the same lines:

Have been negotiating with a young man of the self-made variety that we hope will be able to do the observing with the new telescope and do other assisting about the Observatory. He lives in Kansas, has made some small telescopes (reflectors) for himself and seems not afraid of the work. From his letters it looks as if he has a real interest in astronomy and particularly planetary matters. He is willing to come on a few months trial so it seems we will do well to have him come and see what he can do. His training and schooling is limited to high school, but he has since done a good deal of reading in astronomy and seems to have helped himself also to some considerable knowledge and experience with instruments. He is very eager and enthusiastic to get into observatory work. It seems to me that we would probably get more real assistance from the young man described above than we would from the highly trained variety for the reason that the latter care only to take up new pieces of work for themselves rather than help us with lines the Observatory has been doing.⁵

In another two weeks, the negotiations had been completed and Slipher informed Putnam: "The young man from Kansas is expected to come on next week. I hope he proves to be good help around the Observatory and that after a time he will be able to make exposures with the 13-inch...."⁶

Putnam's reaction to all this was that Slipher's plan was "for the moment ... excellent. The fact of having some one who wants to come, and who is willing to come on a trial basis is thoroughly worth while," he wrote. But, he added, "I don't feel that it answers all our personnel requirements—that is, I think we want to keep our eyes open all the time for a young scientist."⁷

The "young man from Kansas" was 22-year-old Clyde William

Photograph courtesy of C. W. Tombaugh

Clyde William Tombaugh, aged 22, displays his home-made 9-inch reflecting telescope at his family's Kansas farm in 1928.

Tombaugh, who arrived in Flagstaff January 15, 1929,[8] and who, in a little more than a year, would become one of only three men in recorded history to discover a major planet.

Tombaugh was born February 4, 1906, near Streator, Illinois, the son of Muron and Della Tombaugh. He attended rural schools and Streator High School for two years before the family moved to Kansas, where he graduated from Burdette (Kansas) High School in 1925. "Since then," as he wrote shortly after the announcement of Pluto's discovery, "my summer seasons were given to farming, my winters to constructing telescopes."[9]

Tombaugh, incidentally, would remain on the staff of the Lowell Observatory for fourteen years, as an assistant observer until 1938 and then as an assistant astronomer until 1943 when he became an instructor in science at Arizona State Teachers College in Flagstaff. In 1945, he taught at the University of California, Los Angeles, and then worked as a scientist at the White Sands, New Mexico, missile testing range until 1955 when he joined the faculty of New Mexico State University in Las Cruces. While still on the Lowell staff, he received his A.B. and M.A. degrees from the University of Kansas in 1936 and 1939 respectively, and in 1960, he received an honorary Doctor of Science degree from Flagstaff's Arizona State College (now Northern Arizona University) in recognition of his discovery of Pluto and his subsequent planetary studies.[10]

During his first few weeks in Flagstaff, young Tombaugh quickly justified Slipher's hopes for him, as Slipher reported to Putnam on February 7. "Mr. Tombaugh, the young man from Kansas, has been here since the middle of January," he wrote. "We think he is going to develop into a useful man. He has several good qualities that are going to make up for his meager training. He is looking forward to the arrival of the new objective and is eager to be working with it. He has a good attitude: careful with apparatus, willing to do anything to make himself useful and is enthusiastic about learning and wants to do observing...."[11]

In the weeks that followed, while the 13-inch lens was being installed and tested, Slipher's opinion was further confirmed. "Mr. Tombaugh is very much interested in the new instrument and the work ahead of it, and is eager to work with the instrument," he advised Putnam. "He has made a few exposures and I see no reason why he will not be able to get good results with it...."[12]

Tombaugh, parenthetically, had no idea whatsoever of the "work ahead" when he reported to the observatory to take up his duties. Slipher, in his letters to Kansas, had sought assurances of his "good physical health," but had said nothing specific about the kind of work he would be doing. "Only after arriving in Flagstaff," Tombaugh has recalled, "did I realize that I was to join in the search for a new planet, and what it is like to work all night long in an unheated dome in winter at 7,000 feet altitude." The work, most certainly was different from that on a Kansas farm.[13]

Slipher's carefully thought-out plans for the third Planet X search required not only that the new observational survey of the ecliptic be thoroughly systematic, but that the observing techniques be fully standardized.* This was basic, for without complete standardization, plates made of the same star fields could not be meaningfully compared on the observatory's "blink" microscope comparator and Planet X, if it really existed, would not reveal itself by blinking on and off in slightly different positions as it moved slowly among the unblinking fixed stars.

From preliminary tests of the 13-inch, for example, it was determined that one-hour exposures reached stars of the seventeenth magnitude, considered a practical limit in view of Lowell's belief that X's magnitude was 12–13, and this became the rule in the search, although Tombaugh soon realized that under poor seeing conditions, exposures had to be extended by 15 or 20 minutes to reach the required magnitude for the plates.[15] Pairs of plates of fields centered on the same guide star, usually one close to the ecliptic and of about the seventh magnitude, also were to be made within a week, and under similar conditions of seeing and atmosphere whenever possible, and the same chemical developer and development times were to be used in the photographic processing of the plates. Again, Tombaugh soon adopted the practice of taking "three good plates per center within a few nights of each other (preferably all three in a week). The two best matching plates were compared in the blink microscope," he noted, "and the third was available as an immediate check on any suspected moving object."[16]

It was also necessary that the plates themselves be of the same quality, and Slipher tested plates furnished by a number of manufacturers before selecting one firm to supply 14-by-17-inch plates for the search. In requesting test plates, he stressed, "It is important that the plates used for the work be the same or as nearly uniform as possible."[17]

In testing plates, however, it was found that the surfaces of even the best of them were slightly concave, up to three-sixteenths of an inch, varying from plate to plate and seriously affecting the comparability of the plates. To counter this, as well as to extend the overall field of sharp definition on the plates, some special equipment was devised. "We have put together a testing table for determining the concavity of the large plates for the work with the new telescope," Slipher explained at length in a letter to Lampland at Prince-

---

*The procedures set up primarily by Slipher and used in the third Planet X search by Tombaugh have been described in several publications, first by Putnam and Slipher in an article in *Scientific Monthly* in January 1932, and in a number of subsequent papers by Tombaugh, most notably his Astronomical Society of the Pacific *Leaflet 209* in 1946, which was republished in Harlow Shapley's widely read *Source Book in Astronomy 1900–1950*, in 1960; his "Reminiscences of the Discovery of Pluto," published in *Sky and Telescope* that same year to mark the thirtieth anniversary of the event; and "The Trans-Neptunian Planet Search," a chapter in the third volume of G. P. Kuiper's and Barbara Middlehurst's series, *The Solar System*, in 1961.[14] These accounts are excellent summaries and agree generally on the pattern and course of the new search except in minor details. They are brief, however, and undocumented, and here they will be supplemented by material from the Lowell archives and from other sources.

ton in March. "Of these plates the variation in this regard is very marked. For example plates from Seeds that were used in our first focal adjusting were so much more concave that when we started using fresh plates from Cramers [sic] the focus was hopelessly out, the Cramer plates being much more nearly flat. In this respect the Cramer plates appear exceptionally good. It is hoped," he went on, "that with the testing device and a suitably designed plate holder that it will be possible to lay the loaded holder on the testing table and set up pressure screws in the back of the holder so as to bring the center of the plate and at least four other points (the mid points of the sides and ends) always to the same adjustment by springing individual plates to a chosen average shape. We have made tests and find these large plates spring rather easily: also that it would not be feasible to try to use such plates without some way of determining and sorting them for curvature of surface, and so discarding those deviating too much from the average, before making the exposures. Otherwise there would be a considerable loss from poor focusing.... In other words," he added, "that operation will be we hope so simple that one will never be tempted to omit this plate adjustment. And as the focal surface of the lens is somewhat concave it will we hope be possible to get a better focal adjustment over the whole plate."[18]

The special plate holders and testing table were subsequently used with great success in the X search, although on occasion, as Tombaugh soon discovered, plates broke in two when they became chilled in the cold telescope dome.[19] A broken plate, however, was not necessarily a bad plate and once repaired it could be blinked on the comparator.

As noted earlier, with instrument and plates finally in adjustment, Tombaugh formally began the third search for Percival Lowell's Planet X on April 6, 1929. The A. Lawrence Lowell telescope he used in the search was a 13-inch, Cooke type, three-element, astrographic refractor with a focal length of 66.5 inches which yielded a field of 12° by 14°, or 168 square degrees of the sky, on its 14-by-17-inch glass plates. The scale of these plates was 122 seconds of arc per millimeter. The one-hour exposures reached the seventeenth magnitude and commonly recorded from 50,000 to 500,000 stars, depending on the region of the sky being photographed. In the case of the very dense fields in the Milky Way, such as those in Scorpio and Sagittarius, a million or more stars appeared on each plate.[20] It was indeed a superb instrument with which to pursue Lowell's trans-Neptunian exploration. As Putnam and Slipher later wrote: "The performance of the [13-inch] lens exceeded our expectations and we have been led to regard it as possibly the best search instrument in the world today."[21]

Guiding was done with a 7-inch refractor attached to the 13-inch and within two months a 5-inch Brashear camera, loaned for the purpose by W.A. Cogshall, was also put into operation, providing simultaneously exposed 8-by-10-inch plates that were used as checks on the 13-inch plates for objects brighter than the sixteenth magnitude. The Cogshall camera was first used in the search on June 7, 1929.[22]

Tombaugh's first plate, as noted earlier, was of a field in the constellation of Cancer, but two nights later, on April 8, he moved into Gemini which, although low and sinking in the western night sky and thus not in the best position for observations, was nonetheless one of the two most likely locations for Planet X as indicated by Lowell's trans-Neptunian planet memoir. As Putnam and Slipher later noted, "We began the first plates of the Planet X search series in order to get plates that season of Dr. Lowell's predicted favorable region in Gemini...."[23]

During the first month of the search, Tombaugh made ten plates of fields in Gemini—four centered on the star δ Geminorum and three each on 36 and 85 Geminorum—as well as plates of fields in Virginis, Leo and Taurus.[24]

Incredibly enough, his tenth plate, exposed April 11 and centered on δ Geminorum, and a companion plate exposed April 30, contained faint images of what subsequently proved to be Pluto. The plates, Lampland reported after the discovery of Pluto, "were not as good quality as the [later] photographs, both as to sharpness of definition and freedom from troublesome small specks that in some cases had the appearance of faint images. The image of Pluto was not identified on the 1929 plates until the positions were calculated from our preliminary orbit [of Pluto]"[25]

These plates had not been examined immediately after they were obtained, but only some months later and then only cursorily. Plate No. 10, in fact, "broke in two," as noted in Tombaugh's log.[26] Tombaugh has since written that "the reason the 1929 images of Pluto were not recognized was that the region of Gemini was low in the western sky and the stellar limit [of magnitude] was too poorly matched to permit exhaustive blinking. Pluto was expected to be two magnitudes brighter. When you search for something two magnitudes fainter, the increase in the number of stars to be reckoned with is overwhelming." Moreover, "differential refraction badly distorted the star images except in the central portion of the plates containing the guide star. This effect is severe when photographing near the horizon. Slight variations in sky transparency from night to night, made it virtually impossible to reach exactly the same magnitude limit—especially sensitive in low regions of the sky. Unless the stellar limit is the same within 0.2 of a magnitude, the pairs of plates are unblinkable. This is why Pluto was not picked up in the spring of 1929."[27]

On April 15, barely a week after the new X search had begun, Slipher reported optimistically on its progress in a long letter to Lampland. "The 13-inch has been giving some nice plates recently," he wrote. "We were of course eager to begin the ecliptic as far west as possible and we have in consequence been observing much farther from the meridian than should be done for best images over the plate. It will no doubt be found to be advantageous if not necessary to repeat the fields at the same hour angles of the original plates. But we have wanted to get all we could even if not under good conditions, for it seemed that once the weather was more settled and things got going properly that it ought to be no trouble to keep up with the march of the sky westward."[28]

The new search seemed to hold promise for other reasons, too: "It looks as if the survey when completed ought to be very valuable for general information as well as for the special search for Planet X," he confided to Lampland. "The plates contain a lot of material, even those far from the Milky Way. An hour seems to get all that Wolf's photographic charts show with considerably longer exposures; and too the larger plates and wider field are very striking advantages. We have run the series now as far as Beta Virginius and so have touched the spiral nebulae. Some very striking objects have been noted among the spirals and which have apparently not been described from reflector plates. It will be easy from these negatives to make up a list of promising nebulae for the 42-inch." And as to the observatory's new assistant, he added:

Tombaugh seems to hold pretty tight on the guiding eyepiece, and shows carefulness about details, besides he takes pride in and gets pleasure in making faultless plates. New things bother him a bit but when he gets a little more experience with apparatus he will be able to get over some obstacles that bother him now. He likes the instrument and program and is going to be very efficient in making the observations.[29]

The weather cooperated fully with the new search during the spring of 1929, and not only was Tombaugh able to get "a good number of plates"— about a hundred—before the summer rainy season set in early in July, but Slipher was able to obtain a few more spectrograms of nebulae and star clusters, none of which, however, showed any remarkable velocities.[30] The X search, incidentally, and his observations of aurorae, night sky light and zodiacal light, would soon end his work on these deep space objects and in November he advised Knut Lundmark that "our time has been largely taken up with other matters so that we have not been able to devote much of it to nebular researches."[31]

Early in May, Slipher sent Lampland a sample of the photographs Tombaugh was producing with the 13-inch telescope. Lampland, not one to use exclamation points lightly, was enthusiastic. "The superb print of the negative taken with the new lens arrived the other day," he replied. "I cannot tell you how glad I am to see the splendid performance of the new instrument. Today I found the Wolf-Palisa chart to make a comparison and the superiority of the 13-inch stands out very clearly. Has the comparator been put in shape for examination? What a lot of good material should come out of an extended series of plates with the new instrument!"[32]

The comparator, in fact, was in shape but it was used only sparingly during the first few months of the X search. Tombaugh initially was assigned to the observational aspect of the work, to the making of plates, and the examination of the plates was to be accomplished by other, more experienced

Lowell Observatory

The 13-inch Lawrence Lowell photographic refractor mounted in its dome at the Lowell Observatory atop Flagstaff's Mars Hill. The instrument has since been moved twelve miles southeast of the Observatory to eliminate fogging of its long-exposure plates by the night lighting in Flagstaff.

members of the staff. However, he has recalled, "my senior colleagues were busy with other things," and his plates kept accumulating. Thus, with the advent of the summer monsoons, which temporarily halted night photography, Slipher directed him to take over the examination of the plates with the blink microscope himself.[33]

Tombaugh found his examination of the first pair of plates "very laborious because the regions were rich in stars," and his tedious work soon turned up a number of other problems as well. The distortion of star images, caused by the low angle of the observations, and the difficulty of matching the magnitude limits of pairs of plates, both because of the low observation angle and the intrusion of haze during the long exposures, were among these.[34] Also, over the course of the search, the "chance aggregation of silver grains" in the emulsion of the plates "gave rise to thousands of planet suspects.... Each had to be checked, for the risk could not be taken of letting the long-sought planet slip by. Many suspicious images turned out to be faint variable stars whose minima were fainter than the plate limit."[35]

Photograph courtesy of C. W. Tombaugh

Clyde W. Tombaugh at the guiding eyepiece of the 13-inch Lawrence Lowell telescope, 1931.

Fogging of plates by the night light of the sky and by moonlight was still another difficulty, and "even a crescent moon above the horizon could not be tolerated," he has noted. Each month, a week on either side of new moon was used in making exposures at the telescope. The moonlit part of the month was spent in blinking the plates obtained during the previous two weeks.[36]

But a more serious problem—"even more appalling"—was the many asteroids that turned up on his plates. "How was I to distinguish them from Planet X?" he wondered. The difficulty here was that as the earth revolves around the sun it passes each exterior planet or asteroid and causes it to appear to retrograde, that is to move backward or westward against the background of stars. This retrograde motion is directly proportional to the distance, greater for nearby objects and very small for distant objects. Moreover, as a relatively nearby planet or asteroid reverses its apparent movement, it appears to be stationary in the sky for several days, and its apparent motion near this stationary point is so slow as to simulate the motion of a far more distant object.[37]

Finally, about 90 percent of all the planet suspects that appeared on his plates during the search were too faint to be checked for confirmation on the plates made with the 5-inch Cogshall camera.

All of these problems, some of them of course insoluble, combined to frustrate young Tombaugh as he conscientiously scanned pairs of the early search plates. "I just could not reconcile myself to conducting a planet search on such a hit-and-miss manner. It was apparent that I was pretty much on my own. I painfully realized that I would have to think out procedures to greatly refine the methods and techniques, if this laborious planet search was to be worth doing. I spent the rest of the summer rainy season studying these problems. I had to start all over."[38]

Young Tombaugh did some productive thinking while he was blinking his plates that summer. For one thing, he realized that he must pay particular attention to the sky transparency and steadiness of the atmosphere during his observations so that the magnitude limits of the plates were matched as closely as possible, and thus that pairs of plates would be blinkable.[39] For another he realized that the fields photographed should be at or very near to the opposition point, that is, directly opposite from the sun relative to the observer.

"The key strategy," he has since said, "was to take the plates within 15 degrees of the opposition point, so that the daily parallactic motion of the earth [in its orbital revolution around the sun] would cause all bodies external to the Earth's orbit to retrograde (apparent motion) thru the star field—the more distant the object the smaller the angular displacement during the interval between the dates the plates were taken. This solved the asteroid problem beautifully."[40]

In addition, he has pointed out, because the opposition region is on the meridian at midnight, the plates could be taken when the regions were high in the sky, rather than low and close to the horizon, and this yielded sharper images on the plates. "Also, I would duplicate the plate regions at the same hour angle—thus avoiding image distortion and magnitude loss. It proved to be much easier to match the plates of a pair for exhaustive blinking. I initiated the practice of taking three plates of each region within the same week, and the spurious images causing hundreds of planet suspects could be readily checked for confirmation—down to the 17th magnitude."[41]

Parenthetically, the prevailing opinion in astronomy regarding the whole question of trans-Neptunian planets seems to have contributed to Tombaugh's frustration during the summer of 1929. "I remember that the attitude of nearly all astronomers in the decade prior to 1930 was one of considerable skepticism concerning the existence of more planets," he recalled. "Pickering's wild papers probably contributed to this. In fact, I was admonished by a visiting astronomer to Flagstaff in June 1929, privately, that 'Young man, I am afraid that you are wasting your time. If there are any more planets, they would have been discovered long before this.' But I remarked that I felt that the new telescope was an opportunity, and I was going to give it [the project] the utmost effort and care. I would have liked to have seen the expression on his face when the discovery of Pluto was announced nine months later!"[42]

Tombaugh began photographing fields on the ecliptic again early in September in the constellations of Aquarius and Pisces, some 120° from Lowell's predicted position for Planet X. He had now caught up with the opposition point, and his attitude now "was to thoroughly search the entire Zodiac, putting all these refinements into practice. All of this paid off, and a thorough and exhaustive search could be made with definite yes or no answers, instead of 'maybe that suspect was Planet X.'"[43]

From Pisces, he moved into Aries and then Taurus where, he has written, the number of stars per plate increased from 50,000 in the Aquarius-Pisces regions to 500,000 in eastern Taurus and western Gemini.[44] In "the very rich regions of Scorpius and Sagittarius, where each plate contained a million stars, I encountered another type of planet suspect. There were several cases, where two very faint images near each other were popping in and out of view in alternate fashion [under the blink microscope]—a perfect planet suspect. When I checked these with the third plate, they turned out to be two independent, short-period Cepheid variable stars with the maxima out of phase and the minima phase below the limit of recording on the plate."[45]

The summer of 1929 proved to be unusually cloudy in Flagstaff and in mid-October, in a letter to Putnam, Slipher reported: "Tombaugh is getting along fairly well with the Planet X work. He keeps up with the observing work fairly well although the summer weather has been and still is rather troublesome as clouds have made it difficult to get plates at the times when most necessary or advantageous."[46]

The economic weather, incidentally, took a disastrous turn on October 25, 1929—"Black Friday" on the New York Stock Exchange and the start of the Great Depression—but did not seriously affect the observatory's projects, including the Planet X search. "I suspect with all this financial news in the newspapers, that perhaps you were getting worried about the financial status of the Observatory," Putnam wrote Slipher a week after the market broke and the panic set in. "Of course, in the market break, we have lost large amounts of paper profits, but, on the other hand, even at present lows, almost all the securities we own still show us a profit. Our income is going to be just as great as ever, and, in fact, we have been able to buy some stuff at their low points this week. In other words, the Observatory's finances seem to be in apple pie order in spite of the drastic reaction."[47] The extent of the economic catastrophe, of course, was not yet fully appreciated, but the observatory, under Putnam's experienced management, was able to ride out the storm.

By early November, Tombaugh was beginning to encounter the denser star fields of the Milky Way and this slowed the work of examining his plates. "No remarkable objects have yet been found on the Planet X search plates," Slipher advised the trustee on November 8. "But the examination of the plates takes more time than one would have supposed. On some of the plates there are as many as 100,000 stars and to pass over in the eyepiece so many star points requires several hours. And in fact a considerable number of the plates will have at least this number of stars. It will I hope be possible to get more of the examination work off in the future. Maybe it can be speeded up a little, and perhaps Miss [Constance] Brown [then secretary at the observatory] can do some of it."[48]

By January 1930, when Tombaugh was again photographing in western Gemini, more than 400,000 star images were being recorded on his plates and the examination work fell behind, requiring several days for each pair of plates.[49]

On January 21, Tombaugh came back to the star field centered on the star δ Geminorum, exposing "Negative No. 161" for exactly one hour, starting at 10:30 p.m., Mountain Standard Civil Time. The plate, he noted in his log book, yielded fuzzy images for while the sky transparency was good, the steadiness of the seeing was 0 (very poor).[50]

Two days later, on January 23, he made a second plate of the δ Geminorum region, with the hour-long exposure beginning at 9:25 p.m. MST. This was a good plate, obtained when the transparency and the steadiness of the atmosphere were very good. And on January 29, from 9:13 to 10:13 p.m., he exposed still a third plate—"Negative No. 171"—under almost similar conditions.[51]

Because of the quality of the January 23 and 29 plates, and the six-day interval between them, Tombaugh considered them a fairly good pair for blinking. But he could not begin their examination until about February 15, and then he could devote only part of his time to the work. By late in the afternoon of February 18, 1930, only about one-quarter of the pair had been examined.[52]

But then two faint images began blinking on and off on the plates only about three and one-half millimeters apart. "On Feb. 18, 1930 at 4 p.m.," Tombaugh noted in his earliest written description of the event, "planet X (Pluto) was discovered on the comparator. The following ¾ hour was spent investigating the amount of shift or apparent motion of Pluto, also its direction, and genuineness of images in their respective positions on the 5-inch Cogshall camera plates which were made simultaneously with the 13-inch discovery plates (No. 165 of Jan. 23 and No. 171 of Jan. 29). A 3rd plate of Jan. 21, also showed the planet in correct position with respect to the other

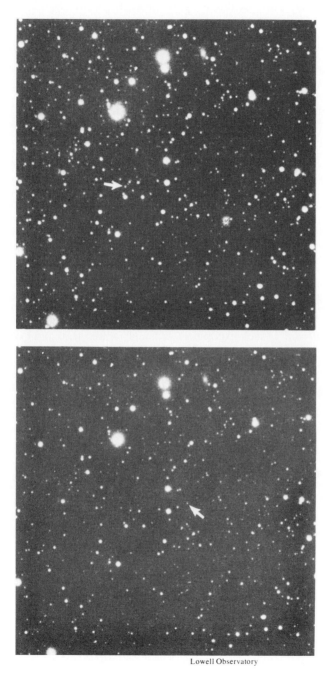
Lowell Observatory

Portions of the discovery plates of January 23 and January 29, 1930, showing images (arrows) of the planet Pluto.

dates. At 4:45 p.m. the discovery was made known to Dr. C. O. Lampland first, and then to Dr. V. M. Slipher and then to Mr. E. C. Slipher. I shall never forget the intense interest the object caused among the rest of the staff."[53]

Thirty years later, Tombaugh recalled these historic moments somewhat more dramatically:

> In February, 1930, after struggling through the Taurus plates, I skipped over to those in eastern Gemini when the stars were less thickly packed. The entire length of the latter constellation had been photographed by the end of January that year. I chose three plates centered on Delta Geminorum, taken Jan. 21st, 23rd and 29th respectively, but bad seeing made the first of these unacceptable for blinking.
>
> I placed the other two in the comparator and began blinking the east half from the south end. By 4:00 p.m. on February 18th, one-fourth of the plate area had been blinked. Upon turning to a new eyepiece field two-thirds of a degree east of Delta, I suddenly spied a 15th magnitude object popping in and out of the background. Just 3½ millimeters away another 15th magnitude object was doing the same thing, but appearing alternately with respect to the other, as first one plate and then the second was visible through the eyepiece.
>
> "That's it!" I exclaimed to myself. The change in position—only three or four millimeters in six days—was much too small for an ordinary asteroid near opposition. But were the images real or spurious? At once I laid out the 8-by-10-inch plates that had been taken by our Cogshall camera simultaneously with the 13-inch exposures. Although nearly at its limit of visibility, there were the images exactly in the same respective positions!
>
> With mounting excitement, I got out the January 21st plates and quickly checked them with a hand magnifier. Even though the 13-inch plate was a sorry one, there was the image displaced about one millimeter east of the January 23rd position, and it was confirmed on the 5-inch exposure. Any possibility of the phenomenon being a pair of variable stars was now ruled out. Next I measured the displacement approximately with a millimeter scale. The object was retrograding about 70 seconds of arc per day. This seemed to be it![54]

When he informed Lampland and Slipher of his discovery, he now recalled, "the air was tense with excitement," and Slipher's first reaction was to declare that no announcement would be made until observational confirmation was completed. But the sky was overcast that night, and the excited astronomers had to postpone further observations of the object with the 24-inch or 42-inch telescopes until the following night. Bachelor Tombaugh whiled away the enforced delay by going to the movies in Flagstaff, viewing, somewhat inattentively one suspects, Gary Cooper in "The Virginian."[55]

On February 19, the skies cleared a bit, although not as much as could be desired, as patches of haze and clouds still interfered with observations. Tombaugh's plates for February 19, 21, and 22 recorded the new object again, despite seeing conditions that were less than optimum. On February 20, the object was located with the 24-inch refractor, but showed no visible disk, and Lampland began photographing it with the 42-inch reflector, getting good images with exposures of only seven minutes. In addition, Tombaugh has recalled, "Lampland made a pair of one-hour exposures, reaching to magnitude 19, in an unsuccessful search for possible satellites. His photographs through blue and yellow filters indicated that the planet was yellowish, resembling the terrestrial planets and unlike Uranus and Neptune which are bluish."[56]

Slipher, in his published report of the discovery with Putnam two years later, added some details of the immediate post-discovery period. "As soon as possible," he wrote, "we pointed the large refractor [24-inch] upon it with one question uppermost on our minds—will it show a disk?—It did not and that was disappointing.... Under the most careful scrutiny, it showed no sign of any cometary features.... Mr. Lampland photographed it with the large reflector to see what such plates might show; and, in particular, to follow its path and to get accurate positions for finding its orbit. The observations continued to find it night after night in its expected place."[57]

The new object's magnitude was fainter than had been expected for Planet X, and Slipher noted that "Dr. Lowell's estimate of the magnitude was too high, doubtless in part because he naturally was influenced by the very high albedos [reflectivity] of Uranus and Neptune." But the object's "drift from day to day, after a few weeks, showed that its distance was definitely much greater than Neptune's and between 40 and 43 times the distance the earth is from the sun. Its drift rate was the distinguishing test for distance, and from the discovery plates on, continued to show that the object was trans-Neptunian."[58]

Thus, a "young man from Kansas," Clyde William Tombaugh, just two weeks into his twenty-fourth year, became the only American to discover a major planet, an event which had occurred only twice before in history. While he has been officially and justly credited with the discovery, it must be noted that he has in subsequent years consistently acknowledged that the many years of work put in by Lowell, Slipher, Lampland and others on the long Planet X search prepared the ground for him.

"In searching for the new planet I was carrying out a systematically arranged program," he wrote within a week of the announcement of his discovery, "and was fortunate to be assigned to this work with the splendid new Lowell photographic telescope."[59] Thirty years later, he declared: "A number of people shared the many years of effort that led to this achievement. I came upon the scene on the eve of the fruition of the enterprise." In addition, he pointed out that he had had the advantage of using the Lawrence Lowell telescope in the search, "the best suited instrument yet brought to bear on the problem."[60] Tombaugh had, in fact, been offered an unprecedented opportunity, but it can also be said that he not only made the best of it, but he improved it.

On February 23, Slipher wrote Putnam a long letter which began with a discussion of the observatory's affairs and the state of the stock market, and ended with comments about the Mars book project and Howard Russell Butler's paintings of Mars. Sandwiched between these items was written:

"Mr. Tombaugh has picked up what looks to be a very interesting object on the 13-inch plates. It is of about the 15th magnitude, and so far has a drift among the stars not much too large for what Planet X would have." It was "not far east of Dr. Lowell's predicted place," he added, but it was "too soon yet" to draw final conclusions.[61]

To Putnam, Slipher's information was "most thrilling and exciting" and apparently he passed the news along to A. Lawrence Lowell, for on March 1 the Harvard president wrote him about "the great news if it turns out to be true. To put Percy's name with that of Leverier [sic] and Adams will be a great and deserved honor, and it distinctly would be a glory for the observatory."[62]

On March 3, Putnam telegraphed Slipher: "Don't be too cautious about publication." He followed his telegram with a letter urging Slipher not to delay "too long" in publishing the news.[63] But the conservative Slipher was well aware from Lowell's controversial career of what can sometimes happen when conclusions too hastily drawn are too hastily promulgated, and he wanted to be absolutely certain that the object was indeed trans-Neptunian and indeed a planet before putting himself on the public record. He was also aware that once the discovery was announced, other observatories with larger telescopes would immediately begin observing the object, and issuing announcements regarding it, and that there would be a race to see who would compute the first accurate orbit. Slipher was hoping, as will be seen, that the Lowell Observatory would provide the first orbit and thus establish the object's planetary status.

Thus, on March 4, he wired Putnam: "Have discussed with men question of announcing object, and find we ought to get some more information first. This because when announced world's largest telescope will immediately observe it and report findings with high authority which would soon amount to eclipsing Lowell Observatory. We appreciate your viewpoint and interest and want to hurry this matter too."[64]

Putnam was agreeable, but worried, and so he "took the bull by the horns" and talked to President Lowell. "His instinct," he confided to Slipher on March 6, "was rather similar to yours, that we had better wait until we are sure before we announce because he felt the danger of someone else's discovering it was slight." At Lowell's suggestion, Putnam also had "a very pleasant talk with Shapley ... he was naturally all for saying something. If I can speak to the faults of your staff, and his, they are both ones of publication—his, too soon, and yours, too late."[65]

A few days later, Putnam wrote Slipher again. "I can appreciate your desire to get some real information before Mt. Wilson is turned loose on it.... Even as it is we are protected on priority from our exchange of letters and my talk with Shapley who is very much interested."[66]

Observations of the new object for position and magnitude continued both with the 24-inch refractor and the 42-inch reflector whenever observing conditions permitted, which was apparently not too often.[67] The examination of the

X search plates stopped abruptly with Tombaugh's discovery, and the blink microscope was turned over to Lampland for his position measurements. It was not until May 26, incidentally, that Tombaugh resumed his examination of plates 165 and 171—the Pluto discovery plates—and it was not until June 9 that he completed the task. In listing the results of his examination, he noted that the plate had yielded eight asteroids, seven variable stars, twelve temporary objects that might be variable stars, novae or defects on the plates; no comets, adding: "Planet 'X' (Pluto) at last found!!!"[68]

While the world was still kept in the dark regarding Tombaugh's discovery, Slipher informed a few people close to, but still outside the immediate observatory circle about it in letters and telegrams dated March 8, 1930. These confidants included importantly John A. Miller, his old astronomy professor now at Swarthmore College's Sproul Observatory; John C. Duncan, who began the photographic search for Planet X back in 1905 and was now at Wellesley College; and Cogshall at Indiana University, another longtime Lowellian.

Miller responded quickly. "I ... appreciate your confidence more than I can express." He added his congratulations.[69] In writing Duncan, Slipher broached the problem of computing an orbit for the object, and in wiring his reply, Duncan advised him: "I could not spare great deal time for computation but might compute preliminary orbit. Prefer positions referred to mean equinox 1930.0. Telegraph three positions representing ends and middle of observed arc and one or two others for check."[70] Slipher did not send any positions to Duncan, but within days, he had invited Miller to come to Flagstaff to supervise the orbital computations there and Miller, as will be seen, obliged.[71]

To Cogshall, Slipher wrote: "In our recent work on the Planet X search there has appeared on the plates of the new photographic telescope what appears to be a remarkable object.... It is too early yet to write anything definite about it, but we hope in the near future to be able to publish something definite. For the present we are of course saying nothing about it except to close friends. You will be interested to know that it was also recorded by your 5-inch Cooke.... So far its apparent motion seems to fit fairly well that of an object beyond Neptune, and it is not very far from one of Dr. Lowell's two predicted positions for Planet X."[72]

Cogshall did not reply until more than a month after the discovery was announced. "I have waited a long time to say anything about your big find that you kindly sent me advance information about," he wrote finally. "Our papers have been full of it of course and everybody talking—some sensibly and some otherwise.... I suppose you saw the article ... attributing credit to everybody for 25 years back. I don't know where it originated but I can guess. [Kenneth] Williams [who had searched briefly for Planet X in 1907] disclaimed it when our local paper played up what was said about him."[73]

Putnam also advised Lowell's widow in advance that an object had been found in the course of the Planet X work, but he had not gone into details. On

March 9, Mrs. Lowell wrote to Slipher and Lampland "about the intensely interesting observation ... that it may be Planet X I pray." She then declared that "Zeus" would be her choice for the name of the new planet.[74]

On March 13, Slipher was too busy talking with newspaper and wire service reporters and answering telephone calls and telegrams to make the trip down from Mars Hill to Arizona State Teachers College (now Northern Arizona University) to present the Lowell Prize to the school's outstanding mathematics student, one Alice Ferrell Miller. But Lampland had the time, and in the course of presenting the award, he announced that a ninth planet had been discovered at the Lowell Observatory.[75]

For late in the evening of March 12, Slipher had sent a telegram reporting the discovery to the Harvard College Observatory, a clearing house for the distribution of important astronomical information. On March 13, the seventy-fifth anniversary of Percival Lowell's birth and 149 years after William Herschel's discovery of Uranus, not only the students at the Flagstaff college but the world learned about the new planet.[76]

Slipher's telegram to Harvard had been brief:

Systematic search begun years ago supplementing Lowell's investigation for Trans-Neptunian planet has revealed object which since seven weeks has in rate of motion and path consistently conformed to Trans-Neptunian body at approximate distance he assigned. Fifteenth magnitude. Position March twelve days three hours GMT [Greenwich Meridian Time] was seven seconds of time West from Delta Geminorum, agreeing with Lowell's predicted longitude.[77]

In an observation circular issued March 13, he expanded somewhat, albeit cautiously, on the discovery:

Some weeks ago ... Mr. C. W. Tombaugh, assistant on the staff ... found a very exceptional object, which since has been studied carefully. It has been photographed regularly by Astronomer Lampland with the 42-inch reflector, and also observed visually by Astronomer E. C. Slipher and the writer with the large refractor.... Besides the numerous plates of it with the new photographic telescope, the object has been recorded on more than a score of plates with the large reflector, by Lampland, who is measuring both series of plates for positions of the object. Its rate of motion he has measured ... at intervals between observations with results that appear to place the object outside Neptune's orbit at an indicated distance of about 40 to 43 astronomical units.... In its apparent path and in its rate of motion it conforms closely to the expected behavior of a Trans-Neptunian body, at about Lowell's predicted distance. There has not been opportunity yet to complete measurements and accurate reductions of positions of the object requisite for use in the computation of the orbit, but it is realized that the orbital elements are much to be desired and this important work is in hand.

In brightness, the object is only about 15th magnitude. Examination of it in the large refractor—but without very good seeing conditions—has not revealed certain indication of a planetary disk. Neither in brightness nor apparent size is the object comparable with Neptune. Preliminary attempts at comparative color tests photographically with the large reflector and visually with refractor indicate it does not have the blue color of Neptune and Uranus, but hint rather that its color is yellowish, more like the inner planets. Such indications as we have of the object suggest low albedo and high density. Thus far our knowledge of it is based largely upon its observed path and its determined rates of motion. Those with its position and distance appear to fit only those of an object beyond Neptune, and one apparently fulfilling Lowell's theoretical findings.[78]

"While it is thus too early to say much about this remarkable object," Slipher concluded, "and much caution and concern are felt—because of the necessary interpretations involved—in announcing its discovery before its status is fully demonstrated; yet it has appeared a clear duty to science to make its existence known in time to permit other astronomers to observe it while in favorable position before it falls too low in the evening sky for effective observations."[79]

CHAPTER 10

PLUTO: THE EARLY CONTROVERSIES

V. M. Slipher's announcement of the discovery of a trans-Neptunian body "apparently fulfilling Lowell's theoretical findings" caused an immediate sensation, and the news, embellished far beyond Slipher's cautious statement, appeared on the front pages of newspapers around the world.

Almost immediately, too, sensation bred controversy—over the nature of the object, its orbit, and whether or not it was indeed one of the two Planets X that Percival Lowell had postulated mathematically fifteen years before. W. H. Pickering's 1919 and 1928 predictions for his Planets O also surfaced briefly in these initial debates,[1] and for the public and the press, at least, the matter of a name for the new planet—for such it was instantly and widely assumed to be—was of overweening interest.

Most of these questions were settled within the first few months after the discovery of what proved to be the solar system's ninth planet, Pluto. But the identity of Pluto with Lowell's Planet X, despite a remarkable similarity of orbits, has remained a longer controversy and has been argued for nearly fifty years by astronomers specializing in celestial mechanics.[2] The point at issue—the validity of Lowell's prediction of an object that *ex post facto* appears to be unpredictable—did not become wholly moot until the discovery of a Plutonian satellite at the U.S. Naval Observatory's Flagstaff station in July 1978, more than forty-eight years after the discovery of Pluto itself.

Slipher obviously timed his announcement telegram to the Harvard College Observatory so that it would be distributed to astronomers and the

Lowell Observatory

Vesto Melvin Slipher, who directed the final search for Percival Lowell's Planet X, as he appeared in the 1930s.

press on Lowell's birthday, a date that Lowell's widow did not allow the astronomers at Flagstaff to neglect in her lifetime. The circumstance, as the journal *Science* opined, showed "a peculiar sense of the fitness of things" in view of Lowell's long theoretical and photographic trans-Neptunian exploration.[3]

Slipher's carefully worded message was turned into large headlines in the evening newspapers of March 13 and the morning newspapers of March 14, and precipitated an avalanche of telegrams and letters at Flagstaff as astronomers and laymen alike sent congratulations and offered suggestions, and the press posed questions and sought more details about the discovery and its possible implications. "We have been much surprised at the great interest in the discovery taken by everyone," Slipher wrote Mrs. Lowell three days after the announcement, "and are so completely swamped with messages of congratulation that we have not had time to write even you since the announcement on Dr. Lowell's birthday."[4]

Among those sending congratulations, it may be noted, were former Planet X searchers Kenneth P. Williams and Thomas B. Gill.[5]

While both Slipher and Putnam realized that Tombaugh's discovery would stir some general interest, neither was prepared for the scope and intensity of the reaction, particularly "the tremendous, instant, and persistent demand from the public press for information" as they later recalled,[6] that followed the March 13 announcement. But Putnam, at least, was concerned that the conservative Slipher might understate the case and not take full advantage of the opportunity presented to draw favorable attention to the observatory and its work. Thus, in telegraphing his own formal congratulations, he advised Slipher: "You will undoubtedly be bothered by the Associated Press, but do not be reticent. This is a wonderful piece of work for you and Dr. Lowell."[7]

The impact of the discovery on the general public is well illustrated in an incident recounted by Harvard Observatory director Harlow Shapley in a lengthy letter to Slipher two days after the announcement. "It seems that we got the Lowell Observatory into the papers," he began, explaining that he had gone to Philadelphia on March 14 to give a lecture on astronomy for the Jayne Foundation. "After the announcement of the planet the Jayne Foundation was bombarded with requests for space, and I was bombarded with inquiries concerning the planet.... They changed the lecture place ... when it became evident that about a thousand more people would try to get into the ballroom than it could accommodate." Shapley also changed his lecture to include some reference to Lowell and the new discovery. "When I ended up that part of the performance with a slide showing Percival Lowell sitting at the business end of the 24-inch refractor, he got a 'hand' as great or perhaps even bigger than he ever received before," he added.[8]

But Shapley had a more important point to make in this letter. "I also hope that positions and motions [of the newly discovered object] are soon forthcoming," he noted. "There are many requests and astronomers may not be able to understand very long why the Lowell Observatory does not give accurate positions. Old plates cannot be satisfactorily examined until we have a provisional ephemeris." To underscore this, Shapley drew Slipher an analogy. "The new comet just discovered [probably Comet Schwassman-Wachmann (1930d)] has been found on some old Harvard plates. Instinctively we should like to hold these positions and compute the orbit ourselves. Ethically we cannot do so, and we are firing away the precious old Harvard positions to Berkeley and Copenhagen where they will probably beat our men in computing the orbit and getting the credit."[9]

Slipher, of course, was well aware of the importance of computing a provisional orbit as quickly as possible for this, he felt, would provide confirmation of the planetary status of the new object and, he hoped, firmly establish it as Lowell's long-sought Planet X. He detailed his position to Mrs. Lowell:

"We see no other interpretation for the object's behavior than that it is a Trans-Neptunian body and apparently Dr. Lowell's Planet X. We did not intend that Harvard Observatory should announce it as Trans-Neptunian planet for we did not feel that we had proof enough to establish that just yet. The object is still ... behaving just like a planet beyond Neptune and in Dr. Lowell's predicted place and about his distance. How much we do want this to come up to indications and expectations! You can not imagine how much this means to us all, for we are so anxious to be able to demonstrate that this is Dr. Lowell's Trans-Neptunian Planet X. After our minds are at ease on that matter then we shall be glad to see all you good people choose a good name.... There is credit enough for Dr. Lowell and to spare, and enough left for everybody who has labored with him...."[10]

Slipher was also determined that the first orbit should come from the Lowell Observatory itself and, because neither he nor the other astronomers at Flagstaff were experienced in such work, he had, as noted earlier, sought assistance in the calculations from Miller and Duncan even before the announcement of the discovery. But other astronomers were also eager to be first to compute an orbit, and soon a lively competition developed.

From the outset, the pressures on Slipher to release the discovery and other pre-announcement positions of the new object were intense. One of the first telegrams that reached Flagstaff after the announcement came from astronomer A. O. Leuschner of the Students' Observatory of the University of California at Berkeley, which had acquired a reputation for computing the orbital elements of asteroids and comets. "Would appreciate by air mail accurate positions of trans-Neptunian object for orbit computation," Leuschner tersely wired the Lowell Observatory on March 13, neglecting, it may be noted, to include any word of congratulation.[11]

A week later, Leuschner made a second request for the positions, this time with effusive congratulations along with an explanation of a statement he had issued to the press which seemed to cast doubt on the trans-Neptunian status of the new object. "Immediately after the release of your announcement the

Sproul Observatory

John A. Miller, director of Swarthmore College's Sproul Observatory, and long-time friend and mentor of Lowell Observatory astronomers.

San Francisco papers made inquiry whether or not we were going to compute the orbit, as has been our practice on practically all comets and other interesting objects," he wrote. "We prepared a statement ... [that] has been much misquoted." This statement, a copy of which he enclosed, concluded: "Until such [orbital] computations should have been made there remains a chance that the observed positions and motions of the new body may find their explanations in an orbit other than should be characteristic of a transneptunian body."[12]

Other early requests for positions came from the U.S. Naval Observatory, from Seth B. Nicholson of the Mount Wilson Observatory, and from George Van Biesbroeck of Yerkes Observatory who wrote on March 18: "The newest addition to our solar system has been photographed here on March 16 and 17 (U.T.) and from the measures I made a rough calculation leading to the

Artist Peter Van Valkenburgh's sketch of Armin O. Leuschner, director of the Students' Observatory of the University of California, as inscribed to astronomer W. W. Campbell, long-time director of the Lick Observatory.

distance of the order of 50 a.u. We are anxious to have an orbit out by means of which the object could be located on prediscovery plates...."[13]

Still other astronomers, in Europe as well as in the United States, appealed to the Harvard Observatory for the positions and on March 15, Shapley sent Putnam a telegram urging him to get Slipher "to release accurate positions for the new planet ... so that preliminary orbital calculations can be made."[14] And as a consequence of his wire, Putnam telegraphed Slipher: "Many requests come for release of accurate positions for the new planet for preliminary orbital computation. Believe this should be wired for distribution. Every one is hot on the subject. Let us feed them what they want."[15]

Slipher was willing to give them what they wanted, but in his own time. On March 16, he had wired the trustee arguing the desirability of having Lowell Observatory issue the first orbit, explaining the need for time to obtain

Pluto: The Early Controversies 203

expert assistance in the computations, and asking permission to invite Miller to Flagstaff to preside over the work. On March 17, Putnam telegraphed his approval of the "confidential plan with Miller." He added: "Hope [he] can come and wish I could too."[16]

A few days later, Slipher, in a long letter to Putnam, explained:

"Our chief concern has been that this matter turn out well and bring to Dr. Lowell's name the honor that we all thru many years have been hoping to see come to him. We are glad to say the object is still continuing to behave as if it were the object he predicted from his mathematical work. Before the announcement we could see no other interpretation for the behavior of the object and we do not now see how it can be anything else than his pla.[net] X. We have had a letter from Yerkes Observatory giving two nights' observations which appear in full confirmation. We are so eager to get out computations of the object's orbit that we are neglecting other things. The orbit will clear up some things about it that we are not sure of yet."[17]

This apparently mollified Putnam, for he replied:

"I begin to think I am too hasty. When I first got Shapley's reports, I thought that we were playing everybody a dirty trick, but I have calmed down, and getting your letter calmed me still further. I am sure you are doing the right thing. As you say, it is not a commet [sic]. In the first place, one is not found every few years. In the second place, it is moving slowly, and is going to stay there a long time."[18]

The suggestion, refuted here by Putnam, that the new object might be a comet apparently had first been publicly made by astronomer John Jackson of the Royal Observatory in Greenwich, England. In a March 23 telegram to Slipher, *The New York Times* noted that Jackson had published a letter in the London *Times* arguing that Slipher's discovery was a comet, not a planet. "Could you oblige us with a dispatch citing evidence it is a planet and adding any reply to Jackson you may care to make?"[19] Slipher's answer to this inquiry, if he made one, apparently is not preserved in the observatory's archives nor in the files of *The New York Times*.

Slipher, now with Putnam's full acquiescence, continued to studiously ignore requests for positions and his silence on the subject, as Shapley had warned, began to generate some resentment among astronomers. But one astute young man, at least, came up with a valid pre-announcement position for the Lowell Observatory object by a quite unorthodox method.

"Perhaps you still await positions of Planet X from the Lowell Observatory," Shapley wrote Putnam pointedly on April 5, three weeks after the discovery was announced. "I can give you one. On March 1, probably at about 12 hours Greenwich Mean time, the planet was in position $7^h\ 16^m\ 12^s$ [right ascension], $+22°\ 5'\ 30''$ [declination]. No," he added, "that did not come from the Lowell Observatory. One of the young men at the Harvard Observatory who was at the talkies the other night saw a newsreel in which Dr. Slipher was pointing at a photograph telling where the planet was on

March 1. He rapidly plotted the field and position, compared with star charts when he got back to the Observatory—or rather with photographs of the region—and handed me the position. Possibly the first newsreel position of an astronomical object that has been obtained!"[20] The position, it may be noted, was right on the mark in relation to the early positions which were finally released a week later by Slipher along with the Lowell Observatory's provisional orbit.

With Putnam's agreement to his "confidential plan with Miller," Slipher wrote to his former professor, appealing to him to come to Flagstaff to supervise the computation of a preliminary orbit for the new object. "The nervous strain of risking an institution's good name on the mind's interpretation of a wholly unique wandering of a fifteenth magnitude star among its fellows for a few weeks is not what we care to experience the second time," he confided. "Wish you could assist to get out orbit. Could you and Mrs. Miller come out for this work if expenses are paid?"[21] Although Miller was pressed for time, he was also eager to undertake the task and quickly agreed, wiring Slipher however: "Must return to Swarthmore not later than April 10. If you believe so short a stay unprofitable wire me at once."[22]

Miller arrived in Flagstaff on March 22 and he and the Lowell astronomers began their calculations. "I remember the occasion very well," Tombaugh has recalled, "seeing Dr. John A. Miller, the two Sliphers and Lampland computing with logarithms at a large table day after day during the spring of 1930."[23]

While the Lowell staff labored over their unfamiliar task, Leuschner and two of his graduate students at Berkeley—Ernest C. Bower and Fred L. Whipple—were also busy. Despite Slipher's refusal to furnish pre-announcement positions, they proceeded to work out no less than six different orbits for the the new object, two of them parabolic and thus typical of cometary motion, from observations made at Lick Observatory that spanned only nineteen days, from March 16 to April 4.

These were announced April 5, and described in a *Harvard Announcement Card* dated April 7 and a week later in a *Lick Observatory Bulletin*.[24] The announcement declared that the work "has resulted in a group of solutions giving a fairly well determined distance from the earth of approximately forty-one astronomical units," an inclination of 17° and a longitude of ascending node of 109°. It also noted that "no definite conclusion concerning the eccentricity and period can be drawn without observations covering a considerably longer interval," and that "at present it is not even possible to decide whether the orbital motion of the object is directed toward or away from the earth." And it added that it was "improbable but not impossible that computations based on more extended observations will materially change the distance, inclination or node."[25]

Putnam telegraphed Leuschner's announcement to Slipher in Flagstaff, and gave his own reactions to the Bower-Whipple orbits in a letter the same day. "It seems to me very interesting, but with your figures going back six weeks earlier I presume that you and Professor Miller will be making some announcement soon."[26]

On April 8, Harvard Observatory promulgated the elements of a circular orbit for the object computed at the Cracow Observatory in Poland, which gave a distance of 37 a.u., an inclination of 13° 5′, and a longitude of node of 108° 42′.[27]

Four days later, on April 12, the Lowell computers completed their work, and Slipher telegraphed their new orbit that same afternoon to Putnam, who promptly forwarded it to Shapley at Harvard. Slipher's telegram reported: "Preliminary orbit of Planet X has been computed by Lowell Observatory staff with collaboration of Dr. John A. Miller of the Sproul Observatory, using positions January 23, February 23 and March 23, determined from Lowell plates by Lampland...." It then listed the orbital elements derived, gave the positions of "Planet X" on the three dates and on March 30, a date used to check the computations. The telegram ended with a brief caveat: "However, because of the very short arc available and the object's extremely small latitude, considerable revision of some elements, especially eccentricity, is not unexpected...."[28]

As it turned out, the Lowell computers might well have paid more attention to Bower and Whipple's reservations concerning the eccentricity and period of the new object, although they probably felt that their observations, extending over two months rather than only nineteen days, provided a sufficient interval to attempt to estimate these elements. At any rate, while the Lowell orbit agreed well enough with Bower and Whipple's results for the distance, inclination and longitude of node, it assigned the new object a near-parabolic eccentricity of .909, and a period of more than 3,000 years—very unplanetary parameters indeed![29]

Parenthetically, Slipher may not have intended this orbit to be distributed immediately to astronomers and the press, and Putnam clearly misconstrued its negative implications as far as the object's planetary status was concerned, for on April 12 Putnam, in wiring his congratulations on the orbit, advised Slipher that he had given Slipher's telegram to Shapley.[30] On April 14, Putnam explained to Slipher that the wording of his orbit announcement "was such that I presumed it was for publication, and passed a copy of it on to Shapley. I didn't study it over carefully—in fact, I had forgotten the definition of 'eccentricity,' and thinking it was the ratio of the major and minor axes, was not surprised to have it come out .909. Shapley called me up Sunday afternoon, and told me he had a telegram from [James] Stokely [a *Science Service* writer] asking for a check on this eccentricity.... It was then that I realized that I had the definition of 'eccentricity' wrong." He added that Shapley "was tremendously excited; he felt that you had found an entirely new class of object, that while it was not exactly a major planet, no one could call it a comet—in fact, he didn't know what to call it."[31]

Slipher's telegram had been sent late Saturday afternoon, too late for the Sunday newspapers. But the news of the new orbit appeared in the papers of

Monday morning, April 14. "'Planet X' Orbit Raises More Doubt," *The New York Times* headlined the widely published *Science Service* report, presumably written by the aforementioned Stokely, that morning. "Lowell Observatory Estimates Put Trans-Neptunian Object in Asteroid or Comet Class—Course Is Long Ellipse," a second headline proclaimed over the story which began:

> Doubts that the new member of the solar system discovered beyond Neptune is the ninth major planet predicted by Percival Lowell have been raised by computations of its orbit made at the Lowell Observatory.... It is now thought it may prove to be a unique asteroid or an extraordinary comet-like object.[32]

This report relied heavily on Leuschner as the source for these statements, quoting him as saying: "The Lowell result confirms the possible high eccentricity announced by us on April 5. Among the possibilities are a large asteroid greatly disturbed in the orbit ... or a bright cometary object.... High eccentricity and small mass would seem to eliminate the object as being Planet X predicted by Lowell and [make it thus] simply an unexpected discovery." Leuschner, incidentally, offered one other curious but intriguing possibility as to the nature of the object, suggesting: "It may be one of many long-period planetary objects yet to be discovered...."[33]

This and other similar reports in the press, some of which also quoted Leuschner, caused considerable concern at Flagstaff, and Putnam hastened to reassure Slipher that the doubts now widely expressed about the nature of the object did not detract from the importance of its discovery. "Don't worry too much about news," he telegraphed Slipher on April 15. "We have had our day with popular news and the facts will take care of scientific reputation. If your new object is not exactly planetary it could have caused the perturbations and is really a more interesting discovery than just another planet."[34] And the same day, he wrote in a letter, "I don't somehow or other feel the worry about any news Stokely may now put out. The facts are gradually going to establish the scientific interest in the object whether planetary or not, and what Stokely says is not going to seriously affect scientific judgment. We certainly have had our day with the popular newspapers, more than any of us, I think, expected, and what Stokely says there will not affect us, either.

"The new object," Putnam continued, "whatever it is, is a most interesting discovery. If it is not strictly planetary, it comes into a wholly new class, which is, in some ways, more exciting. Altogether I think it is a great piece of work that everybody has done. One of the things of which I am anxious to get an idea is the approximate size. I don't suppose the orbit can be figured accurately enough to give the mass from that, and the only way is to estimate from brightness, which, of course, must be hazy."[35]

The announcement of the Lowell Observatory orbit brought a flood of new inquiries from the press, and on April 14 Slipher sent Miller, who was still en route to Swarthmore, a night letter to brief him on developments.

"Wired orbit Saturday afternoon, answered excited press today that our orbit and observations show definitely object is Trans-Neptunian and afford strong evidence planetary nature. Does not look cometary. Understand Leuschner suggests asteroid or comet...."[36] Miller's telegraphic reply to this, sent after his arrival at Swarthmore April 16, was: "Believe in orbit. The newspapers here had seen Leuschner's comments. Impossible to avoid interviews."[37]

Slipher, in fact, had not lost faith, and expressed his thinking at length on April 19 in long letters to both Miller and Putnam. To Miller, he wrote: "Evidently the type of orbit has greatly impressed some astronomers. And as was to be expected in the eyes of some it looked to change the whole nature of the object. Leuschner seemed to think it an asteroid or comet—just the sort of thing he is working with every day! ... Earl [Slipher] has pointed out the fact that some plates of Halley's Comet got about the first of June 1911 after the c)met had got a long way out from the sun, still show only a hazy spot of light that in no sense resembles the starlike image of X. There seems to be no doubt that the term planet to the best of our knowledge best fits the object. Of course if Leuschner likes he might call it a giant minor planet, but it will go right on being just what it is in spite of all we think or say about it or the names we call it. Still," he added, "it is very evident that it was important that we could get out the first general orbit here.... We shall always look back with pleasure on having you here to help get the baby settled and made as much at home [as] could be done under the circumstances. While there have been anxious moments it does look now as if the infant will survive."[38]

And to Putnam, he wrote: "I hope and believe the object will prove to be of very great interest and importance even if it should finally not quite fit Dr. Lowell's predictions. It was clearly his work and courage in predicting it that were directly responsible for the find. And the seeming willingness on the part of some, like Leuschner, to belittle the matter in quite off-hand suggested interpretations appeared uncalled for if not unfriendly. He was doubtless peaved [sic] that we did not send him early positions with which to determine its orbit. Am sending him a letter that will explain the situation as we saw it but it may not quite satisfy. Presumably he was one of those that [sic] was appealing to Shapley for positions."[39]

Slipher had, in fact, written to Leuschner the day before, enclosing "some early positions of the Trans-Neptunian object, apparently Lowell's Planet X. Also," he added, "I wish to explain the peculiar situation in which we found ourselves with regard to the matter of positions and the preliminary orbit, and to apologize sincerely for not having written you long ago in explanation.... [A]ll criticism for the policy followed belongs on my shoulders." He then explained at length that "it seemed best for the observatory and those most deeply concerned for us to find it out and make it known if the object were less important than it had appeared. Lowell Observatory had put so much into this problem during the last quarter century as to justify the policy.... We felt we should proceed very carefully and that we have tried to do...." To further justify his stand, Slipher also made a curious and perhaps apocryphal statement that may refer to Miller but is not supported elsewhere in the available record. "A number were asking for positions for orbit purposes and among them one claim was advanced that positions had been promised him by Dr. Lowell (if Planet X were found) that he might be the first to compute the orbit, and we could not question the sincerity of the claim," he wrote. "[Thus] perhaps our early decision to do a preliminary orbit here was after all as fair a policy as any that could have been followed. We were for a long time concerned only with rates of motion and not with positions."[40]

Slipher's apology clearly satisfied Leuschner, who may not have been as peeved as Slipher thought, and who again sent his congratulations, pointing out: "I trust that you realize that even if the object should not be found to conform to the requirements of Lowell's predicted Planet X, its discovery is of equal if not greater astronomical importance."[41]

Parenthetically, the problem with the extreme eccentricity of the Lowell orbit seems to have been two-fold, for Leuschner later attributed it to "a slight and unavoidable inconsistency in the Jan. 23, 1930, position from a small scale plate, amounting to less than three seconds of arc in each of the coordinates." Tombaugh, writing in 1978, suggests that there is yet another factor which has not been known. "The images on the 13-inch plates exhibit a tiny core image slightly off center of the total image," he has recalled. "When Lampland was measuring the exact positions of the Pluto images on the discovery plates, he nervously confided to me: 'How does one decide the true center of the image in this situation.'"[42]

In his April 19 letter to Putnam, Slipher candidly set down his thoughts concerning the new object. "If it is a comet it is different from all other known comets and if it is an asteroid the same is to be said of it, if it is a planet it does appear to have the most exceptional orbit of all, and much the most exceptional. On the other hand, we know that in the outer satellites of Jupiter we have a similar change in the satellites as regards size and orbit both as compared with the four large satellites, and so we might find that there is a somewhat similar change in the planetary bodies of the sun's system between Neptune and Planet X....

"Your question as to size and mass," he continued, "is not easy to answer; in fact as you say, its magnitude is our present most available indication. The indication is that the quality of light is like that of an inner planet rather than like Uranus's or Neptune's [and] is evidence for a rather low albedo [reflectivity] and a body of high density. When the orbit is more accurately determined a year or two hence it will be possible to see how much it has perturbed the other planets, and in this way get a measure of its mass.... So Planet X could still be some thousands times more massive than all the asteroids rolled together. Hence only in apparent size does it compare with asteroids, owing to its enormous distance. If Planet X had the albedo of

Mercury or Mars it could with its present star magnitude still have a mass a few thousand times the aggregate of all the asteroids and comets combined.... Hence the term planet is clearly to be preferred to asteroid or comet. It took a long time to find Planet X," he reminded Putnam, "and so it may take a lot more time and work in determining clearly its real nature. Even in the minds of many astronomers it is easy to translate without thinking a 15th magnitude [object] into a trivial body. We would," he added in conclusion, "have wished for a brighter object and an orbit that would seem to fit better what was expected, but this, whatever may be said of it, is causing more thinking than anything that has been found in a very long time."[43]

Putnam acknowledged this long letter April 25, noting particularly Slipher's analogy between the outer satellites of Jupiter and the outer planets of the sun. "One thing in your letter which strikes a familiar chord in my mind was Dr. Lowell's feeling that an exterior planet might have an orbit with a large inclination, and he was arguing just as you point out, from the outer satellites of Jupiter. I thought later that he gave this up, feeling that Jupiter's outer satellites that behaved so queerly were, perhaps, captured comets, and Planet X's behavior makes it look as though possibly there was something generic about them after all."[44]

A few days later, Slipher received some undoubtedly welcome support both for his decision to withhold the early positions of the object until an orbit could be computed at Flagstaff and his efforts to convince astronomers and laymen alike that the object was indeed trans-Neptunian, planetary in nature, and, in effect, Lowell's Planet X. This came from Indiana University's W. A. Cogshall who wrote: "Some people do anything to get into the papers—I think your announcement and other statements since are exactly right—Harvard, nor anyone else have [sic] no right to demand your observations. They wouldn't give anybody a thing they could use themselves. Let them stew." Then, noting a new press report speculating that the Lowell object might be a "comet" of a "very rare variety," he added: "Why not say that nobody ever heard of such a 'comet' before and this can only be a comet if a comet can be something else. Of course this may not be Lowell's planet at all—It may even have a hyperbolic orbit but even so it is the only thing of the kind ever found in all astronomical history."[45]

Late in April, as doubts about the planetary nature of the Lowell Observatory object continued to appear in the press and in astronomical publications, Slipher started to draft a second observation circular regarding the discovery which not only included the positions and orbital elements derived by Miller *et al* but presented for public consumption some of the arguments he had outlined to Miller and Putnam.

In this two-page printed circular, which was dated May 1, 1930, Slipher declared that the early Lowell positions for the object, along with more recent ones, "will, it is hoped, offer an opportunity for others interested to determine an improved orbit." The Lowell orbit, he wrote, "was, of course, based upon the best available observations, but these covered only an exceed-

Indiana University photograph

Wilbur A. Cogshall of Indiana University's Kirkwood Observatory, photographed in the 1930s.

ingly small fraction of the indicated orbit. Hence our knowledge of the orbit's shape and size—its eccentricity and semi-major axis—must be regarded as probably subject to considerable modification when more extended positions are available."[46]

Slipher pointed out, too, "that if Mars were removed to the distance of 41.3 astronomical units his stellar magnitude would not be greater than Planet X. That is, this new body may be comparable with Mars in size and mass." He conceded that visual observations had not shown a recognizable disk for the object, but declared that "the object's faintness doubtless raises the question of size of minimum detectable disk" and that "how large the disk may be and escape detection remains to be determined."[47]

Moreover, he noted, repeated photographic observations with the 42-inch reflector and visual observations with the 24-inch refractor "have revealed no cometary features about the object. Besides, comets have been observed at nothing like this distance from the sun. As for asteroids, Planet X would outshine 100-fold the largest of the asteroids, Ceres (diameter 480 miles) if removed to the same distance." Finally, he declared, "the orbit is no sufficient ground upon which to venture a decision as to the nature of such an object," and here he drew "an obvious analogy between the Sun's increased family of planets and the satellite system of Jupiter." Thus, he concluded:

> This then appears to be a Trans-Neptunian, noncometary, non-asteroidal body that fits substantially Lowell's predicted longitude, inclination and distance for his Planet X. Lowell considered his predicted data as only approximate, and a one to one correspondence between forecast and find would not be expected by those familiar with the problem.... This remarkable Trans-Neptunian planetary body has been found as a direct result of Lowell's work, planning and convictions and there appears present justification for referring to it as his Planet X.[48]

Slipher's argument here was soon to receive strong support from other quarters and, indeed, shortly after it was published the respected A. C. D. Crommelin in England, long one of Percival Lowell's friends, announced still another orbit to the British Astronomical Association that not only led to the full confirmation of the object's trans-Neptunian planetary status, but indicated, as will be seen later, that it did in fact appear to be Lowell's postulated planet.

Crommelin had written Slipher on April 16 his congratulations "to you and your staff (particularly Mr. Tombaugh) on the great discovery of a Trans-Neptunian Planet."[49] A week later, Slipher sent him some of the early positions along with an explanation of the delay in releasing them. "Before announcing the finding of what we have had reason to believe is Lowell's Planet X," he wrote, "it seemed to me—and the other staff members heartily agreed—that we here should determine a preliminary orbit.... Dr. Lowell's extensive work and the Observatory's long search devoted to the problem was some justification of this plan; as have been some of the remarks and suggested interpretations that have since appeared." To reconcile the high eccentricity of their recent orbit determination with Lowell's original work, Slipher cited the analogy with the Jovian satellite system, and also the careful qualifications that Lowell had put on his conclusions in his trans-Neptunian planet memoir. "You will recall his pointing out near the end that the 'actual as against the probable errors of observation might decidedly alter the result,'" he reminded the English astronomer.[50]

On May 9, Crommelin announced to a meeting of the British Astronomical Association that an image, apparently that of the Lowell object, had been found on a plate exposed January 27, 1927, at Belgium's Royal Observatory at Uccle. This, of course, considerably extended the time span of observations of the new object, and using the Uccle position, Crommelin now computed an orbit that was clearly planetary in its characteristics. The distance, inclination and longitude of node were in close agreement with the earlier provisional orbits, including the Lowell one, but the eccentricity was now 0.287, still high in relation to the other known planets, but nevertheless valid as a planetary parameter, and the period was a more reasonable 265.3 years.[51]

Crommelin's orbit, and the provocative conclusions he drew from it, will be discussed in greater detail in the following chapter. Here, it is only necessary to note that it provided the first confirmatory evidence of the planetary nature of the Lowell Observatory object and led directly to further confirmations. After the promulgation of Crommelin's elements, this status was not seriously challenged again, although early in June, Van Biesbroeck, by doubling the orbital arc used by Miller and Slipher in computing their orbit, came up with hyperbolic elements for the object.[52] By this time, however, its planetary status was already widely accepted and Van Biesbroeck's hyperbolic orbit, where it was noted at all, was not taken seriously.

By June, too, Milton Humason of Mount Wilson Observatory had found spectrographic indications of the object's planetary character. Late in May, Mount Wilson's director Walter S. Adams advised Slipher that Humason would arrive in Flagstaff in a few days to make some more seeing tests in connection with the 200-inch project, noting in passing and without elaboration that Humason had obtained a low dispersion spectrogram of the object shortly after the discovery announcement.[53] Slipher replied immediately that he would cooperate in the new tests. "It will naturally interest us much to hear of Mr. Humason what the spectrogram showed," he wrote. "Its light never appeared to me to be of cometary quality, but with an object so faint it is not possible to see well enough visually to decide such a question."[54] Humason duly arrived on May 30 and, Slipher reported to Miller: "Their spectrogram ... shows a solar type spectrum."[55] This, Miller opined, "is a good argument for its being a planet." And in discussing the early orbits, including the one of Van Biesbroeck, he added: "Of course if the Euclla [sic] position is a true one, then Crommelin's orbit would carry more confidence than any that has been derived."[56]

Crommelin's orbit greatly facilitated the search for images of the object on old plates and this search, pursued at observatories around the world, quickly produced results. Among the first pre-discovery images found, as noted earlier, were those recorded on March 19 and April 7, 1915, on Lowell Observatory Planet X search plates, but these were also among the last to be announced. Earl Slipher had telegraphed word of their discovery June 12, 1930, to his brother, then in Indiana to receive an honorary doctoral degree from Indiana University, and suggested the immediate release of their approximate positions.[57] But the elder Slipher again was reluctant and opted for delay, in part certainly to assure that the positions when released would be accurate. At any rate, despite some pressure from Putnam during the summer,[58] the discovery of the 1915 images did not become general knowledge among astronomers until December, when Lampland described them in a paper read before a meeting of the American Astronomical Society in New Haven, Connecticut,[59] and their positions were not published until late the following February.[60]

Of more immediate importance in the post-discovery controversies regarding the new object were four images found a few days earlier by S. B. Nicholson and N. U. Mayall at Mount Wilson Observatory on plates exposed there December 27–29, 1919, by Humason with a 10-inch Cooke photographic triplet during a systematic search largely inspired by W. H. Pickering's Planet O prediction earlier that year. The two astronomers had relied on orbits by Bower and Whipple and by Crommelin to locate the images on the star-crowded plates, and then had used their positions to compute both heliocentric and barycentric* orbits that were in good relative agreement with Crommelin's orbit and further confirmed the planetary nature of the object.[61]

Only four days after Nicholson and Mayall announced the 1919 images, astronomer Frank Ross at Yerkes Observatory reported that images of the new planet had also been found on two plates exposed by E. E. Barnard January 29, 1921, and on a plate he himself had made January 6, 1927.[62] Subsequently, five additional pre-discovery images were turned up, two recorded at Germany's Königstuhl Observatory and one at Harvard in 1914, and two at Mount Wilson in 1925. By September 1931 Bower, in publishing still another orbit for Pluto, could list no less than fifteen such observations, not including the April 1929 images recorded shortly after the Lowell Observatory had resumed its long-delayed Planet X search.[63]

These pre-discovery images, extending as they did the observational time span over sixteen years, permitted progressively refined orbits, by Bower and Whipple in August 1930, by F. Zagar in November 1930, and finally by

---

*That is, with reference to the center of mass of the solar system.

Bower in 1931, which firmly established the Lowell Observatory object as a trans-Neptunian planet.[64]

If the determination of the true nature and orbit of the Lowell Observatory object was the first order of post-discovery business for astronomers, the matter of its name held top priority for the public and the press from the beginning. From the day of Slipher's announcement suggestions poured in, many of them duplications and most arguing the appropriateness of the choice, until the list of proposed names numbered nearly one hundred. Suggestions ranged from "Atlas," one of Putnam's early favorites,[65] to "Zymal," urged by one W. E. D. Stokes Jr., as "the last word in the dictionary" for "the last word in planets."[66] Other names proposed included, for example, "Artemis, far ranging goddess of the hunt,"[67] "Perseus, as being the nearest name to Percival, in honor of Dr. Lowell, at the same time being a name which has classical significance,"[68] "Vulcan," the name of Leverrier's postulated 1859 intra-mercurian planet,[69] and "Tantalus, as a very appropriate designation for the somewhat tantalizing Planet 'X'."[70] A Pittsburgh, Pennsylvania, man telegraphed to "P. Lowell" to ask: "Why have only one lady in our planetary system?" He suggested "Idana, because she is so distant."[71] Newspapers, too, joined in with *The New York Times*, for example, promoting "Minerva—the planet that sprang to human view full panoplied from the mind of man."[72]

Naming the new planet, indeed, became a game, a lively competition for many and while this probably would have happened in any event, Putnam made it certain through his post-announcement remarks to the press. "Apparently I got quoted in the Boston papers as saying we wanted suggestions for a name," he wrote Slipher apologetically two days after the announcement. "I didn't mean to give any such impression to anybody, and have told everyone since that the naming should properly come from you and your associates."[73] And in another two days he telegraphed Flagstaff: "Popular clamor here is all about a name for the new wonder. Many have been suggested, both modern and classical, and to end foolish and undignified newspaper talk in Boston I think a name should come from you soon. Best suggestion that has come to me is Cronus, the youngest of the Titons [sic], son of Uranus and Father of Neptune. Real certain Dr. Lowell would wish a classical name."[74]

The suggestion of "Cronus" may have come originally from Putnam's mother, Percival Lowell's sister, Mrs. William L. Putnam, and apparently she also made it directly to Slipher at Flagstaff as well. For on March 21, he replied: "Your suggestion is excellent, and it is too bad that Dr. [T. J. J.] See ... had used the name Cronus, so he tells us in a recent letter, for a planet he had predicted some years ago, and hence, of course, it would not do to apply it to Dr. Lowell's planet. Dr. See would then claim we had only discovered his planet and used his name. Unfortunately we have been so rushed off our feet for some weeks with this matter that we have not found time to consider the name."[75]

Shapley at Harvard had been busy, too, but he found time to make some suggestions. "I write this note while I am in the midst of considerable perturbation arising from Planet X," he confided mischievously to Slipher on March 20. "An hour ago we first determined the photographic magnitude on Harvard plates. Two hours ago we received the first precise positions from the Yerkes Observatory. Three hours ago, and before and after, letters come and reporters come with inquiries and suggestions. Is the name to be Osiris, or Bacchus?" Other names proposed, he added, included "Apollo" and "Erebus."[76]

If the astronomers at Flagstaff had any names to suggest in the weeks between Tombaugh's discovery and its announcement, they left no record of them. But Mrs. Lowell, as noted earlier, had suggested "Zeus" in a letter March 9. "Mr. Putnam asked me if I had any thoughts about the name," she had written Slipher then. "He said he had thought of Diana.... Zeus is my choice. Some people believe the name to be one and the same as Jupiter but of what I have read it is not so—Jupiter is a Roman god—Zeus is a Greek god...."[77]

But after Slipher's announcement and the stir it caused in the press, Lowell's widow changed her mind. On March 14, she telegraphed Slipher: "In eastern newspapers and at luncheon today unanimous demand that the planet be named Percival, and we hope that you at the observatory who made possible its finding will be in sympathy with the appropriateness of this name. They think as I that the gods of the past are worn out. I find from [A.] Lawrence [Lowell] and Roger [Putnam] what you all have to say is going to have great weight."[78] The following day she sent news clippings to Slipher and Lampland, noted that the name "Constance"—her given name—had also been suggested, and asked: "Are you willing to have the Planet named Constance?"[79] Slipher apparently maintained a discreet silence on this question, at least as far as the observatory archives show.

Many people, in fact, proposed the names "Percival" or "Lowell" for the new planet in the first days after its discovery was revealed.[80] But soon the idea of having a classical name gained in the popular mind. Arizona's United States senator, Henry Fountain Ashurst, reflected the feeling of many when on March 31, after congratulating the observatory with typical grandiloquence, he wrote: "When the time shall have arrived to name this planet, if you do not name it Percival, I would humbly suggest that you return the Olympian nomenclature...." Ashurst's choices were "Athenia" or "Minerva."[81]

Just a few days earlier, with no recommendations for a name forthcoming from Flagstaff and suggestions continuing to pour in, Putnam had penned an urgent appeal to Slipher for a decision, declaring in a letter on March 27 that "apparently you and I are passing the buck back and forth about this naming. I feel very strongly that you in Flagstaff are the ones to pick the name, and I hope very much that you all won't wait too long to pick it because if you do, some one else will name it, and the name will stick. The three best suggestions to my mind so far are Cronus, Minerva, and Pluto," he added, noting that his "only objection" to Pluto stemmed from its possible association with "Pluto Water," a mild, mineral water laxative widely advertised in the 1920s and 1930s, and that his "only strong feeling" was that the name should be classical. "I do hate, though, all this attempt to run competitions and general newspaper publicity, which is why I wish a name could be picked out promptly.... It really is amazing to me what a tremendous amount of popular interest has been taken in what I would consider a wonderful scientific achievement."[82] The following day, after presumably hearing from his mother that Cronus had been preempted by See, Putnam wrote that "to my mind, that leaves only Pluto and Minerva, with Minerva ranking ahead in my choice. I don't think Pax or many of the other suggestions are really properly in keeping...."[83]

But Slipher continued to procrastinate, even after the aforementioned Stokely of *Science Service* telegraphed him early in April: "Interest in the new planet still continues at high pitch among newspapers we serve, especially regarding name...."[84] Through the month, suggestions kept coming in, but no definitive word came from Flagstaff as to what the name would be.

Not until May 1 did Slipher formally propose the name "Pluto" and the planetary symbol "♇"—the superposed initials of Percival Lowell's name—and publicly announce the choice in a Lowell observation circular.[85] By then, to his consternation, the word was already out. "The announcement by the United Press of the naming of the new planet discovered at Lowell Observatory was unfortunate and premature," he declared in a statement to the press. "It came out through the personel [sic] of a printing office where a bulletin was to be printed. We regret very much the announcement of our choice of a name getting into the newspapers before it could be communicated to the American Astronomical Society and the Royal Astronomical Society as it appears inconsiderate to those bodies...."[86]

Putnam's thinking on the subject at this point is contained in an undated, typewritten statement in the Lowell archives apparently prepared for the press sometime after the announcement of the name. "We felt in making our choice of a name for Planet X that the line of Roman gods for whom the other planets are named should not be broken and we believe that Mr. Lowell whose researches led directly to its discovery would have felt the same way," he began.

There have been many suggestions which have been weighed and sifted and suitable ones narrowed down to three—Minerva, Cronus, and Pluto. The discovery of this planet is so preeminently a triumph of reasoning that Minerva, the goddess of wisdom, would have been our choice if her name had not for many years been borne by an asteroid.... Cronus, the son of Uranus and the father of Neptune, would have been appropriate but so also is Pluto, the god of the regions of darkness where "X" holds sway. Pluto's two brothers, Jupiter and Neptune, are already in the heavens and it seemed particularly appropriate that the third brother should have a place. Now one is found for him and he at last comes into his inheritance—in the outermost regions of the sun's domain."[87]

The preempted "Minerva," however, proved to be by far the most popular choice, with nearly half of all the letters received by Putnam and at the observatory suggesting the name.[88]

Slipher's circular announcing the choice of the name Pluto had noted that the symbol "♇" would "be easily remembered because [of] the first two letters of the name," and that this symbol would not be "confused with the symbols of the other planets."[89] In his subsequent statement, Putnam added another argument for this symbol, declaring that the use "of the initials of Percival Lowell's name would be a fitting memorial to him in being given to his planet."[90] Somewhat later, in a formal article with Slipher on the discovery of Pluto, he expanded on this somewhat, writing that the "symbols of all the existing planets, except that of Neptune, are so conventionalized and similar that we felt that the proper symbol for Pluto would be the letters 'PL' in the form of a monogram which would be easy to write, and to the layman bring Pluto to mind. The fact that these were Percival Lowell's initials added weight to the thought."[91]

"Pluto" was, in fact, one of the very first suggestions received at the Lowell Observatory following the announcement of the discovery of the new planet, and within weeks it was in quite common use among astronomers in Europe at least, as Slipher later noted.[92] Slipher, in announcing the name, credited the suggestion to a "Miss Venetia Burney, aged 11, of Oxford, England," adding that it had been "kindly cabled" to Flagstaff by Oxford's Professor H. H. Turner.[93]

The name and symbol proposed by the Lowell Observatory was quickly accepted both within astronomy and by the lay public. By the end of July 1930, Miller could assure Slipher: "I note from publications that Pluto has been very well received and such astronomers as I have spoken to have agreed that the name was well chosen. Laymen often quote its name so I think the general public has also given its approval," he wrote.[94] There were at least two notable exceptions to this, however.

For one, Lowell's widow retained her belief that the old gods were "worn out," confiding ruefully to Slipher on New Year's Day 1931: "I wish the Planet could be called Planet X and not Pluto. I do not like Pluto for the name."[95]

And the aging planet-hunter William Henry Pickering, while at first favoring the name and indeed claiming it as his own, later contended with some bitterness that it was inappropriate for such a small planet and more properly should have been reserved for his postulated Planet P.

Pickering had received Slipher's observation circular announcing the discovery of the new object somewhat belatedly at his private observatory at Mandeville, Jamaica, in the British West Indies, where he had continued to work following his retirement from the Harvard observatory staff in 1924.[96] Even before sending a scribbled postcard of congratulations to the Lowell staff,[97] he had dispatched an article, dated April 10, 1930, to *Popular Astronomy* on "The Trans-Neptunian Planet" in which he proposed that the planet be called Pluto.[98] But by January 1931, in another article in the same journal, Pickering declared that because Pluto penetrates Neptune's orbit for a brief period during its revolution around the sun, it "should be called Salacia ... or Amphitrite," after the wife of the Roman Neptune or the Greek Poseidon. "This would have saved the name Pluto," he wrote, "... to be given to the large but still unknown planet [P] revolving outside of Neptune at a distance of about 70 [astronomical] units, as determined by both planetary perturbations and cometary aphelia."[99]

A few months later, again in *Popular Astronomy*, he fumed petulantly that when he had first recognized the importance of Planet P "some twenty years ago, I mentally reserved for it the name Pluto.... Pluto should be named Loki, the god of thieves! A suitable name for P will now indeed be difficult to find when that planet is discovered."[100]

Pickering, be it noted, apparently eventually became reconciled to the name Pluto and its symbol "♇" for after his death in 1938 at the age of eighty, noted Harvard astronomer Annie J. Cannon remembered his remarking: "That's a good name—Pickering-Lowell!"[101]

CHAPTER 11

THE LONGER CONTROVERSY

The question of whether the Lowell Observatory's trans-Neptunian object was actually the Planet X predicted by Percival Lowell in his "Memoir of a Trans-Neptunian Planet" in 1915, however, remained the subject of a longer and more complex controversy that lasted almost fifty years.

To argue *for* the proposition, as did V. M. Slipher and others from the moment of Slipher's announcement of the discovery, it was necessary to ignore certain facts which soon became evident and which seemed to make it impossible for anyone to have predicted the orbit of Pluto with any degree of accuracy from the perturbations of Uranus. Yet one of Lowell's suggested orbits for his Planet X, based on such apparent perturbations, bear what a number of astronomers were quick to admit was a "remarkable" resemblance to those of Pluto. Thus to argue *against* the proposition, it is necessary to concede one of the most incredible sets of coincidences in the history of science.

This is not to say that the discovery of Pluto was an accident in the usual sense of the word. Pluto was discovered as the result of a long, deliberate and systematic search—a search inspired by Lowell, carefully planned by Slipher, and conscientiously carried out by young Clyde Tombaugh. The observers at Flagstaff were looking specifically for a trans-Neptunian planet postulated by Lowell and they found a trans-Neptunian planet with elements surprisingly similar to Lowell's Planet X_1* within 6° of longitude of where he said it would be. If this was simply fortuitous, as some very eminent celestial mechanicians contended in subsequent years, the coincidence was compounded by the fact that Pluto was also found within 6° of the position posited by William Henry Pickering for his 1919 Planet O.

Moreover, when reasonably accurate orbits were finally computed for Pluto some months after its discovery, Lowell's elements for his Planet X_1 were found to be extraordinarily close, given the inherent uncertainties of the problem, to those of the new planet, as first England's A. C. D. Crommelin and later America's Henry Norris Russell, among others, pointed out in print. The overall agreement, in fact, was far better than that of Leverrier's and Adams's postulated elements for a trans-Uranian planet with Neptune. Furthermore, some of Pickering's elements for his 1919 and 1928 Planets O turned out to be also close to the mark, as shown in the following table, which has been adapted from one prepared and published by Crommelin in 1931:[1]

Elements		X_1 (Lowell)	O (Pickering)	Pluto (1930)
a	(mean distance)	43.0	55.1	39.5
e	(eccentricity)	0.202	0.31	0.248
i	(inclination)	10°±	15°±	17°.1
Ω	(longitude, node)	—*	100°±	109°.4
ω	(longitude, perihelion)	204°.9	280°.1	223°.4
μ	(mean annual motion)	1°.2411	0°.880	1°.451
P	(period, in years)	282	409.1	248
T	(perihelion date)	1991.2	2129.1	1989.8
E	(longitude, 1930.0)	102°.7	102°.6	108°.5
m	(mass, earth = 1)	6.6	2.0	<0.7
M	(magnitude)	12–13	15	15

*Not predicted

In this, it is evident that Lowell's elements for the smaller of his two orbits for Planet X are in fairly good agreement with Pluto's, relatively speaking, for the distance, period, eccentricity, and date and longitude of perihelion, that is to say, for the size and shape of its elliptical orbit and the direction of its major axis. But Pickering's 1919 prediction for O was closer for mass and magnitude, and he was very close indeed in suggesting a longitude of ascending node and an inclination of the orbit—the elements which together define the position of the planet's orbital plane in relation to the plane of the ecliptic. And also in 1928, Pickering had assigned his second Planet O a mass of only

---

*"Planet X_1" will be used throughout this chapter to designate one of the two orbits postulated by Lowell for Planet X.

0.75, very near the first post-discovery estimates of Pluto's mass of <0.7 earth masses. Furthermore, in 1928 Pickering, alone among planetary prognosticators, had predicted the unprecedented circumstance of one major planet penetrating the orbit of another, as Pluto briefly penetrates Neptune's path.

That Pluto's discovery position was close to Pickering's as well as Lowell's planets was recognized immediately after Slipher's announcement. "Is it your 'X' or 'O' planet?" *International News Service* correspondent Lyle Abbott wired Slipher on March 13.[2] "How much does this differ from Pickering's position?" James Stokely of *Science Service* telegraphed the same day.[3]

Two days later, in a long letter to Putnam about the discovery, Harvard's Harlow Shapley warned: "We shall soon be hearing from W. H. Pickering. According to a local computation today his Planet O (1919 orbit) was +5°.6 from the object found and Lowell's planet +5°.9 according to the published 1914 orbit." However, he added, Pickering now "preferred three planets instead of a single planet O."[4]

Shapley's last remark seems to have reassured Putnam, for shortly thereafter he wrote to reassure Slipher: "I don't think I should worry too much about Pickering's predictions. I think he has predicted just about everything —from one planet to three, in varying positions. At any rate, Dr. Lowell predicted it, and you have found it, which is more than Pickering has done."[5]

Pickering was at his private observatory in Mandeville when he learned of the Lowell Observatory's discovery. Always ready, even eager, to comment on major astronomical developments in print, he reacted typically by writing "The Trans-Neptunian Planet," dated April 10, 1930, for *Popular Astronomy*. In this, he reviewed the various attempts over the years to locate a trans-Neptunian planet, including Lowell's 1915 Planet X work, and then discussed at some length his own 1919 prediction for Planet O. His elliptical orbit then, he wrote, "differed entirely from Lowell's, and would locate the planet in an entirely different part of the sky." But, he added, "by a curious coincidence at the time the planet was publicly announced in 1930.2 the two positions very closely agreed. Lowell's orbit placed the planet in longitude 102°.2, the writer's placed it in longitude 102°.8, or about a moon's diameter distant, and the planet was actually found in longitude 107°.5."[6]

To Pickering, this situation was analogous to the discovery of Neptune and the fact that while Leverrier and Adams had been uncannily close in predicting the planet's position, they had been considerably wide of the mark in predicting its orbital elements.

To justify this analogy, Pickering relied on the argument Sir John Herschel had used to explain this fact—that even though the elements of an unknown planet cannot be determined with any precision, its location can be found from the perturbations it causes in its neighbors' paths.

"Until the discovery at Flagstaff," Pickering declared, "the fact is that no astronomer had the means to compute an accurate elliptical orbit for planet

O. All that he could do was to make an approximation of its position, which was also true in the case of Leverrier and Adams. As we have now seen both Lowell's and my positions turned out to be reasonably accurate.... As in the case of Leverrier and Adams, the computed orbit I published at that time [1920.0] was all wrong, but the location was nevertheless satisfactory. ... Evidently the search for O was well worth making and could have profitably been made ten years earlier, by any observatory having the suitable equipment."[7]

In this article, Pickering made no direct claims, although here and in subsequent papers he consistently, if casually referred to the new found object as his Planet O. Five days later, when he dispatched a "Supplementary Note" to *Popular Astronomy,* he was even less proprietary. While briefly questioning the efficacy of analytical methods in trans-Neptunian planetary prognostications, he warmly praised his one-time friend and colleague, Lowell, at some length.

Lowell's analytical search, he now wrote, "is undoubtedly a very 'elegant' mathematical investigation.... but it seems to me that mathematics is out of place, and should be replaced by common sense. In the case of the discovery of Neptune the perturbation was enormous in proportion to the accidental errors, and the mathematics fully justified. Such was by no means the case with planet *O,* when the largest perturbation on which Lowell's investigation was founded, did not exist at all, and the remote ones on which he places considerable reliance, going back to 1690, are of doubtful value...."[8]

Pickering continued:

I do not wish it thought that my criticisms of his work are made in a spirit of rivalry. What I have stated is explicit and clear. If the investigation of others prove [sic] that I am mistaken ... I shall be only too glad to acknowledge my error. Lowell and I were at one time more or less personal friends and throughout his life I backed him in his Martian work far more fully, I believe, than did any other professional astronomers, because I believed that he was largely right. It was only because I was unwilling to back him in all his theories, and because I said so, that we later, to my regret, became more or less estranged. He was glad and appreciative of my favorable comments on his work, but hurt by those of my criticisms that were unfavorable. I wish to point out that all of us who are interested in the new planet, and myself in particular, owe Lowell a distinct debt of gratitude. Had he not established his observatory at a very considerable expense, and had his former associates not continued his work of searching for the unknown planet, it is highly probable that it would not have been discovered for another hundred years, that is until it again perturbed the planet Uranus.... If the only astronomical work that Lowell had done was to build his observatory, and plan that his former associates should look for the unknown planet, that should be sufficient to establish his place in astronomical biography. In closing I therefore say all honor to the late Dr. Lowell.[9]

These articles certainly do not suggest that Pickering was vigorously claiming credit for predicting the new planet, and his congratulatory postcard to the Lowell Observatory, dated after his first article but before his "Supplementary Note," reinforced this impression for the Lowell astronomers at least.[10] In subsequently commenting on a report that "Pickering was about to claim his ephemeris for X against Lowell's," Slipher advised Putnam that "a card came since from Pickering himself which indicates that he has a peaceful attitude of mind."[11]

Slipher was also reassured by what Pickering did not say in his articles, as he pointed out to Putnam. "He carefully avoids telling the reader that he had the Harvard Observatory more recently photographing for his planet O near R.A. [right ascension] nine hours as I recall here a couple of years ago, which shows that he himself abandoned this 1919 prediction he had made." And in a more general vein, he added: "I suppose we must let some of these things die of themselves as probably most of them will. At least it is hoped that after a while these unfavorable details will be lost and forgotten and the main thing remain, i.e. that Lowell predicted the planet and it was found at his Observatory in carrying on the search he had inaugurated."[12]

Pickering's initial article and his "Supplementary Note" were written before he learned of Bower and Whipple's series of orbits, or of the Lowell Observatory's own orbit which indicated that the new object's motion resembled that of a comet, and he, like just about everyone else, had jumped to the conclusion that it was a planet. But when word of the preliminary orbits reached him, he quickly jumped from one conclusion to another, penning still another *Popular Astronomy* article, dated April 30, 1930, in which he flatly declared that the new object was not a planet at all.

"I feel as if there could be no doubt, based in part also on my own computations of possible planetary orbits, but that this object is simply a comet," he wrote. "Had it been announced when discovered, as is customary, we should doubtless now know its orbit very completely, but such unfortunately was not the case, and we shall probably have to wait another year before we know the semi-major axis and the argument of perihelion [*] with certainty. In the meantime the comet must be growing fainter, but can probably be photographed for a few years longer,—at least we hope so." The object was nonetheless "a most interesting body on account of its unusual brilliancy" and the fact that "no previous comet has been followed, much less discovered, that is more remote than Saturn." But it was "not Lowell's so-called Planet X," nor was it now his Planet O. "There is no reason to imagine that this comet differs from any other comet except that it would appear to be somewhat more massive, and all the excitement about a new planet seems to be a case of much cry and very little wool."[13]

Only twenty days and less than fifty pages of *Popular Astronomy* separate Pickering's assumption that the new object was a trans-Neptunian planet—his

---

*The argument of perihelion is the angle between the ascending node and the perihelion point of an orbit, measured along the arc of the orbit.

Planet O—from his equally hastily drawn conclusion that it was "simply a comet," and these precipitous plunges into print certainly did not enhance his credibility on the subject. Astronomer Philip Fox, for example, opined that Pickering "appears to me to be like a man sitting on the end of a limb and sawing it off between himself and the trunk."[14]

Pickering's abrupt demotion of the Lowell object to cometary status was also self defeating. For not only did his 1919 prediction for Planet O agree quite closely with Pluto's discovery position, but even the first preliminary orbits by Bower and Whipple and Miller and Slipher gave values for the inclination of orbit and longitude of ascending node very close to those Pickering had suggested in 1919. Crommelin, in commenting on the early orbits, wrote Slipher: "I was rather impressed by noting that W. H. Pickering in his 1919 forecast gave the node as 100 degrees and the inclination as 15 degrees, both quite near the true values; it is difficult to think that this agreement is entirely fortuitous."[15] And later, after subsequent orbits sustained these similarities, Crommelin confessed "some surprise that Pickering so quickly adopted the cometary view, in face of the fact that so many of his predictions were satisfactorily fulfilled."[16]

It is one of the lesser ironies surrounding Pluto's discovery that Crommelin, in announcing the discovery of the 1927 Uccle position and his resulting orbital computations that led to confirmation of the object's planetary status, added new fuel to the fires of controversy over the prediction of Pluto, although the controversy was clearly distasteful to him. "I think it is a mistake," he had written to Slipher in mid-April, "to try to stir up rivalry between the partisans of Lowell and Pickering as regards their claims; we all know that both of them made forecasts that agreed well in longitude with the actual discovery, but that the Flagstaff search was carried on in consequence of Lowell's forecast. From the faintness of the new body its mass is probably a good deal less than either of the predicted values."[17]

But in reporting his new orbit to the British Astronomical Association a month later, Crommelin gave strong support to the validity of Lowell's prediction, while hardly more than mentioning Pickering's forecast. During his computations, he said, he "did not consult Lowell's predicted orbit ... and had forgotten what his elements were." But, he noted, "when my orbit was completed I made a comparison, and was agreeably surprised to find a close agreement between Lowell's and mine. The close agreement in the three independent elements of the eccentricity, direction of major axis, and time of passing perihelion is beyond what can be ascribed to chance; we must conclude that Lowell's forecast was sound."[18]

Crommelin added, notably, that Lowell had "clearly exaggerated the mass, which he gave as six times that of the earth," and on the basis of an estimate of 4,000 miles (6450 kilometers) as the diameter of Pluto made at France's Meudon Observatory, he himself assigned the new planet a mass of less than one-ninth the earth's.[19] Pluto's mass, as will be seen, then and for many years since has been the crux of the controversy.

Crommelin also summarized his conclusions in a letter to Slipher. "I send circular with the orbit I deduced for the new planet assuming that it was photographed at Uccle in 1927," he wrote. "It is in good agreement with Lowell's forecast. He gave longitude of perihelion 203°.8 (Ep^t 1850) = 205° (Ep^t 1930). My value is 216°. He gave perihelion passage 1991; my value 1984.9. He gave period 282^y, my value 265^y. He gave eccen[tricity] .202, my value .287. He gave inclin[ation] 10°, my value 17°. It is very difficult to think that all these points are flukes!" And as a postscript, he declared: "It was not until my orbit was in print that I made the comparison with Lowell's prediction, so that there was no *cooking* in arriving at the agreements."[20]

Not everyone could agree with Crommelin's assessment, however. Even before the English astronomer announced his orbit and its seeming similarities to Lowell's X$_1$ prediction, Ernest William Brown, Sterling professor of mathematics at Yale University and a leading authority on celestial mechanics, was contending before the National Academy of Sciences that it was impossible for anyone to have predicted Pluto at all, and that its discovery was "purely accidental."

Brown's paper, "On the Predictions of Trans-Neptunian Planets from the Perturbations of Uranus," touched off a new and longer-lasting controversy.[21] In this, he dealt entirely with Lowell's analytical investigation, with only passing reference to Pickering's work. In fact Pickering's Planet O never seriously entered into this broader, more complex debate, perhaps because of the admittedly empirical methods he had used. Although Brown did cite Pickering's predictions in subsequent papers, he did so only to dismiss them, and the mild protests that this treatment drew from Pickering himself had little if any effect on the course of the controversy.[22]

In his paper for the Academy, Brown presented a number of points for his thesis that Pluto was unpredictable from the residuals of Uranus and that, despite the relatively close agreements between its orbit and that of Lowell's Planet X$_1$, its discovery had been purely accidental.[23]

That Pluto was found in the region favored by Lowell, he explained at the outset, "suggested a fresh examination of Lowell's work.... It has always been difficult to understand why predictions of an exterior planet by Lowell and his predecessors, Gaillot, Lau and W. H. Pickering, were possible from the very small residuals which the longitude of Uranus exhibits. The definiteness of these predictions appeared to be quite outside the possibilities of the material under discussion and yet it was not easy to point out any fundamental error in the arguments."[24]

Nonetheless, Brown's "fresh examination" now led him to several conclusions critical of Lowell's findings. The oscillations shown in Lowell's curve of the perturbations of Uranus "seem to have periods too short for an explanation on the basis of an exterior planet, and neither of the two hypothetical planets of Lowell seem to account for them," he found. There was also the "accidental circumstance" that the interval of observations used by Lowell was near the synodic period of Uranus and Pluto and as there "must be

Yale University

Ernest William Brown of Yale University.

approximate symmetry about the middle of the interval," this "gives at once a ficticious longitude at epoch and longitude of perihelion which are almost exactly those of Lowell's hypothetical planets." It further appeared that "the actual values he obtains for distance, mass and eccentricity substantially depend on three groups of observations [of Uranus] made before 1783, having large probable errors." Moreover, he added, the residuals of Uranus since the last observation used by Lowell (1910) "bear no resemblance to those which either of his solutions require."[25]

It is unfortunate that, if my analysis is correct, so much careful and laborious work can lead to no result. However, in so far as it has stimulated a search for an outer planet which has proved successful, one cannot regret its completion and publication.[26]

Brown pointed out that Lowell's method, following that of Leverrier for the discovery of Neptune, was the reduction of the squares of the residuals to a minimum, and he conceded that it had been "carried out in great detail and apparently with high accuracy...." But he contended that Lowell had given no probable errors for his results, nor had he weighted the residuals he had used. "The weighting of the material in a problem of this nature is of fundamental importance," he explained, "because the question of the validity of the hypothesis depends almost entirely on the probable error of the unknowns...."[27]

But Brown's most telling point concerned Pluto's apparently small mass which, as the first post-discovery observations strongly indicated and as Slipher had tacitly conceded in suggesting a high density, was clearly less than the 6.6 earth masses that Lowell had assigned Planet X_1. On this point, the Yale mathematician rested his case:

The information concerning the newly found planet at this moment is scanty but it appears to be sufficient to prove that it could not have been predicted from its effect on Uranus. On the most favorable assumption as to albedo (0.06) and density (5.6), the mass cannot be greater than that of the earth or one-seventh of that predicted by Lowell. At a distance of 40 to 43 units the perturbations of such a planet on Uranus are not within the range of detection from the material available. Among the numerous solutions given by him, only those with an eccentricity of at least 0.5 appear to be possible. ... Such a solution, however, puts the planet more than 30° away from the place it was found.... We must therefore regard the fact that it was found near the predicted place as purely accidental.[28]

Crommelin, however, stuck to his guns and in formally publishing his new orbit he defended the validity of Lowell's work. After listing his own elements for Pluto and again noting their close agreement with Lowell's, he asked: "Is Lowell's forecast then, wholly illusory? I find it hard to admit this in the face of so many accordances...."

He then sought to justify Lowell's use of the early prediscovery observations of Uranus in his solutions, and to explain Lowell's overestimate of X's mass. "Now there would have been a fairly close approach to Uranus in 1719, at the last perihelion passage of Pluto," he pointed out, and "it was at about this time that Lowell located his largest perturbation; my suggestion is that Lowell correctly located the maxima and minima of the curve of perturbations, but exaggerated their height considerably."

In a direct reference to Brown's thesis, he declared flatly: "Prof. Brown has expressed an opinion adverse to the soundness of Lowell's work, but the proof of the pudding is in the eating and the fact that the elements of the orbit were predicted with considerable precision overrides *a priori* arguments against the possibility of such an achievement."[29]

Slipher's initial reaction to Brown's conclusions, incidentally, were quite superficial, if nonetheless interesting. To Putnam he wrote that from the beginning "the view that the great credit due Dr. Lowell should be safeguarded" had been "uppermost in my mind, and it seems for the most part the result has been very satisfactory, excepting for the Yale attack, but I just supposed it is in their nervous systems to tackle a Harvard man when they see him carrying the ball, even in the heavens."[30]

Lowell himself almost certainly would have responded publicly and with alacrity to Brown's criticisms. But the cautious Slipher, aware of Brown's prestige and his own lack of expertise in the field of celestial mechanics, published no reply, preferring to await further developments and to rely on the similarities between Lowell's X_1 orbit and that of Pluto to make the case for Lowell's prediction. Clearly he found justification for this course of inaction in subsequent events.

Late in June, for example, he advised Putnam:

> The 1927 Uccle position of Pluto used by Crommelin now looks to be authentic as some other orbits, especially by Nicholson at Mt. Wilson and by Leuschner, seem to check very closely with Crommelin's in all regards. In other words, these latter orbits which could utilize a much longer arc have come back surprisingly near to Dr. Lowell's eccentricity etc.! These recent orbits are of course becoming much more dependable, and are to my mind a sufficient answer to Prof. Brown and defense of Lowell's work. I am very sorry our orbit got such a high eccentricity, but it seemed to be in the positions used owing to the exceedingly short arc available, and it was no doubt better for us to have done the orbit than to have left it to others to have got the same result from the meager data. However that may be it is very gratifying to have the most favorable turn of the matter: The close agreement with Dr. Lowell's orbit, and the finding of Pluto on the earlier plates here and elsewhere, which are getting Pluto established much more promptly than seemed possible a few weeks ago....[31]

A month later, after the Astronomical Society of France had awarded its Medal of Honor to Lowell Observatory for Pluto's discovery, Slipher again confided to the trustee: "This kind of thing does a lot of good and makes it easier to accept such view points as Brown's.... It is very satisfactory to see how much the recent orbits resemble Dr. Lowell's. The agreement is no doubt much closer than could really be expected."[32]

Putnam, replying to this, agreed and evoked the "happy accident" controversy over Neptune. "The history of the discovery of Neptune is certainly being repeated in a remarkable way when we think that as eminent a mathematician as Benjamin Pierce [sic] acted just the way Brown is going now, and we know that neither of them is the type that acts from rancor, but only from conviction." But, he also pointed out, "we know what the ultimate judgment of history has been, and I am sure it is being repeated now."[33]

The news continued to be encouraging through the summer, and Putnam, after attending a meeting in Chicago in September, advised Slipher: "I saw Brown, and he is still convinced of what you might call his 'coincidence theory,' chiefly now because the mass of Pluto is presumably so slight, but I found in talking with other people ... that they felt there was too much to be coincidence—in fact, everybody was anxious for news, and realized the importance and greatness of what we had done."[34]

While Crommelin had made only slight mention of Pickering in his first announcements of his orbit, in his formal publication he noted that Pickering's 1919 postulated position and his suggested node and inclination were "nearly correct," although his period and longitude of perihelion were "probably less accurate than Lowell's." Pickering's 1928 orbit for O also "made the planet sort of a distant satellite of Neptune, with the same period but a larger eccentricity."[35] But even after this, Pickering still held to the cometary view, citing the 0.909 eccentricity of the Miller-Slipher orbit and Van Biesbroeck's hyperbolic orbit to his point at the June 25 meeting of the British Astronomical Association. It is probable here that Pickering, working at his relatively remote West Indian retirement retreat, simply was not up to date on the fast-paced developments concerning the Lowell object. At any rate, Crommelin was obliged to point out to him that the 1927 Uccle position and the 1919 images reported two weeks earlier by Nicholson and Mayall at Mount Wilson had virtually settled the question of Pluto's planetary nature.[36]

As such new data became available, the similarities between Lowell's predicted Planet X_1 and Pluto did indeed become more and more evident with each subsequent orbital computation, as not only Slipher kept reminding Putnam but as a number of other astronomers now began to point out in print. Certainly the most influential of these was the widely respected, highly articulate Henry Norris Russell, who wrote three articles for popular consumption on Pluto's discovery which were published in the July, November and December 1930 issues of *Scientific American*. The December article in particular provided grist for the Lowellians' mill, and Putnam and Slipher, in publishing what can be considered the Lowell Observatory's "official" version of Pluto's discovery more than a year later, quoted it at some length to their purpose.[37]

Russell's articles, taken together and casually read, probably gave readers the impression that he believed Lowell's X prediction was a valid one, for he stressed the resemblance between forecast and fact, and praised Lowell and his assistants effusively. Nowhere, however, did he actually commit himself on the identity of Pluto with Planet X, the crucial point in the controversy, although this omission probably was missed by most readers.

His July article was entitled "Planet X" and was written from Athens, Greece, in April, while Pluto's planetary status was still being debated and before Crommelin announced his orbit. In this article, the proofs of which had prompted Slipher to come to a decision on Pluto's name, Russell noted that the object's mass and size seemed to be smaller than Lowell had predicted. "The general agreement of its distance of 43 to 44.7 astronomical units shows, however, that this prediction is near the truth." Of the discovery itself, he added that "two tributes must be paid where honor is due":

First, to the skill, assiduity, and devotion of the workers at the Lowell Observatory who have made this important discovery; and second to the memory of Percival Lowell, traveler, man of letters and affairs, and observer of the planets. He did many things in the course of a crowded life and he did them well. The mathematical researches took him outside his other fields of work and into a region full of traps, for the unwary, yet the discovery of this new planet has justified him by his works despite the doubts of many of his contemporaries.[38]

Russell's November article, written from Italy, was simply a popularized explanation of "How Pluto's Orbit Was Figured Out" by the methods of celestial mechanics.[39] But in his December article from Paris, "More About Pluto," he had more to say about the new planet vis à vis Lowell's forecast. "The orbit, now that we know it, is found to be so similar to that which Lowell predicted from his calculations 15 years ago that it is quite incredible that the agreement can be due to accident...." Then, in a passage later quoted by Putnam and Slipher, he added:

Lowell saw in advance that the perturbations of the latitudes of Uranus and Neptune ... were too small to give a reliable result, and contented himself with the prophecy that the inclination, like the eccentricity, would be considerable. For the other four independent elements of the orbit, which are those Lowell undertook to determine by his calculations, the agreement is good in all cases, the greatest discrepancy being in the period which is notoriously difficult to determine by computations of this sort.... The actual accordance is all that could be demanded by a severe critic.[40]

Russell also briefly and somewhat obscurely discussed Brown's contention that Pluto was unpredictable from its effects on Uranus. "If someone," he wrote, "should maintain that a calculation based on wholly insufficient data (and hence physically meaningless) might by some strange bit of luck result in predicting all four of the important elements of the orbit with the relatively small errors shown above, he would still have to account for the fact that the outstanding errors are *so adjusted* that their influences partly counteract one another and bring about the closest agreement of prediction and fact—just at a time when genuine prediction, based on physically significant data, would have behaved the same way!"[41]

Lowell, of course, had stressed in his X memoir that planetary theories could be "legitimately juggled" by a "suitable shuffling of the cards" to come out correctly, and Russell, like Lowell, cited the case of Neptune to illustrate the point. Leverrier and Adams, he pointed out, had assumed too great a distance for their unknown planet from Bode's law, but their calculations also "made the orbit considerably eccentric (although it is really very nearly circular)" and this "spurious eccentricity brought the predicted orbit toward the sun in the region where the planet actually lay at the time, and went far to undo the error of the two original assumptions. History," he concluded, "seems this time to have repeated itself closely, except in one tragic detail—Percival Lowell did not live to see his prediction thus fully confirmed."[42]

In these articles, Russell did not consider directly the specific criticisms made by Brown, but in fact Russell held with the Yale mathematician, however incredible the coincidences might be. Subsequently, in a carefully written appendix to A. Lawrence Lowell's 1935 *Biography of Percival Lowell*, he reviewed the issue again, citing Pluto's apparently small mass and noting particularly Brown's references to the "large probable errors" in the pre-1783 observations of Uranus and Lowell's reliance on them in his computations. "The question arises," he wrote, "if Percival Lowell's results were vitiated in this way by errors made by others more than a century before his birth, why is there an actual planet moving in an orbit which is so uncannily like the one he predicted?"

There seems no escape from the conclusion that this is a matter of chance. That so close a set of chance coincidences should occur is almost incredible, but the evidence assembled by Brown permits no other conclusion.[43]

This circumstance, however, did not deprive Lowell of the honors. "In any event," Russell declared, "the initial credit for the discovery of Pluto justly belongs to Percival Lowell. His analytical methods were sound; his profound enthusiasm stimulated the search; and, even after this death, was the inspiration of the campaign which resulted in its discovery at the Observatory which he had founded."[44]

Russell, it may be noted, did not refer to Pickering or his various Planets O in his articles and, in fact, most of the post-discovery discussion over Pluto's prediction and/or predictability centered on Lowell's work. Pickering himself, after his precipitous dismissal of the Lowell object as a comet in April (and perhaps because of it), had little to say on the subject in print for some months. In October, however, he completed a long, rambling article, tinged with bitterness, which appeared in the January 1931 *Popular Astronomy* under the title: "The Mass and Density of Pluto—Are the claims that it was predicted by Lowell justified?" Not surprisingly, he no longer considered Pluto to be a comet, but he could not resist the speculation that it might "turn out to be a piece of a star."[45]

In this article, Pickering came as close as he ever would to claiming Pluto as his Planet O and the Mount Wilson pre-discovery images figured prominently in his argument. They showed, he pointed out, that in 1919 Pluto had been only 1°.15 away from his predicted right ascension.[46]

Pickering also noted that in 1928 he had reduced O's mass from 2.0 to 0.75 that of the earth, a value he now found compared favorably with his own estimate for Pluto's mass of 0.71, and had shortened O's distance from the sun to bring its orbit within Neptune's—"a fact hitherto without precedent among planetary orbits." From his mass estimate, and the assumption that Pluto's diameter was 4,000 miles, he derived a density of 31, or nearly six times that of Mercury, densest of the known planets, and this suggested to him an interesting, if unusual possibility. "If this conclusion is correct," he opined, "then Pluto is the first planetary body discovered whose internal structure indicates that it had an origin outside the solar system. Its density would thus seem to imply a relationship to those stars described as being of the white dwarf type...."[47]

But as the title of his article indicates, Pickering was not entirely concerned with advancing his own claims and theories. His other major thrust was against "Dr. Lowell's work on this planet," a digression that was necessary, he explained, because of "the surprising and reckless claims to the exclusive prediction of its location put forth by his active adherents and administrators, backed up by extensive newspaper propaganda."[48]

Pickering then referred to Lowell's graph of the perturbations of Uranus in his X memoir and, echoing Brown, declared:

"A glance will show that its irregularities are due wholly to accidental errors of observation. It therefore would be utterly impossible to compute either a position or an orbit for Pluto based on the observed perturbations of Uranus alone, as Dr. Lowell attempted to do. Even as an aid to the perturbation of Neptune, which planet the writer employed, they would be perfectly useless. No astronomer that I can recall has pointed with pride to Lowell's success in locating the planet at a longitude differing by a little over 5° from its actual place in the sky. It was over 6° at the time for which he actually computed it in 1914.5. No one who has carefully studied his work has expressed confidence that even this very moderate success was due to other than mere accident. None has stated that the very excellent and extensive mathematical work that Lowell published had any particular relation to the irregular deviations in the orbit of Uranus now actually observed."[49]

In support of these strangely ambivalent statements, Pickering cited Brown and Bower and Whipple, who had also cited Brown, but with one reservation concerning his own reliance on Neptune's residuals in his 1919 O prediction. "They even doubt if the effect [of Pluto] on the orbit of Neptune would have produced any appreciable perturbation!" he exclaimed. "I think, however, if they look the matter up they will be convinced that some effect is noticeable in that case."[50]

Crommelin, meanwhile, continued to be impressed by the close agreements between Lowell's and Pickering's predictions and the progressively refined orbits of Pluto and unimpressed by Brown's arguments, and he now decided that the credit for Pluto's discovery should be shared. In February 1931, in the Royal Astronomical Society's *Monthly Notices*, he tabulated all the various predictions of trans-Neptunian planets over the years by Todd, Forbes, Lau, See, Gaillot, Pickering and Lowell and compared them with his own elements for Pluto's orbit. From this, he concluded that the honors must be divided between Lowell's orbit and Pickering's of 1919. "Lowell's is superior as regards the period, the date and longitude of perihelion," he wrote, "Pickering's as regards node, inclination, magnitude and mass. Pickering, in his 1928 prediction, took the period the same as Neptune's; this prediction is inferior to that of 1919, except in two points. Its value for the mass appears to be nearer the truth, and it predicts the unexpected fact that the orbit of Pluto penetrates that of Neptune."[51]

Surprisingly, in view of his earlier bitterness, if not his ambivalence, Pickering himself was one who demurred at Crommelin's attempt to divide the honors for Pluto's discovery. For he now changed his mind again and recorded this change with some magnanimity in the *Monthly Notices* in May. "I think we may agree that Lowell was surprisingly successful," he now declared. "Indeed no other computer of unknown planetary orbits has ever done so well.... Lowell's great achievement ... was undoubtedly the successful computation of the size of Pluto's orbit."[52]

Of his own planetary predictions, Pickering noted:

"I wish to point out that my empirical method gives only a position for the unknown planet on the date of maximum perturbation of the known one.... The search should be made as soon as possible after the date of maximum perturbation has been recognized, so that the orbit shall not have time to develop any large error. It may be further pointed out ... that whether the analytical or the empirical method of search is employed, the only thing of importance to determine is the location of the planet and its brightness."

In closing may I remark that I am not claiming to share in the principal credit for the discovery of Pluto with my late friend Dr. Lowell. One or the other of us should have primary credit for it alone.... If it was a mere accident, as some claim, that his position was no near the actual one, then it was certainly a very fortunate accident for the finding of the planet. Had it not been for his bequest it is quite probable that the planet would not have been found for many years, and he should receive therefore full credit for this very important part in the discovery....[53]

With this, Pickering and his planet O predictions disappear from the post-discovery controversies over Pluto for all practical purposes, although, as will be seen, they were considered briefly again some ten years hence in the context of the larger debate over Pluto's predictability.

In this larger debate, Brown's position continued to gain support, now from Lick Observatory's Ernest Clare Bower who, in the course of publishing a durable orbit and ephemeris for Pluto in September 1931, discussed Lowell's investigation at some length noting, incidentally, that Lowell himself had clearly placed "no great confidence" in his results. This discourse, Bower added, surely with Crommelin's conclusions in mind, was prompted by the fact that the "extraordinary agreement between the elements of X_1 and of

Pluto has led a few astronomers to accept Lowell's work as a trustworthy prediction with merely an overestimation of the mass."[54]

Bower however, although relying on Brown's analysis to a considerable degree, did not go so far as to say, with Brown, that Pluto was not predictable at all from the residuals of Uranus, but only that under the circumstances that prevailed in Lowell's investigation, it could not have been predicted. Pluto's mass again was crucial here. Bower pointed out: This datum "cannot be satisfactorily determined gravitationally at present" from Uranus' motion, and "is now, and for many years will be, indeterminate from Neptune's residuals." Nonetheless, from his "best assumptions" as to Pluto's magnitude, albedo and density, he set an upper limit for its mass of 0.3 that of the earth, adding: "Until a disk is actually seen, however, the most probable value for *Pluto's* mass may be taken to be about 0.1 Earth's," a value close to Crommelin's earlier estimate.[55]

Two months later, in the *Monthly Notices,* Brown gave a detailed, updated exposition of his argument,[56] and followed this with a somewhat less technical summary in an address as retiring president of the American Astronomical Association delivered in Washington, D.C., December 29, 1931, in which he stressed the difficulty of describing and understanding the problem.[57]

To simplify matters, Brown had devised a mathematical transformation which, he noted, enabled a comparison of residuals with a minimum of computation. "The transformed observations will, in certain cases, show whether a given set of residuals can be accounted for by the existence of an unknown planet or not," he explained. "It is then seen that, if the hypothesis appears valid, a first approximation of the place of the unknown planet can sometimes be immediately obtained."[58]

When this transformation was applied to the residuals of Uranus used by Lowell, Brown declared, "It is at once seen that the slightly systematic character they appear to possess cannot be due to the action of an exterior planet. The investigation stops here because if nothing observable is present any prediction based on the observations cannot have any value."[59]

Only a few days after Brown spoke, Putnam and Slipher published in the *Scientific Monthly* what was perhaps—until the publication of this book—the most detailed, if still quite brief, description of Pluto's discovery under the title: "Searching Out Pluto—Lowell's Trans-Neptunian Planet." In this largely narrative account, they made no reference whatsoever to the arguments of Brown and others, although toward the end they urged the view that Pluto was Lowell's Planet X. They used quotations from Crommelin's May 1930 letter to Slipher, and from Russell's December 1930 *Scientific American* article to emphasize the close agreement between more recent orbits and what Lowell had predicted.[60]

Brown, of course, in contending that any similarity between Pluto and Lowell's Planet X was purely accidental, was questioning the assumptions that Lowell had made in attacking the problem, and in doing so he himself made some assumptions. But there were some astronomers, now and later, who also questioned Brown's assumptions, and who pointed out that Lowell's assumptions just possibly might prove to be right. A. O. Leuschner, the eminent if impatient computer of planetary orbits, was one of these.

Leuschner entered the broader controversy in June 1932, making a review of the events that followed the discovery of Pluto the subject of his address as retiring president of the Pacific Division of the American Association for the Advancement of Science.[61]

At the outset, Leuschner had a word of caution for the Lowellians: "Until the identity of *Pluto* and 'Planet X' is really established, *Pluto* should not be referred to as 'Planet X'." But Brown and his supporters were not to have it all either. "The astronomical world is still divided on the question of whether *Pluto* actually was predicted by Lowell or whether its discovery was an accidental by-product of the remarkable search instituted at the Lowell Observatory for 'Planet X'," he noted. "The scientific arguments pro and con appear to be equally valid and we are not in a position to render a final verdict at the present time."[62]

Then, after discussing the preliminary orbits and the various pre-discovery observations of Pluto, he turned to Brown's thesis.

"E. W. Brown of Yale has explained that, when the interval of observation is nearly the same as the synodic period, this method produces a symmetry in the finally outstanding residuals, the use of which would lead to ficticious elements of a suspected body, and would produce results given by Lowell, whether or not the sought for body really exists. Brown therefore considers the discovery of *Pluto* an accidental by-product of the Lowell search. I venture to observe here that the accidental feature of Lowell's discovery might be interpreted to be that the circumstances outlined by Brown fortunately prevailed for a correct prediction on the part of Lowell. This, of course, would imply a reality of the available residuals of *Uranus,* including that of the 1795 [1715?*] position, doubted by Brown, Bower and others. Any possible error in the theory of *Uranus,* of short- or long-period term, may completely mask the admittedly small effect of *Pluto* on *Uranus.* The point I wish to make is that, although different solutions may give residuals within the errors of observation, a solution including a correct assumption may give more correct results in spite of the seeming indeterminateness of the problem.... Thus, in the rejection of Lowell's orbit of "Planet X" and its identity with *Pluto,* undue emphasis has been placed on the fact that the perturbations are masked in the residuals of *Uranus* during the period of observation. By making the assumption that the residuals of *Uranus* include the action of an extra-Neptunian body, Lowell proceeded along logical lines and the similarity of the orbit of his "Planet X" with that of *Pluto* may be the result of the fact that his assumption may prove correct..."[63]

---

*The year "1795" is probably a typographical error. Leuschner had been discussing Neptune's residuals at length, referring frequently to Lalande's May 8 and 10, 1795, pre-discovery observations. Here, of course, he is referring to observations of Uranus, specifically the earliest pre-discovery ones used by Lowell, including those of Flamsteed in 1715.

Leuschner's somewhat inconclusive assessment was the last word, for all practical purposes, on the problem of Pluto's predictability for nearly ten years. The new planet, while shining brightly in the reflected glare of worldwide publicity in the months immediately following its discovery, soon lost its luster in the public eye, although astronomical interest in it remained high for a somewhat longer period. By the time Leuschner spoke, however, most astronomers recognized that the issues of Pluto's predictability and mass would be resolved, if they were to be resolved at all, only with the passage of considerable time.

The next major development concerning Pluto came in 1940, and it passed virtually unnoticed within astronomy, in part certainly because of the disruption attendant upon World War II. That year, amid the repressive atmosphere of Nazi-occupied France, Victor Kourganoff, a doctoral candidate at the Paris Observatory, completed an exhaustive mathematical reinvestigation of the Pluto controversy, including the work of Lowell, Pickering and Brown, in which he systematically refuted Brown's criticisms and assigned primary credit for the prediction and discovery of Pluto to Percival Lowell.[64] Four years later, Kourganoff summarized his findings in *Ciel et Terre,* but again they remained almost entirely outside the ken of astronomers.[65] Not until 1951, when Gibson Reaves of the University of California's Berkeley astronomy department translated Kourganoff's thesis and published an outline of his work in the *Publications* of the Astronomical Society of the Pacific did his conclusions gain a wider audience among astronomers.[66]

Reaves' outline, by which Kourganoff's work is still known in the main, is important because in some quarters at least it has left "little doubt" that the discovery of Pluto "was not an accident" and that both Lowell and Pickering "had detected genuine evidence of the planet's existence years before its discovery."[67]

Moreover, his summation, which he stressed was not a critical review, is excellent, although it omits a point or two here and there which may modify Kourganoff's conclusions for some readers. For example, in the introduction to his thesis, Kourganoff reveals a bias, which Reaves does not mention, when he declares that Brown developed his thesis by "picking out the slightest details" in Lowell's work "with an animosity which can be compared only with that displayed a century before by the depreciators of Leverrier," and by characterizing Brown's contention that Pluto's discovery was "pure chance" as an "insinuation."[68] Even Putnam, as we have seen, did not consider that Brown was criticizing Lowell's prediction out of rancor.

It is important to keep in mind that in refuting Brown's conclusions, Kourganoff consistently emphasized the adjective "pure" or its equivalents or modifiers. His point was that the discovery of Pluto was not the result of chance alone, but of chance in the form of logical assumptions which later proved to be viable, combined with careful observation and the effective application, particularly by Lowell, of the methods of celestial mechanics.[69]

In his introduction, Kourganoff severely criticized the data base of

Brown's investigation and contended that the agreements between Lowell's prediction and the actual fact of Pluto were too many and too close to be easily dismissed as coincidence without adequate explanation. "Indeed, if Brown has discovered any mistakes in Lowell's computations, and such was not the case, it would seem almost incredible that Lowell succeeded in determining by 'pure accident' the elements of the orbit so accurately in spite of their number," he wrote. "Brown's criticism is characterized by the lack of direct analysis of Lowell's calculations, and by the absence of the use of the elements of Pluto.... [Brown] tries to resolve the problem as it were from outside, an elegant method which leaves unresolved a good deal of questions and does not bring a crucial test; we attempt to show the intimate mechanisms of the solutions of Lowell and of Pickering, and compare them, item by item, with data given by the observations of Pluto."[70]

Kourganoff divided his study into four major parts, devoting the first to an extensive recalculation of the perturbations of Uranus, using the laborious classical method and the best elements of the two planets available to him at the time and checking the results of his computations with a modification of Leverrier's interpolative method and plotting them on graphs.

He found, among other things, that the pre-discovery perturbations of Uranus, which enabled Lowell to make his prediction, were markedly larger than the modern ones; that these ancient observations were in better accord with theory than had been suspected, certainly by Brown; that the modern residuals of Uranus, i.e. since 1781, were indeed extremely small, but that a determination of Pluto's mass was nonetheless possible and yielded a value about equal to that of the earth. Interestingly, he also found that extremely small residuals of Uranus remained even after accounting for the influence of an earth-sized Pluto and concluded: "Whether these are due to systematic errors or to another trans-Neptunian planet is for the future to decide."[71]

Kourganoff next took up Lowell's and Pickering's work in succession. Lowell, he pointed out, improved on Leverrier's work by treating the mean distance from the sun of his unknown planet as a variable, by using least squares reductions almost exclusively, by carrying his solutions to the squares of the eccentricities, and by using the percentage of the reduction of the residuals of Uranus to evaluate his solutions. To demonstrate the closeness of Lowell's prediction, Kourganoff diagrammed the orbit of Pluto with those of Lowell's Planet X_1, Uranus, Neptune and Leverrier's predicted orbit for Neptune, showing among other things that the conjunctions and nearest approaches of Pluto and Uranus do not always coincide, as they had in the early eighteenth century where Lowell had found his largest perturbations, and that successive conjunctions were not equivalent as a result of the high eccentricity of Pluto's orbit.[72]

Of Pickering's Planet O predictions, Kourganoff found that his "semi-graphical" method "was a little difficult" to follow through Pickering's numerous publications, but that while much of his work was "crude" and "had an empirical or arbitrary character," he nonetheless made some contributions to the problem, notably in regard to his "original" use in 1919 of Neptune's residuals which "confirm the success of Celestial Mechanics."[73]

Kourganoff also credited Pickering with removing the duality of Lowell's solution, although Pickering, while he had been content to predict only a single position for his planets, had long recognized after Leverrier, Lau and others that the perturbations could be explained by symmetrical positions on opposite sides of the sun.[74] In this regard, Kourganoff plotted the 1920.0 positions of Lowell's two Planets X, Pickering's 1909, 1919 and 1928 Planets O, and Lau's and Gaillet's two trans-Neptunian planets to illustrate how the predicted positions were clustered in two opposing groups.[75]

After systematically comparing the work of Lowell and Pickering, Kourganoff came to a key conclusion which Reaves reported as follows: "...while the errors of Lowell's prediction are due primarily to inadequacies in the given *data*, the errors of Pickering's work are due primarily to the inadequacies of his *method*. So, then, to Lowell goes first credit for the 'theoretical' part of the discovery of a trans-Neptunian planet, and to Pickering, second place for removing the indeterminacy of Lowell's double solution and for confirming Lowell's work by the use of Neptune as well as Uranus."[76] Kourganoff also noted that Pickering, in his 1928 Planet O memoir, had given up his 1919 conclusions.[77]

In his discussion of Brown's criticisms of Lowell's prediction which make up the fourth part of his thesis, Kourganoff made a point-by-point refutation of four of the Yale mathematicians major arguments.

Of Brown's contention that the unweighted pre-discovery observations of Uranus on which Lowell relied so heavily contained "large probable errors," Kourganoff found that his "numerical examination of this question" sufficed "to refute Brown's opinion."[78] As Reaves summarized his point: "Of course Lowell was not able to know, as we do now, how unusually accurate the old observations were. Lowell regarded the assumption of high precision for the old observations as a 'working hypothesis.' No solution would have been forthcoming without this assumption."[79]

A numerical examination of the recomputed residuals also sufficed to refute Brown's point concerning the interval of the modern observations that Lowell had used in his computations. The "accidental circumstance" that this interval closely paralleled the synodic period of Uranus and Pluto, and produced a symmetry around the middle of the interval which led to a fictitious if coincidentally accurate longitude was not germane to the question. Kourganoff's point, Reaves notes, was that "because one may achieve Lowell's result by incorrect reasoning, it does not follow that Lowell's reasoning was incorrect." Kourganoff showed the invalidity of Brown's argument mathematically by demonstrating that if Brown were correct, Lowell's extension of the interval of observations beyond 1903 to 1910 should have increased his predicted longitude for X by 15° when, in fact, it had decreased it by 4°. The error in Brown's argument here, Kourganoff felt, was that he considered only the modern residuals and had confused the angular conjunctions of Pluto and Uranus with their minima of approach.[80]

Kourganoff also examined the argument of Brown, Bower and others based on "the presumptive littleness of Pluto's mass and on the great disproportion between this mass and that of Lowell's solution" and concluded that, as Brown himself had pointed out, the discovery of the longitude of an unknown planet is, to some degree, independent of its mass, noting that both Leverrier and Adams had derived masses for Neptune twice the actual value, but that this had not affected their longitude prediction.[81]

Finally, Kourganoff rejected Brown's contention that the modern residuals of Uranus do not show the characteristic form to be expected from the action of an unknown exterior planet, and that thus "if nothing observable is present any prediction based on the observations cannot have any value." Brown had relied on his newly developed transformation for this concluson and Kourganoff, while conceding the value of this transformation in some work, found that it was not adequate for the case of Pluto because it was not directly applicable to the isolated pre-discovery observations of Uranus, and because Brown, in deriving it, had assumed a circular orbit, a "poor assumption for the orbit of Pluto which has an eccentricity of 0.25." Moreover, Brown applied his method only to the modern residuals of Uranus which admittedly showed only small perturbations. Kourganoff modified Brown's transformation and reworked the problem, using both the ancient and modern observations to find that for both Uranus and Neptune, the transformed residuals did show the characteristic form expected from an exterior planet.[82]

In commenting on his overall results, Kourganoff declared: "We think we have now established that the thesis of 'pure chance' is completely untenable, and we are happy to have satisfied Leverrier's view that to credit only the myth of chance in great scientific discoveries is to risk profoundly discouraging those who devote themselves to the progress of science."[83]

Kourganoff's general conclusions were as follows:

> Deduction, observation, and chance: these are the three essential factors, indissolubly linked, in the discovery of Neptune, Pluto and, in fact, in all great scientific discovery.
> To attribute the discovery of Pluto to chance alone is as contrary to the strict truth as to attribute the discovery of Neptune solely to Celestial Mechanics....
> In the case of Pluto, the part played by chance was unquestionably greater than in the case of Neptune; but our researches prove: that it is only through Celestial Mechanics that chance could exert its beneficial role. This is because it appears in the effect of the residuals of Uranus and Neptune and not in some circumstance related to the photographic searches for the star. Celestial Mechanics played a direct part in the discovery of Pluto: it is that which transformed the idea of the existence of the star from a gratuitous hypothesis into an absolute certitude; it is that which permitted its location in the 'vast

reaches of the sky.' We have seen, finally, that the zeal of the observers and the incomparable search instrument with the "blink microscope" also played their part in the success of the Lowell Observatory.

In the problem which has occupied us, one cannot adopt a too limited point of view without entirely distorting the truth.

The thesis of "pure chance" is absolutely untenable. Pluto was "discovered" in 1915 by Lowell, and "re-discovered" in 1919 by Pickering, by the methods of Celestial Mechanics.[84]

But Kourganoff notwithstanding, most astronomers who concerned themselves with the problem of Pluto over the years remained skeptical that it was predicted by anyone, and this skepticism was primarily based on the uncertainty of Pluto's mass. That Pluto's mass is small relative to the earth's has not been seriously disputed since its discovery; the question, rather, has been: How small?

The difficulty of determining Pluto's mass with the precision that astronomers consider desirable was for many years compounded by several factors. The lack of a satellite was one, for since Lampland first searched unsuccessfully out to the nineteenth magnitude for possible companions shortly after the planet's discovery, none was found until 1978. The perturbations of Uranus were another, for they proved to be virtually useless in the problem. Nor have Neptune's residuals been of any help, for this planet has still not completed a full revolution around the sun since its discovery in 1846, and attempts to determine Pluto's mass from Neptune's motion have necessarily been largely dependent on the prediscovery observations by Lalande in 1795. The value of Pluto's mass in such analytical attempts, modern investigators have noted, "is critically dependent on the time span of the data used."[85]

Still, through the years, celestial mechanicians have considered Neptune their best hope. Seth B. Nicholson at Mount Wilson Observatory was apparently the first to analyze Neptune's residuals for a clue to Pluto's mass. In 1931, he derived a value of 0.94 that of the earth, with an error factor of 25 percent, a figure well above the first estimates of 0.11 by Crommelin and by Bower, but still far short of the 6.6 earth masses that Lowell had assigned for his Planet X.[86] Nicholson's approximation stood for some years and received support from the 1942 work of Lloyd R. Wylie of the U.S. Naval Observatory who, again using Neptune, came up with a figure of 0.91 earth mass for Pluto.[87]

Some astronomers, however, were not convinced by Nicholson's finding. "I felt quite skeptical about this determination," Yale University's highly respected Dirk Brouwer wrote in 1950 in assessing the "Current Problems of Pluto" on the twentieth anniversary of its discovery. "The remarkable thing is, however, that all that has since been done on the problem—which is a good deal—had substantially confirmed the result obtained by Nicholson." Brouwer here cited the work of the Dutch-born astronomer G. P. Kuiper who

Dirk Brouwer of Yale University.

in 1949, with the 82-inch reflector at McDonald Observatory in Texas, had attempted to measure Pluto's diameter and had obtained a value of 0.4 second of arc, corresponding to a linear diameter of 0.8 that of the earth, or about 6,400 miles (10,300 km.). "Hence Pluto is a much larger body than the moon or mercury," Brouwer noted, concluding: "According to the best evidence available the mass of Pluto is 0.9 or 0.8 times the mass of the earth, and the density nearly the same as the earth's."[88]

The following year, Brouwer, with colleagues W. J. Eckert and G. M. Clemence, came to a similar conclusion in their "Coordinates of the Five Outer Planets 1653–2060."[89] In summarizing their work in *Sky and Telescope*, Brouwer and Clemence pointed out "a curious situation with regard to Neptune ... the modern observations of this planet and the prediscovery observations of 1795 [by Lalande] are compatible only if the attraction of Pluto with a mass comparable to the mass of the earth is included."[90]

Dutch-born American astronomer
Gerard P. Kuiper.

By this time, however, Kuiper had already produced new evidence that Pluto's mass was not comparable to the earth's, and that Crommelin's and Bower's early estimates were nearer the mark. In March 1950, even as Brouwer's summary of Pluto's problems appeared in print, Kuiper obtained a new measure of Pluto's angular diameter, this time working with the 200-inch reflector on Palomar Mountain in California. Now he discerned a tiny disk for Pluto only 0.23 arc seconds in width, corresponding to 0.46 the diameter of the earth, or about 3,600 miles (5800 km). Assuming the same density as earth, this figure yielded a mass for Pluto of only 0.10 that of the earth.[91] Some astronomers subsequently suggested that what Kuiper might have measured was merely a specular reflection from the smooth, frozen, spherical surface of the planet, rather than its full disk, but others quickly pointed out that even if this were so, Pluto's mass could not be appreciably larger. It is fair to add that such measurements as Kuiper attempted are subject to rather large systematic errors.[92]

Kuiper's second measurement, however, subsequently received confirmation of sorts in 1965 when, on the night of April 28–29, a near occultation of a

The Longer Controversy   243

faint star by Pluto occurred and was observed at the U.S. Naval Observatory's Flagstaff station and at Kitt Peak National Observatory near Tucson, Arizona, and at the Lick Observatory in California. A minimum separation of 0.125 second of arc was measured as Pluto passed just south of the star, presumably fixing an upper limit for Pluto's diameter of 3,600 miles (5,800 km), Kuiper's earlier figure. The investigators concluded that with the probable errors of the measurements, the odds were nine-to-one (9–1) that the diameter of Pluto was less than 4,225 miles (6,800 km) which, again assuming equal densities, yielded a mass of less than one-seventh (<0.14) that of the earth.[93]

Further evidence that the mass of Pluto might be of this order came in 1968 when R. L. Duncombe, W. J. Klepczinski and P. K. Seidelmann of the U.S. Naval Observatory found that the use of Wylie's 1942 value of 0.91 for Pluto's mass in existing theories of Neptune's motion gave relatively large residuals. They reported that Newcomb's 1899 theory of Neptune, adjusted for Wylie's mass, misrepresented the actual observations of Neptune up to 1938 by more than five seconds of arc, and that for the 1950 theory of Eckert, Brouwer and Clemence, again with Wylie's mass, residuals from the observations amounted to nearly 4 arc seconds. "The apparent failure of the two theories to represent the observations of Neptune's longitude over extended periods," they noted, "indicated the possibility that an adjustment to the mass of Pluto incorporated in the theories might be required." In other words, Neptune's motion was beginning to get out of whack with theory so that "a suitable shuffling of the cards," in Lowell's earlier, more colorful phrase, now seemed advisable. The Navy astronomers concluded that a mass for Pluto of 0.18 that of the earth best fulfilled the requirements of the observations. This value too, they found, apparently resolved another problem. For Wylie's mass of 0.91, combined with Kuiper's 3,600-mile estimate for Pluto's diameter, produced an "unacceptably large value for the mean density of Pluto of at least 40 g/cm^3 [grams per cubic centimeter]." Taking 4,000 miles (6450 km) as an upper limit for Pluto's diameter, they pointed out that their new value for its mass reduced the planet's density to about 1.4 times that of the earth.[94]

The problem of the observations of Neptune vis à vis theories of its motion came up again two years later in 1970 and produced a somewhat more intriguing result. Astronomer Dennis Rawlins used the value of 0.18 for Pluto's mass to again reduce the 1795 Lalande prediscovery observations of Neptune and to "suggest that we may have, via Lalande's data, the largest meaningful residuals in the tables of the planets since the removal of the anomalies in Uranus' orbit in 1846." Rawlins's computation indicated that the 1795 residual should be $-7''$, and he concluded that it "seems possible that this discrepancy is due to an alien perturbation."[95]

In 1971 the value for Pluto's mass was reduced again—in fact, brought back down to the original 1930 estimates of Crommelin and others. Seidelmann, Klepczinski, Duncombe and E. S. Jackson of the Naval Observa-

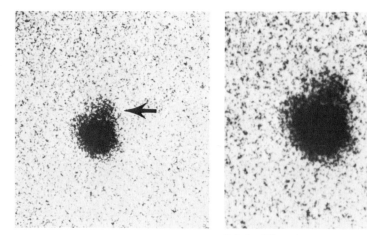

U.S. Naval Observatory

Pluto's satellite, Charon (arrow), *left*, and in an enlarged print, *right*. Specks are emulsion grains.

tory, from a new analytical determination from Neptune's residuals, reported that 0.11 earth mass was "the most likely value based on the observational data currently available" for the mass of Pluto. And again assuming 4,000 miles (6450 km) as an upper limit for Pluto's diameter, they derived a density of 0.88 that of the earth.[96]

Pluto's mass was destined to be reduced even further. In November 1976, astronomers Dale P. Cruikshank, Carl B. Pilcher and David D. Morrison of the University of Hawaii, announced photometric evidence that Pluto reflects sunlight from a surface of predominantly frozen methane (CH_4), and thus may have a higher albedo, or reflectivity, than had been generally supposed. If so, they suggested, Pluto's diameter may be smaller than that of earth's moon (2,160 miles or 3480 km) and might even be as small as 1,750 miles (2820 km). Again if so, and assuming a mean density of 1 to 2 g/cm^3 which would be likely if the planet were composed primarily of frozen volatiles, its mass might be only "a few thousandths that of the earth...." Such a mass, they pointed out, is "much less than would be required to perturb the motions of Uranus and Neptune measurably" and thus:"If this train of logic is basically correct, it appears that Tombaugh's discovery of Pluto in 1930 was the result of the comprehensiveness of the search rather than the predictions from planetary dynamics."[97]

That Pluto's mass *is* only "a few thousandths that of the earth" was dramatically demonstrated in July 1978 with the announcement that photographs of Pluto, made with the 61-inch reflector at the U.S. Naval Observatory's Flagstaff station, had revealed a Plutonian satellite. The tiny moon was

The Longer Controversy 245

U.S. Naval Observatory

U.S. Naval Observatory astronomers James W. Christy, *seated*, who discovered Pluto's satellite, Charon; and Robert Harrington, who first computed the mass of the Pluto-Charon system.

immediately, and appropriately, named Charon for the boatman in Greek mythology who ferried the souls of the dead across the River Styx to Hades, Pluto's underworld realm.[98]

The discovery, which startled the astronomical world, was made in June by astronomer James Christy in Washington, D.C., while examining plates exposed April 12–13 by astronomer Anthony Hewitt at Flagstaff in the course of a routine program designed to further refine the elements of Pluto's orbit. Additional plates, made by other astronomers early in July both at Flagstaff and at Cerro Tololo in Chile under superior observing conditions, confirmed Charon's existence.[99]

The discovery of a Plutonian satellite, of course, permitted the first direct and reasonably accurate determination of Pluto's mass since the planet's discovery forty-eight years before. Careful measurements of the plates yielded a separation between Charon and Pluto of 0.8 to 0.9 second of arc, the equivalent of about 10,500 miles (17,000 km), and from this, using Kepler's third law, Navy astronomer Robert Harrington quickly found that the mass of the Pluto-Charon system is only about two-thousandths (0.002) that of the earth. Pluto is thus a very, very small planet indeed, only about 20 percent as massive as the earth's moon, and with a probable diameter of only 1500 miles (2420 km), about two-thirds that of the earth's moon.[100]

The discovery of Charon, and the consequent determination of Pluto's very small mass, made it certain that Pluto could not be Percival Lowell's predicted Planet X, and thus the long controversy over this question ended.

It also makes certain that Pluto cannot have caused the perturbations in the motion of Uranus that Lowell, Pickering, Kourganoff and others found over the years, and this circumstance may prove to have even more important implications for the future of planetary discovery.

CHAPTER 12

MORE PLANETS X?

Inevitably, the discovery of Pluto has stimulated new speculations on the part of some astronomers over the years that still other unknown planets may be orbiting the sun in the remote vastnesses of trans-Neptunian or trans-Plutonian space. Vesto Melvin Slipher was perhaps the first to consider the possibility; certainly he was the first to act on it. "The thought that has been in my mind from my first look at it," he confided to Lowell Observatory trustee Roger Lowell Putnam barely a month after announcing Pluto's discovery, "was that there were probably more like it, and so the search is to go on...."[1]

Thus, when the initial excitement and controversy over Pluto's discovery began to subside late in May 1930, young Clyde Tombaugh once again took up his tedious duties at the Lowell Observatory's blink comparator to resume the systematic search for trans-Neptunian planets that had been so abruptly interrupted by success some three months before. He began where he had left off on that historic February afternoon, by carefully scrutinizing the unblinked portions of the January 23 and 29 Pluto discovery plates of the star-rich Gemini region. The task took him a full two weeks.[2]

For the next thirteen years, the photographic search continued, probing the entire sky visible at Flagstaff from 50° south to the North Pole, most of it to magnitudes sixteen and seventeen and in some selected regions out to the eighteenth magnitude. But although nearly 90 million images of some 30 million stars over more than 30,000 square degrees of the sky were photographed and examined, no more planets were found. Tombaugh felt that the care and thoroughness of this survey justified the conclusion that "no unknown planets brighter than the sixteenth magnitude exist and that any planet between magnitude 16 and 17 had a good chance of being discovered."[3]

Nevertheless, as Slipher later explained, the search "needed to be made and the conditions could not be more favorable than then ... for Mr. Tombaugh's recent success made him willing and eager for the long and laborious program of painstakingly examining plates for years in the hope that among the millions of stars there might be found one that would possess that tell-tale shift which is necessary and sufficient proof that here is the outermost member of the sun's planets...." Moreover, such a systematic observational search was the only practical approach, Slipher felt, because the faintness of Pluto "indicated that such planets as might exist in orbits still more distant from the sun would be too small and weak in their gravitational perturbative action on the other members of the solar system to expect aid from mathematical analysis of the problem."[4]

The cost of continuing the hunt, parenthetically, was apparently of some initial concern. For Putnam soon sought outside funds to support the work, writing to Henry Norris Russell in September to inquire "how we can get a grant from the National Research Council, or some such body, to carry our systematic search around the heavens. Before Pluto was discovered," he noted, "this seemed like a foolish thought which was too much in the nature of an experiment, but now that we have proved that such things can be found this way, and we are going to find many valuable side-lines which we haven't time, ourselves, to work up, it seems a very proper thing for some financial body to undertake."[5]

Russell agreed, replying: "Heartily in sympathy with your suggestion.... I should put it very high indeed on any list of projects, and will back it to the best of my power."[6] No National Research Council grant was forthcoming, however, although the Lowell Observatory did subsequently receive two small grants for the search from the Penrose Fund of the American Philosophical Society.[7]

Again, parenthetically, one of the first things that Tombaugh looked for when he resumed his search was something called the "Ottawa object," a short-lived trans-Neptunian planet suspect that was reported on April 22, 1930, at the height of the excitement over Pluto, by R. M. Stewart, the director of the Dominion Observatory in Ottawa, Canada.[8] Stewart announced that the object had been found during a search for pre-discovery observations of Pluto by C. F. Henroteau and Miss M. S. Burland on plates taken at Ottawa in 1924. He added that the orbit of the object was uncertain, but the two positions on the 1924 plates could "be satisfied by two circular orbits at trans-Neptunian distances and one at an asteroid distance."[9]

Two weeks later, Crommelin issued a British Astronomical Association *Circular*, giving a distance of 39.82 a.u., a longitude of ascending node of 280°29'.4, and an inclination of 49°.7 for the object.[10] On May 8, 1930, he advised Slipher at Flagstaff: "The Ottawa object may be another trans-Neptunian, but unless more images are found its orbit must remain rather uncertain."[11] As of 1979 nothing more concerning this object has apparently ever been reported.

For the first two years of the extended search young Tombaugh worked virtually alone, making one-hour exposures with the 13-inch refractor and the 5-inch Cogshall camera of strips of the sky paralleling the ecliptic, and generally following the procedures that had proven to be effective in the early phase of the third Planet X survey. From 1933 to 1935, while Tombaugh was completing his A.B. degree at the University of Kansas,[12] the work was taken over by Frank K. Edmonson, who came to Flagstaff under a brief revival of the short-lived Lawrence Fellowship program initiated in 1905 by Percival Lowell to staff the first photographic search for his Planet X. Tombaugh did the blinking of the plates during the summer. By 1936, Tombaugh was back at the observatory and devoting his full time to the examination of plates while Henry L. Giclas, who had joined the Lowell staff in 1931, made most of the exposures at the telescope. With two men working, the survey progressed rapidly for the next two years.[13]

In 1939 and 1940, Tombaugh himself made a series of two-and-a-half-hour exposures of fields along a segment of the ecliptic, identifying stars on his plates out to magnitude 18.6. But the examination of these plates, he found, became "prohibitively tedious" because of the great number of stars they recorded. In some cases, nearly a month was required to conscientiously blink a single pair. In the early 1940s, the observatory briefly considered acquiring a 25-inch f/4 Schmidt camera to carry the extended search out to magnitude nineteen, but this plan was frustrated by the advent of World War II. Such a search would indeed be an arduous undertaking, Tombaugh felt, particularly in the regions of the Milky Way. As the work of examination would increase roughly in proportion to the number of stars on the plates and "since the star density in the Milky Way between 17 and 20 mag. would increase 10–20 fold, the work would become truly prohibitive."[14]

As it was, Tombaugh's labors in the extended search were considerable. "Generally," he has recalled, "the blink examination of 30,000–60,000 stars for planetary motion was a good day's work.... Three to six hours a day were all that one could blink with efficiency." He has estimated that in the entire survey, he spent a total of 7,000 hours, or about 3.3 years assuming 40-hour weeks, at the Lowell comparator, examining plates and checking planet suspects. In addition, he spent several thousand hours at the telescope.[15]

With the war, the Lowell Observatory's extended observational search for new planets virtually came to an end. Only a few unexamined plates were blinked during the late summer and fall of 1945. Its negative result, as far as the primary objective was concerned, has since stood the test of time and of subsequent predictions of new major planets that supposedly lay within the fields and magnitudes that Tombaugh and his colleagues covered.

Photograph courtesy of C. W. Tombaugh

Clyde W. Tombaugh, age 32, examines planet-search plate in 1938 with Lowell Observatory's Zeiss "blink" comparator.

There were some other negative results as well: No zones of large asteroids beyond Saturn were found, for example, nor was anything analogous to the Trojan asteroids of Jupiter detected at the equilateral points in the orbits of Saturn, Uranus or Neptune although, as Tombaugh has noted, such asteroids would have to be very large to be visible at the distances of Uranus and Neptune.[16]

The positive results of the search—the "very valuable side-lines" that Putnam had mentioned—included the discovery of a new globular cluster, five galactic star clusters, and a cloud of some 1,800 galaxies. Only one comet, of magnitude 9-10, was found and this on plates blinked a year after they were made, but more than 1,800 variable stars, most of them presumably of short period, were found on the plates, and some 4,000 asteroids were recorded. Of the latter, approximate positions and magnitudes were determined and published for 744, with 145 of these turning out to be new objects. In addition, a total of 29,548 extra-galactic nebulae were counted, several thousand of which were large enough to permit classification with Hubble's system.[17]

Arizona Daily Sun

During a visit to the Lowell Observatory in 1978. Clyde W. Tombaugh, age 71, checks the "blink" comparator he used to discover Pluto.

Parenthetically, this exhaustive trans-Neptunian planet search is proving valuable to astronomy in still another way. Since 1957, Lowell astronomer Henry L. Giclas has been duplicating the earlier program with the 13-inch for use on a comparative basis not in the search for new planets, but for stars with large proper motions, that is, stars with relatively fast angular movements across the sky. By 1980, summary catalogs of more than 14,000 stars fainter than magnitude 8 with proper motions larger than 0".26 arcsecond in the northern hemisphere, and 0".19 in the southern, were complete, along with a catalog of red and white dwarf stars with very small proper motions. Positions had been published for 820 minor planets, and five comets had been discovered.[18]

But while Slipher and the Lowell Observatory opted for a strictly observational search, a few other astronomers in the years since Pluto's discovery have sought a possible trans-Plutonian planet theoretically—a truly formidable task. Not surprisingly, given the dearth of data bearing on the problem and the uncertainties of what little data there are, they too have been unsuccessful.

The idea that there might be a major unknown planet more distant from the sun than Pluto, of course, was some fifty years old when Pluto was discovered. In fact, almost all predictions of one or more hypothetical trans-Neptunian bodies, beginning with Todd and Forbes in 1880, have placed them at distances considerably greater than Pluto's mean of 39+ a.u. The major exceptions to this have been Pickering's 1928 Planet O and his post-Pluto postulations for Planets T and U which concerned, it will be recalled, either Neptunian or intra-Neptunian bodies.[19] Lowell himself predicted that his Planet X would be somewhat more distant than Pluto turned out to be, and he also suggested a second trans-Neptunian planet at about 75 a.u., although he concluded it would be futile to search either theoretically or observationally for such a body.[20] A distance of this order has been favored for a hypothetical tenth planet by a number of post-Pluto planetary prognosticators, including Pickering. For what it is worth, the distance also fits Bode's law.

Serious efforts to discover a trans-Plutonian planet analytically have been few and far between, and those few who have undertaken this onerous problem have preferred to rely on analyses of the motions of periodic comets, rather than the dubiously small residuals that remain in the motions of the known outer planets.

A singular exception to this, however, is the work of France's M. E. Sevin who in 1946 suggested a tenth planet existing beyond Pluto at a distance of almost 78 a.u. Sevin first derived a period for this hypothetical body by a curious, empirical method in which he divided the known planets, along with the erratic asteroid Hidalgo, into two groups of inner and outer bodies thusly:

Group I	Group II
Mercury	?
Venus	Pluto
Earth	Neptune
Mars	Uranus
Asteroids	Saturn
Jupiter	Hidalgo

He then added up the logarithms of the periods of the horizontally paired planets in each group, finding that log T (Group I planet) + log T (Group II planet) $\cong$ 7.340, a number which he concluded was a constant of sorts. When a possible trans-Plutonian planet is paired with Mercury in this simple equation, its period turns out to be 247,275 days, or about 677 years. Sevin was apparently encouraged by this work for he subsequently worked out a full set of elements for a hypothetical "planet 10" at a distance of 77.8 a.u. with a period of 685.8 years, an eccentricity of 0.3, and a mass of 11.6 earths.[21] His prediction, however, stirred little interest among astronomers.

Indeed most astronomers have apparently felt, as had Lowell, that the distance alone which must be postulated for a trans-Plutonian body was sufficient to make its detection from the motions of the known planets improbable, if not impossible, for at least many decades to come. Nor can the problem be effectively minimized by increasing the size and/or mass of the hypothetical planet, or by juggling its other orbital elements to counter its remoteness, for both theory and observation place limits on these parameters. Furthermore, in any analytical solution to the trans-Plutonian planet problem which relies on planetary residuals, Pluto's mass must be taken into account, and Pluto's mass, despite progressive downward refinements in its value, remained an uncertain quantity through these years and until the discovery of Pluto's satellite, Charon, in 1978.

One prediction made soon after Pluto's discovery was that astronomical interest in the new planet would intensify during the 1960s and thereafter, and indeed it has. For Pluto then not only began to approach a conjunction with Uranus, but it began to swing rapidly sunward toward its perihelion passage in 1989, when it will be only some 29 a.u. from the sun. From 1979 to 1998, because of the high eccentricity of Pluto's orbit, Neptune will be the outermost known planet of the solar system. The relative nearness of Pluto in the final decades of the twentieth century, in fact, should provide astronomers with their best chance to acquire new definitive knowledge of the planet since its discovery and for several centuries to come. Indeed, Pluto's relative nearness was a contributory factor to the 1978 discovery of the Plutonian satellite, Charon, made at a time when the planet was only about 30 astronomical units from the sun.

Parenthetically, the eccentricity of Pluto's orbit and the fact that it penetrates Neptune's orbit very early raised an interesting question: How close do the two planets come to each other in the course of their mutual motions through space? Crommelin, after completing his elliptical elements for Pluto in May 1930, addressed himself to this question and concluded that Pluto's high orbital inclination prevented a very close approach of the two planets, and that their most proximate distance would be no less than four astronomical units.[22] In more recent years, this minimum distance has been considerably extended—to at least 18 a.u.—and Pluto's orbit has been found to have been stable for at least 120,000 years.[23] So there is apparently no possibility of a cataclysmic encounter in Neptunian space.

If the residuals of Neptune and Uranus were poor indicators of Pluto's mass, they have been equally unproductive as indicators of a possible trans-Plutonian planet, or planets, and post-Pluto planet-hunters necessarily have had to look elsewhere for a data base in their prognostications. Generally, they have turned to analyzing the motions of the periodic comets, and such analyses have been carried out with widely varying degrees of mathematical rigor.

Pickering in 1931, of course, used cometary motions in combination with his empirical graphical method to revise the orbits of the pre-Pluto Planet P, assigning it a distance of 75.5 a.u. with a mass of fifty earths, and Planet S, which he now placed at 48.3 a.u. with a mass of 5.3 earths. In both cases, he relied on the occurrence of cometary aphelia to derive their mean distances for his hypothetical planets.[24] Comets are still popular with planet-hunters. As recently as June 1975, Gleb Chebotarev, director of the Leningrad Institute of Theoretical Astronomy, announced that a mathematical analysis of cometary orbits over the past century indicated that undiscovered planets exist at 54 and 100 a.u. from the sun, and his announcement made headlines far from the Soviet Union.[25]

There have been only a few other such predictions, the most notable perhaps, although still obscure, being a series beginning in 1950 with the work of Professor K. Schütte of Munich who used data on eight periodic comets to suggest a trans-Plutonian planet at a distance of 77 a.u. from the sun.[26] Four years later, H. H. Kritzinger of Karlsruhe, using the same eight comets, extended and refined Schütte's work, finding the supposed planet to be somewhat nearer the sun at 65 a.u. Other elements of a circular orbit he derived included a period of 523.5 years, an orbital inclination of 56°, and an estimated magnitude of 11.[27]

In 1957, Kritzinger reworked the problem, pairing data for two of the eight comets and computing a circular orbit for a planet at 75.1 a.u. with a period of 650 years, an inclination of 40°, and a magnitude, assuming it to be similar to Neptune, of 10.[28] And finally, two years later, after an unsuccessful photographic search for this planet, he worked over the problem once again, dropping two of the eight comets from his computations and coming up with elliptical elements for a planet at Schütte's original distance of 77 a.u., with a period of 675.7 years, an inclination of 38°, and an eccentricity of 0.07, a planet not unlike Sevin's "planet 10," and in some ways similar to Pickering's final Planet P.[29] But needless to say, no such planet has as yet been found.

Of greater interest, perhaps, have been the sporadic attempts to use Halley's comet to locate a second trans-Neptunian planet. This justly famous comet, which first came into the planetary prediction picture in the pre-Neptunian speculations of 1835, reappeared again in 1942 when astronomer R. S. Richardson sought to determine Pluto's mass from anomalies in the comet's motion. The difficulty with Halley's comet, whose periodic returns to the sun every 76 years have been recorded for two millenia, has been that its perihelion passage has consistently lagged several days behind predictions. Something out there was perturbing it, Richardson decided, and he thought it might be Pluto. It was not, he found, but in the course of his investigation he determined that an earth-sized planet in an orbit 36.2 a.u. from the sun, or one astronomical unit beyond Halley's aphelion, would delay its perihelion passage about half a day, while a similar planet at 35.3 a.u., a tenth of a unit beyond its aphelion, gave a perihelion delay of six days.[30]

Thirty years after Richardson's work, Joseph L. Brady of the University of California's Lawrence Livermore Laboratory, used the perihelion delay of

Halley's comet to postulate a truly spectacular trans-Plutonian planet, surely as strange a planet as has ever been predicted and certainly one of the most publicized. This massive, erratic body, he found, reduced the residuals of the time of perihelion passage of Halley's comet over seven apparitions dating back to A.D. 1456 by 93 percent, and improved the theories of the periodic comets Olbers and Pons-Brooks as well.[31]

Brady prefaced his 1972 prediction by declaring that there "is no logical reason to suppose Pluto to be the outermost planet in the solar system. Indeed there is a great deal of evidence that it is not, and the literature of the last 100 years records numerous attempts to predict planets beyond Neptune and Pluto, some of which led to telescopic searches." This literature included "the fruitful prediction of Percival Lowell" in 1915, and its "most consistent message is that there must be two trans-Neptunian planets and the second one appears to be about twice the distance of Neptune from the sun."[32]

Moreover, he contended, the evidence for such a body was becoming stronger as observations of Neptune's motion accumulated over the years. "Whether Pluto was actually Lowell's Planet X, or its discovery an extraordinary coincidence as Brown claimed, or fortuitous as Kourganoff indicates in his refutation of Brown, is still an intriguing question," he wrote. "But even more intriguing is the consistent failure of the theories of Neptune's motion to represent the observations. As indicated by Rawlins, the unexplained residual in the orbit of Neptune may be due to an alien perturbation; but years may have to pass before planetary residuals will reveal the position of a trans-Plutonian planet."[33]

Thus for Brady the well-documented anomalies of Halley's comet seemed to offer the most immediate and practical chance for analytically locating such a body. From his computations, Brady concluded that a trans-Plutonian planet existed at a distance of 59.9 a.u. in an orbit with an eccentricity of 0.07, and with a period of 464 years. There is nothing particularly unusual about these parameters in light of earlier trans-Plutonian predictions. But his other postulated elements for what came to be called "Planet X" in both the professional and popular press, raised many an astronomical eyebrow. For this "Planet X" moved in a retrograde direction around the sun, i.e. opposite the motion of all the known planets; its orbital inclination was a startling 120° to the plane of the ecliptic, its mass three times that of massive Saturn, and its magnitude was estimated at 13–14, assuming Pluto's albedo.[34]

The reaction within astronomy to Brady's well-publicized "Planet X" was immediate and negative. Seidelmann of the Naval Observatory, Brian G. Marsden of the Smithsonian Astrophysical Observatory and Giclas of Lowell Observatory quickly teamed up to investigate Brady's prediction and produced a sharp critique. Given the mass and orbital inclination of Brady's planet, they pointed out, "its effect on the latitude of the known planets would be large"—too large to have reasonably escaped detection over the years. And even accepting Brady's magnitude estimate, they found it "rather surprising that a planet with an annual average motion of 46 arc minutes has been overlooked." Giclas, using Brady's predicted position, searched the indicated area of the sky on plates taken by Tombaugh in 1940-41 in the extended trans-Neptunian planet search and plates he himself had made in 1969 during his own subsequent proper motion survey, but found no such body. "We conclude," the three astronomers wrote, "that, while small or very distant unknown planets may exist, there is no planet having the mass, magnitude, mean distance and orbital inclination hypothesized by Brady."[35]

Others came to more or less similar conclusions. Peter Goldreich and W. R. Ward argued "The Case Against Planet X" in print by noting that as Brady's planet would contain most of the angular momentum in the solar system, it would produce serious, and observable effects on the known planets which had not, in fact, been observed. They also pointed out that the residuals of Halley's comet could be otherwise explained in terms of non-gravitational forces such as the jet action of outgassing by the comet.[36] And A. R. Klemola and E. A. Harlan, using Lick Observatory's 20-inch double astrograph, searched an area of the sky 6°.3 square centered on Brady's predicted position, making both blue and yellow plates 24 hours apart and finding no images with the expected motion out to a visual magnitude of seventeen to eighteen. "It appears likely," they concluded, "that there is no planetary body at the ephemeris position to within 3° for the indicated limits of apparent magnitude."[37]

Apart from speculations of possible trans-Plutonian planets and continued efforts to resolve the problem of Pluto's mass, Pluto has been of interest to some astronomers for another important reason—for what it might reveal about the origin and evolution of the solar system. Early in 1932, in publishing the Lowell Observatory's "official" version of Pluto's discovery, Putnam and Slipher made a significant point:

> Thus the finding of Pluto is impressive not merely because it adds another planet and greatly expands the known domain of the Sun, but also because it hints that, out in this new extension of the Sun's domain, planetary affairs may differ markedly from what we might have supposed, influenced by our acquaintance with the previously known planets.[38]

And indeed, in the years since this was written, it has become quite certain that planetary affairs beyond Neptune, to the extent they have become known, do "differ markedly" from what astronomers once had every reason to expect they might be. Lowell, in attempting to predict his trans-Neptunian planet, cannot be faulted for proceeding on the not-illogical assumption that the next planet beyond Neptune would closely resemble Neptune and the other outer planets in its physical condition. Yet Pluto has turned out to be a very unusual, if not entirely unique body that cannot be easily classified among either the outer or the inner, terrestrial planets. In short, Pluto appears to be in a class of its own.

Before adequate orbits for Pluto were derived, it may be recalled, the newly discovered Lowell object seemed to be strange enough to prompt suggestions, such as Leuschner's, that it might be an unusual comet or a large asteroid in an erratic orbit. This latter idea, incidentally, appears as late as 1938 when S. H. J. Wanrooy speculated that Pluto might be a planetoid and a large member of an outer ring of such bodies circling the sun beyond Neptune as a more distant counterpart of the asteroid belt between Mars and Jupiter.[39]

That farther-ranging asteroids or other as yet unclassified objects exist, incidentally, was apparently demonstrated in October 1977 with the discovery, by Charles T. Kowal of the Hale Observatories, of a small trans-Saturnian object which Kowal immediately named Chiron. Preliminary elements computed by Astronomer Brian G. Marsden from seventeen pre- and post-discovery observations of Chiron indicated a mean distance of 13.7 a.u., an eccentricity of 0.3786, an inclination of 6°.923, and a period of 50.7 years. At its closest approach to the sun, calculated for February 1996, Chiron will be a 8.5 a.u., and at aphelion, it reaches a distance of 18.9 a.u., nearly to the orbit of Uranus. Its diameter appears to be of the order of 50 kilometers.[40]

Chiron, in a moment of institutional exuberance and to Kowal's dismay, was first announced as "a tenth planet," but almost immediately it was officially designated as asteroid 1977 UB. Kowal, however, was not at all sure that this designation was correct either, and others suggested that Chiron might be a distant, nearly extinct comet. But until astronomers come up with more precise definitions for the terms, "planet," "minor planet," "asteroid," and "comet," its classification as an asteroid must suffice.[41]

What is more interesting about Chiron's discovery from the standpoint of the history of planetary discovery, however, is that Kowal and some other observers, such as the University of Arizona's Tom Gehrels, have been making extended, systematic surveys of the more remote regions of the solar system since the early 1970s, using the 48-inch "Big Schmidt" telescope on Palomar Mountain, and are quite optimistic that new "planets X" will be found in the course of this work. Such observing programs as Kowal's and Gehrels' are, in fact, the best indication to date that the final chapter of this history cannot yet be written.[42]

One particularly intriguing possibility relating to Pluto, perhaps, has been the proposal that Pluto is not one of the sun's original planets at all, but is rather a renegade satellite of Neptune. This suggestion was apparently first advanced by a Japanese astronomer, Issei Yamamoto, in 1934.[43]

Two years later, astronomer Raymond A. Lyttleton developed the idea in a paper entitled, "On the Possible Results of an Encounter of Pluto with the Neptunian System," in which he concluded not only "that Pluto may originally have been a direct satellite of Neptune" but that "the encounter which gave it an existence as an independent planet also reversed the general direction of the motion of Triton," thus providing a "rather spectacular explanation of the retrograde motion of a true satellite." Lyttleton was vague on the nature of the encounter, however.[44]

Lowell Observatory

The end of an era. In 1967, Roger Lowell Putnam, *second from left,* turned over the trusteeship of Lowell Observatory to his son, Michael C. J. Putnam, *right.* With them on the occasion are John S. Hall, *left,* Lowell director from 1958–77, and V. M. Slipher, *second from right,* still spry at 92. Slipher, the longest-lived of Percival Lowell's assistants, died in 1969 at the age of 94.

In 1956, two decades later, Kuiper also concluded that Pluto must be an escaped satellite of Neptune, contending that because of its unusual orbital features, such as its high eccentricity and inclination, it could not have been an original proto-planet of the sun. If Pluto originated as a satellite of Neptune, it would be expected to have a relatively long period of rotation, on the order of a week, at least if its original distance from the planet was less than 60 planetary radii, the distance of Iapetus from Saturn, where synchronization of rotation and revolution occurs.[45] He pointed out that the problem of Pluto's rotation had been apparently solved photometrically at Mount Wilson and the Lowell observatories by astronomers M. F. Walker and R. Hardie in 1953–54. They had observed periodic light variations of the planet and had derived a rotational period of 6.390 ± 0.003 days, slower by a factor of at least six than the other outer planets whose rotation periods can be counted in hours alone.[46]

This period, incidentally, apparently coincides with the period of revolution of Pluto's satellite, Charon, which thus may be in a synchronous orbit. Navy astronomers, however, shortly after Charon's discovery, noted that Pluto's minimum brightness corresponded with Charon's maximum elongation from the planet. Thus the photometric variations in Pluto's brightness do not appear to be the result of the satellite's motion around the planet.[47]

Kuiper's contention that Pluto is an errant Neptunian moon stirred wide interest in the popular press as well as some intra-fraternity bickering within astronomy. At least one well-known astronomer, W. J. Luyten, of the University of Minnesota, publicly accused Kuiper of plagiarizing and exploiting Lyttleton's 20-year-old hypothesis.[48]

Kuiper, in fact, had referred to Lyttleton's work, had stressed the difference between it and his own, and had flatly rejected Lyttleton's suggestions, particularly as they related to Triton's retrograde motion. "The conclusion so reached regarding Pluto's origin as a satellite of Neptune should not be confused with Lyttleton's (1936) hypothesis that Pluto and Triton were initially both satellites of Neptune and then had a close encounter which caused Pluto to leave the system and Triton to become retrograde," Kuiper had written. "There is no reason to suppose that an encounter between regular satellites has ever occurred; and there are five retrograde satellites other than Triton."[49]

Indeed, Kuiper's hypothesis of the origin of Pluto was only a part of his broader theory of the origin of all the planets. In this, he suggested that sometime after the formation of "proto-planets" from the primordial solar nebula, the sun had become brighter, causing the proto-planets to lose mass through "evaporation." As proto-Neptune thus shrank, then Pluto, as its satellite, progressively moved away from its primary to a point where it finally entered a solar orbit and became a planet itself.[50]

The discovery of the Plutonian satellite, Charon, does not, it may be noted, preclude the possibility that Pluto is a renegade satellite of Neptune. Naval astronomers have pointed out that the cataclysmic event required to eject Pluto from a Neptunian orbit could also have split a large chunk of the planet away, leaving it to orbit Pluto as a satellite.[51] More speculative is the suggestion that the variation in Pluto's brightness is caused by the scar left by such a schism.

In the final analysis, however, whether Pluto is an original planet, a renegade satellite of Neptune, or even an unusual asteroid is simply not yet known.

It would be foolish indeed, in view of what has here been set down and the capriciousness of chance, to say that the history of planetary discovery has run its course, and that there is no other "planet," in the vaguely defined but commonly accepted sense of the word, circling the sun in the farther reaches of the sun's domain. Some astronomers already are convinced that one or more Planets X will be discovered through systematic observing programs such as those that produced Kowal's Chiron and the U.S. Naval Observatory's Charon. It is, they feel, largely a matter of time.

It may also be a matter of number, for undoubtedly the "eye of analysis," in Percival Lowell's phrase, will again be brought to bear on the problem. The progressive reduction of Pluto's mass over the years, culminating with a direct determination of this value from the presence of its satellite, is the significant circumstance here.

If Pluto's mass is only a few thousandths that of the earth as now seems certain, the planet cannot produce any discernible perturbations in the motions of its near neighbors. And if this is so, one must still account for the perturbations that Lowell and others have found, and that Kourganoff at least, after a careful re-analysis of their work, concluded were real. Kourganoff, in fact, found that even an earth-sized Pluto left small residuals in Uranus' motion, although he was unwilling to decide whether these were attributable to systematic errors or to "another trans-Neptunian planet." There are also problems with Neptune's motion, and notably the "great unexplained residual" that Rawlins found after reducing again Lalande's 1795 prediscovery observations on the basis of a Pluto one-sixth as massive as the earth, and which led him to suggest "an alien perturbation."

In the analytical problem, as Pluto's mass is reduced, these residuals must grow larger unless, of course, the theories of the outer planets can be "legitimately juggled"—Lowell's phrase again—to keep these planets in their proper courses.

If theory and observation cannot be thus reconciled, it is a reasonable assumption that still another Planet X may be causing the discrepancies, and the future of planetary discovery may thus be very exciting indeed.

CHAPTER NOTES

Unless otherwise noted, correspondence cited below is contained in the Lowell Observatory Archives, Flagstaff, Arizona.

Notes to Pages 1–12

Introduction: Of Time and Number

[1] P. Lowell, "Memoir on a Trans-Neptunian Planet," *Memoirs of the Lowell Observatory*, vol. 1, no. 1 (1915).
[2] P. Lowell, *The Evolution of Worlds* (New York: The Macmillan Company, 1909), 119–20.

Uranus and the Asteroids

[1] W. Herschel, journal entry, in introduction to *The Scientific Papers of Sir William Herschel*, J. L. E. Dreyer, ed. (London: The Royal Society and the Royal Astronomical Society, 1912), xxix: hereafter cited as the *Herschel Papers*.
[2] *Ibid.*, xxx.
[3] W. Herschel, "Account of a Comet," *Herschel Papers* 1:30–38.
[4] *Ibid.*
[5] *Ibid.*
[6] C. A. Lubbock, *The Herschel Chronicle* (Cambridge: The University Press, 1933), pp. 84–85.
[7] *Herschel Papers*, xxx; Lubbock, *op. cit.*, p. 79.
[8] Lubbock, *op. cit.*, p. 79.
[9] *Ibid.*
[10] *Ibid.*, p. 106.
[11] "A Letter From William Herschel, Esq., F.R.S.," *Herschel Papers* 1:100–101.
[12] Lubbock, *op. cit.*, 82–83; see also J. B. Sidgwick, *William Herschel: Explorer of the Heavens* (London: Faber and Faber, Ltd., 1953), p. 79.
[13] Lubbock, *op. cit.*, p. 86.
[14] W. Herschel, "A Paper to obviate some doubts concerning the great Magnifying Powers

[263]

used," *Herschel Papers* 1:97.
[15] W. Herschel, memorandum, *Herschel Papers*, p. xxxv.
[16] *Ibid.*, p. xxxiii.
[17] *Ibid.*
[18] *Ibid.*, p. xxxiv.
[19] *Ibid.*, p. xxxiii.
[20] A. F. O'D. Alexander, *The Planet Uranus* (New York: American Elsevier Publishing Co., Inc., 1965), p. 25.
[21] E. S. Barr. "The Infrared Pioneers—I. Sir William Herschel," *Infrared Physics* 1:1–4 (1961). Herschel in 1800 used thermometers to measure the radiation of the solar spectrum, finding radiation in the "invisible," or infrared range as well as in the visible range.
[22] W. Herschel, memorandum, *Herschel Papers*, p. xiv.
[23] *Ibid.*, p. xv.
[24] *Ibid.*, p. xxiv.
[25] *Ibid.*
[26] *Ibid.*, p. xv.
[27] Lubbock, *op. cit.*, p. 59.
[28] W. Herschel, memorandum, *Herschel Papers*, p. xxii–xxxiii.
[29] *Ibid.*, p. xxix.
[30] *Ibid.*, p. xxxiii.
[31] *Ibid.*
[32] Lubbock, *op. cit.*, pp. 60–61.
[33] Alexander, *op. cit.*, p. 37.
[34] *Ibid.*, p. 39.
[35] *Ibid.*, p. 37; Lubbock, *op. cit.*, p. 93.
[36] Alexander, *op. cit.*, p. 38.
[37] *Ibid.*, p. 41.
[38] "Boscovich, Roger Joseph," *Encyclopedia Britannica*, 14th edition, 1928. The neglect of Boscovich, perhaps, can be noted in the fact that he is not listed in the 15th edition of the *Britannica*, published in 1974.
[39] Alexander, *op. cit.*, p. 39.
[40] *Ibid.*, pp. 43, 46–47, 49.
[41] E. T. Bell, *Men of Mathematics* (New York: Simon and Schuster, 1937), p. 152.
[42] Alexander, *op. cit.*, pp. 31, 51.
[43] Lubbock, *op. cit.*, p. 93.
[44] *Ibid.*
[45] *Ibid.*, p. 95.
[46] Alexander, *op. cit.*, p. 35, quoting Herschel's letter to Banks, Nov. 19, 1781, as published in the *Occasional Notes of the Royal Astronomical Society* 3:115 (1949).
[47] Sidgwick, *op. cit.*, p. 73.
[48] Alexander, *op. cit.*, p. 35.
[49] Lubbock, *op. cit.*, p. 95.
[50] W. Herschel, "An Account of the Discovery of Two Satellites revolving round the Georgian Planet," *Herschel Papers* 1:312–14; and "On the Georgian Planet and its Satellites," *Herschel Papers* 1:317–26.
[51] W. Herschel, "On the Discovery of Four Additional Satellites of the Georgium Sidus," *Herschel Papers*, 2:1.
[52] "Occultation observations reveal ring system around Uranus," *Physics Today* 30:17 (1977), and J. L. Elliot, E. Dunham and Robert L. Millis, "Discovering the Rings of Uranus," *Sky and Telescope*, 53:412–16 (1977). *See also* B. A. Smith, L. A. Soderblom *et al.* "The Jupiter System Through the Eyes of Voyager 1," *Science* 204:951–71, 1979.
[53] Sidgwick, *op. cit.*, p. 81.
[54] *Herschel Papers, op cit.*, especially "Experiments on the Refrangibility of the invisible Rays of the Sun," 2:70–76 (1800), for his infrared work, and "On the Nature and Construction of the Sun and the Fixed Stars," 1:470, for the sun "as an inhabitable world."
[55] W. Herschel, "On the Diameter and Magnitude of the Georgium Sidus," *Herschel Papers* 1:102–7.

[56] *Ibid.*, p. xxxv.
[57] Lubbock, *op. cit.*, p. 112.
[58] *Herschel Papers, op. cit.*, p. xxxiv.
[59] *Ibid.*, p. xxxv.
[60] *Ibid.*, p. xxxvi.
[61] *Ibid.*, p. xxxvii.
[62] *Ibid.*, p. xxxvi; *see also* Lubbock, *op. cit.*, pp. 119–23.
[63] W. Herschel, "A Letter from William Herschel, Esq., F.R.S.," *Herschel Papers*, pp. 100–101.
[64] Lubbock, *op. cit.*, pp. 119–23.
[65] *Ibid.*, p. 131.
[66] *Ibid.*, p. 123.
[67] Alexander, *op. cit.*, p. 54.
[68] *Ibid.*, pp. 51–53.
[69] M. Grosser, *The Discovery of Neptune* (Cambridge: Harvard University Press, 1962), pp. 124–26.
[70] Lubbock, *op. cit.*, p. 201.
[71] See, for examples, A. M. Clerke, *A Popular History of Astronomy During the Nineteenth Century*, 4th edition (London: Adam and Charles Black, 1902), p. 5; and Sidgwick, *op. cit.*, p. 74.
[72] G. D. Roth, *The System of Minor Planets*, trans. Alex Helm (New York: D. Van Nostrand Co., Inc., 1962), pp. 18–19.
[73] S. J. Jaki, "The Titius-Bode Law: A Strange Bicentennial," *Sky and Telescope* 43:280–81 (1972).
[74] *Ibid.*
[75] F. G. Watson, *Between the Planets* (Philadelphia: The Blakiston Co., 1941), p. 5. Roth, *op. cit.*, gives a formula for Bode's Law as $a = a + 2^n b$, where $a = 0.4$, $b = 0.075$, and n = the number of the planet from the sun.
[76] Roth, *op. cit.*, p. 22; *see also* Clerke, *op. cit.*, p. 72.
[77] Roth, *op. cit.*, p. 24; *see also* Grosser, *op. cit.*, p. 30.
[78] Roth, *op. cit.*, p. 24; Watson, *op. cit.*, pp. 10–11; Grosser, *op cit.*, pp. 30–31; Clerke, *op cit.*, p. 73.
[79] Clerke, *op. cit.*, p. 73.
[80] Grosser, *op. cit.*, p. 31.
[81] *Herschel Papers, op. cit.*, p. xliv; Lubbock, *op. cit.*, pp. 268–69; Grosser, *op. cit.*, p. 32.
[82] Bell, *op. cit.*, p. 259; Clerke, *op. cit.*, p. 74.
[83] Grosser, *op. cit.*, pp. 32–33; Bell, *op. cit.*, pp. 239–42.
[84] Clerke, *op. cit.*, p. 74; Grosser, *op. cit.*, pp. 33–34.
[85] Lubbock, *op. cit.*, p. 269.
[86] W. Herschel, "Observations on two newly discovered bodies (Ceres and Pallas)," *Herschel Papers* 2:197; Grosser, *op. cit.*, pp. 34–35.
[87] Lubbock, *op. cit.*, 271.
[88] *Ibid.*
[89] *Ibid.*
[90] *Ibid.*, p. 220.
[91] W. Herschel, "Observations on two newly discovered bodies (Ceres and Pallas)," *op. cit.*
[92] Lubbock, *op. cit.*, p. 274.
[93] *Ibid.*
[94] *Ibid.*, p. 272.
[95] *Ibid.*
[96] *Ibid.*
[97] Clerke, *op. cit.*, p. 75.
[98] Clerke, *op. cit.*, p. 77; Roth, *op. cit.*, pp. 30, 34; A. C. D. Crommelin, "Minor Planets," *Encyclopedia Britannica*, 14th edition, vol. 15, pp. 575–77 (1928).
[99] Lubbock, *op. cit.*, p. 272.
[100] Watson, *op. cit.*, pp. 87*ff.*

Neptune

[1] A. F. O'D. Alexander, *The Planet Uranus* (New York: American Elsevier Publishing Co., 1965), pp. 39, 42. Alexander cites Boscovich, *Opera* 3:369 (1785); and Lexell, *Nova Acta Petersburgi* 1:69 (1787).
[2] C. A. Lubbock, *The Herschel Chronicle* (Cambridge: The University Press, 1933), p. 127.
[3] Alexander, *op. cit.*, p. 82.
[4] *Ibid.*, p. 41.
[5] *Ibid.*, pp. 85–87.
[6] *Ibid.*, pp. 85–88; *See also* D. Rawlins, "A long lost observation of Uranus, Flamsteed 1714," *Publications of the Astronomical Society of the Pacific* 80:217–19 (1968). Rawlins notes that the 1714 observation may have been made by Flamsteed's assistant, J. Crosthwaite.
[7] Alexander, *op. cit.*, p. 89; *See also* M. Grosser, *The Discovery of Neptune* (Cambridge: Harvard University Press, 1962), p. 41.
[8] Grosser, *op. cit.*, pp. 25–26.
[9] *Ibid.*
[10] Alexander, *op. cit.*, pp. 87–89.
[11] A. Bouvard, *Tables astronomiques publiées par le Bureau des Longitudes de France contenant les Tables de Jupiter, de Saturne et d'Uranus construites d'après la théorie de la Mécanique céleste*, (Paris: 1821), p. vix.
[12] Grosser, *op. cit.*, pp. 42–43.
[13] B. A. Gould, *Report on the History of the Discovery of Neptune*, (Washington: Smithsonian Institution, 1850), 10.
[14] Grosser, *op. cit.*, pp. 45–46.
[15] *Ibid.*
[16] *Ibid.*
[17] *Ibid.*, p. 49. See also A. M. Clerke, *A Popular History of Astronomy During the Nineteenth Century* (4th ed.; London: Adam and Charles Black, 1902), p. 78. Clerke wrote: "The idea that these enigmatic disturbances were due to an unknown exterior body was a tolerably obvious one. Bouvard himself was perhaps the first to conceive it."
[18] G. B. Airy, "Account of some circumstances historically connected with the discovery of the Planet exterior to Uranus," *Monthly Notices of the Royal Astronomical Society* 7:123 (1846).
[19] Airy, *op. cit.*, p. 124.
[20] G. B. Airy, letter to the editor, *Astronomische Nachrichten*, no. 349 (1838).
[21] Grosser, *op. cit.*, pp. 50–51.
[22] *Ibid.*, pp. 51–52. *See also* E. Everett, "On the New Planet Neptune," *Astr. Nach.*, no. 599 (1847); and J. E. B. Valz, letter to the editor, *Astr. Nach.*, no. 600 (1847).
[23] Airy, "Account," *Astr. Nach.* 600:125 (1847).
[24] *Ibid.*, p. 127; *see also* Gould, *op. cit.*, p. 15.
[25] Gould, *op. cit.*, pp. 10–11, quoting F. W. Bessel, *Populaire Vorlesungen*, p. 448 (1840).
[26] *Ibid.*, pp. 14–15.
[27] Grosser, *op. cit.*, p. 55.
[28] Gould, *op. cit.*, p. 13.
[29] *Ibid.*, p. 3.
[30] Airy, "Account," *op. cit.*; and *Astr. Nach.*, nos. 585, 586 (1846).
[31] Gould, *op. cit.*, pp. 3–4.
[32] J. C. Adams, "An Explanation of the observed Irregularities in the Motion of Uranus, on the hypothesis of disturbances caused by a more distant planet; with a determination of the mass, orbit and position of the disturbing body," *Appendix* to the *Nautical Almanac for the Year 1851* (London: 1846), p. 3.
[33] W. M. Smart, "John Couch Adams and the Discovery of Neptune," *Occasional Notes of the Royal Astronomical Society*, vol. 2, no. 11, p. 47 (1947).
[34] Clerke, *op. cit.*, p. 79.
[35] Adams, *op. cit.*, p. 4.
[36] *Ibid.*

[37] Grosser, *op. cit.*, pp. 86ff.
[38] Airy, "Account," *op. cit.*, p. 129.
[39] Grosser, *op. cit.*, p. 88.
[40] Airy, "Account," *op. cit.*, pp. 129–30.
[41] *Ibid.*, p. 128.
[42] *Ibid.*
[43] Smart, *op. cit.*, pp. 56–57.
[44] U. J. J. Leverrier, "Premiere memoire sur la théorie d'Uranus," *Comptes Rendus* 21:1050–55 (1845).
[45] Grosser, *op. cit.*, pp. 58–59.
[46] *Ibid.*, pp. 61–64.
[47] U. J. J. Leverrier, "Théorie et Tables du mouvement de Mercure," *Annales de L'Observatoire Impériale de Paris* 5:102 (1859).
[48] Grosser, *op. cit.*, pp. 65–69, 82, 84.
[49] U. J. J. Leverrier, *op. cit.*; see also "Recherches sur les mouvements d'Uranus," *Comptes Rendus* 22:907–18 (1846); "Sur le planète qui produit les anomalies observées dans le mouvement d'Uranus—Determination de sa masse, de son orbite et de sa position actuelle," *Comptes Rendus* 23:428–38 (1846); and "Recherches sur le mouvement de la planète Herschel (dites Uranus), in *Additions* to the *Connaissance des Temps pour 1849* (Paris: 1846), pp. 3–254.
[50] Grosser, *op. cit.*, p. 23.
[51] *Ibid.*, p. 22.
[52] Clerke, *op. cit.*, p. 80.
[53] Adams, *op. cit.*, p. 5.
[54] *Ibid.*, pp. 5–13.
[55] *Ibid.*, pp. 13–15.
[56] *Ibid.*, pp. 17–18.
[57] *Ibid.*, pp. 18–26.
[58] Airy, "Account," *op. cit.*, 137.
[59] Adams, *op. cit.*, pp. 25–26.
[60] *Ibid.*, p. 31.
[61] *Ibid.*, pp. 28–29.
[62] *Ibid.*, p. 29.
[63] Gould, *op. cit.*, p. 24.
[64] *Ibid.*, pp. 25–26.
[65] *Ibid.*, p. 28.
[66] Airy, "Account," *op. cit.*, p. 131.
[67] Gould, *op. cit.*, p. 29.
[68] *Ibid.*, pp. 30–31.
[69] *Ibid.*, pp. 31–32.
[70] *Ibid.*, pp. 32–33.
[71] *Ibid.*, p. 33.
[72] *Ibid.*
[73] Airy, "Account," *op. cit.*, p. 134.
[74] *Ibid.*, p. 132.
[75] *Ibid.*, p. 136.
[76] Grosser, *op. cit.*, pp. 107–8.
[77] *Ibid.*, pp. 108–10.
[78] S. C. Walker, report to Joseph Henry, *Astr. Nach.*, no. 605 (1847).
[79] Gould, *op. cit.*, p. 21; see also Grosser, *op. cit.*, p. 111.
[80] Airy, "Account," *op. cit.*, p. 142.
[81] Gould, *op. cit.*, p. 19.
[82] J. Challis, "Account of observations at the Cambridge Observatory for detecting the planet exterior to Uranus," *Mon. Not. R.A.S.* 7:145 (1846).
[83] Grosser, *op. cit.*, pp. 113–14.
[84] Airy, "Account," *op. cit.*, p. 137.
[85] H. H. Turner, obituary of J. G. Galle, *Mon. Not. R.A.S.* 71:278 (1911).
[86] *Ibid.*

[87] J. F. Encke, letter to the editor, *Astr. Nach.*, no. 580 (1846).
[88] *Ibid.*
[89] Grosser, *op. cit.*, p. 119.
[90] Turner, *op. cit.*, p. 280.
[91] *Ibid.*
[92] Grosser, *op. cit.*, p. 124.
[93] J. Pillans, "Ueber den Namen des neuen Planeten," *Astr. Nach.*, no. 600 (1847).
[94] "Namen des neuen Planeten," *Astr. Nach.*, no. 581 (1846).
[95] Leverrier, letter to the editor, *Astr. Nach.*, no. 591 (1847).
[96] J. F. Encke, report to the Royal Academy of Sciences, *Astr. Nach.*, no. 588 (1846).
[97] Grosser, *op. cit.*, pp. 127–28.
[98] Challis, *op. cit.*.
[99] Grosser, *op. cit.*, p. 128.
[100] Challis, *op. cit.*
[101] Airy, "Account," *op. cit.*, p. 143. Challis had even suggested the name "Oceanus," implying to the French at least a claim to the right to name the new planet.
[102] Grosser, *op. cit.*, pp. 129, 131.
[103] Smart, *op. cit.*, p. 65.
[104] Grosser, *op. cit.*, p. 134.
[105] Smart, *op. cit.*, p. 65.
[106] *Ibid.*, pp. 66–68.
[107] Airy, "Account," *op. cit.*, pp. 121–22.
[108] *Ibid.*
[109] *Ibid.*, pp. 121ff.
[110] *Ibid.*, pp. 143–44.
[111] *Ibid.*, p. 144.
[112] Adams, *op. cit.*, p. 5.
[113] Smart, *op. cit.*, pp. 75–76.
[114] *Ibid.*, pp. 77–78.
[115] Gould, *op. cit.*, p. 20.
[116] J. C. Adams, presidential address, *Mon. Not. R.A.S.* 36:232 (1876).
[117] W. Lassell, letter to the editor, *Astr. Nach.*, no. 589 (1846).
[118] W. Lassell, letter to the editor, *Astr. Nach.*, no. 611 (1847).
[119] Lassell, *op. cit.*, no. 589 (1846).
[120] J. Challis, "Observations of Le Verrier's Planet taken at the Cambridge Observatory," *Astr. Nach.*, no. 591 (1847); and "Second Report of Proceedings at the Cambridge Observatory relating to the new planet (Neptune), *Astr. Nach.* no. 596 (1847). Encke's division has been reported by a number of astronomers, but many doubt the visual reality of this and additional reported divisions other than Cassini's. Incidentally, the British amateur observer J. R. Hinds in *Astr. Nach.*, no. 589 (1847), wrote that from his observations of Neptune, "The existence of a ring appears to be as yet undecided, though most probable. *Le Verrier* presents an oblong appearance in Mr. Bishop's refractor."
[121] Lassell, *op. cit.*, no. 611 (1847).
[122] W. G. Hoyt, "Reflections Concerning Neptune's 'Ring'," *Sky and Telescope* 55:284–85 (1978).

Beyond Neptune

[1] W. C. Bond, letter to the editor, *Astronomische Nachrichten*, no. 591 (1847); and "Observations of Neptune made at Cambridge Observatory," *Astr. Nach.*, no. 595 (1847).
[2] W. C. Bond to E. Everett, *Astr. Nach.*, no. 611 (1947). Everett was then president of Harvard University.
[3] J. Challis, letter to the editor, *Astr. Nach.*, no. 583 (1846).
[4] J. F. Encke, report, *Astr. Nach.*, no. 588 (1847).
[5] J. Challis, "Second Report of Proceedings in the Cambridge Observatory relative to the new Planet (Neptune), *Astr. Nach.*, no. 596 (1847).
[6] W. M. Smart, "John Couch Adams and the Discovery of Neptune," *Occasional Notes of the Royal Astronomical Society*, vol. 2, no. 11, p. 80.
[7] A. C. Petersen, "Nachsuchtung fruherer Beobtung des Le Verrier'schen Planeten," *Astr. Nach.*, nos. 594, 595 (1847); and E. Everett, "On the new Planet Neptune," *Astr.*

Nach., no. 599 (1847). Petersen's first brief announcement appeared in the issue dated April 3, 1847. *See also,* S. C. Walker, letter to Joseph Henry, *Astr. Nach.,* no. 605 (1847).

[8] Walker, *op. cit.*; E. Everett, *op. cit.*
[9] Walker, *op. cit.*
[10] *Ibid.*
[11] Petersen, *op. cit.*
[12] F. V. Mauvais, "On Two Observations of Neptune May 8 and 10, 1795, in the Histoire Céleste," *Astr. Nach.,* no. 607 (1847). *See also,* B. A. Gould, *Report on the History of the Discovery of Neptune* (Washington, D.C.: The Smithsonian Institution, 1850), p. 44.
[13] W. M. Smart, *op. cit.*, p. 81.
[14] B. Peirce, "Investigation in the action of Neptune to Uranus," *Proceedings of the American Academy of Arts and Sciences,* 1:65 (1847).
[15] *Ibid.*, p. 66.
[16] *Ibid.*, pp. 66–67.
[17] *Ibid.*, pp. 67–68.
[18] *Ibid.*, p. 144.
[19] *Ibid.*
[20] Gould, *op. cit.*, pp. 47–48. *See also* P. Lowell, "Memoir on a Trans-Neptunian Planet," *Memoirs of the Lowell Observatory,* vol. 1, no. 1 (1915), p. 33.
[21] B. A. Gould, *op. cit.*, pp. 48–49.
[22] *Ibid.*, p. 52.
[23] W. M. Smart, *op. cit.*, p. 83.
[24] B. A. Gould, *op. cit.*, p. 52.
[25] *Ibid.*, p. 53.
[26] *Ibid.*
[27] *Ibid.*, pp. 53–54.
[28] *Ibid.*, p. 54.
[29] *Ibid.*, p. 55.
[30] *Ibid.*
[31] *Ibid.*, pp. 55–56.
[32] W. M. Smart, *op. cit.*
[33] *Ibid.*, p. 82.
[34] *Ibid.*
[35] M. Grosser, *The Discovery of Neptune* (Cambridge: Harvard University Press, 1962), pp. 140–41.
[36] F. Baldet, "Le corps céleste transneptunienne," *Bulletin de la Société astronomique de France* 44:228, 1930.
[37] U. J. J. Leverrier, "Théorie et Tables du Mouvement de Mercure," *Annales de L'Observatoire Impériale de Paris,* vol. 5(1859), table of contents.
[38] "Suspected Existence of a Zone of Asteroids Revolving Between Mercury and the Sun," *Monthly Notices of the Royal Astronomical Society* 20:24–26 (1859).
[39] U. J. J. Leverrier, *op. cit.*, p. 102.
[40] *Ibid.*, pp. 396–97.
[41] *Ibid.*
[42] "A Supposed Interior Planet," *Mon. Not. R.A.S.* 20:98–100 (1860).
[43] U. J. J. Leverrier, *op. cit.*, p. 398.
[44] Radan, M. R., "Future observation of the supposed new planet," *Mon. Not. RA.S.* 20:195–97 (1860).
[45] "Lescarbault's Planet," *Mon. Not. R.A.S.,* 20:344 (1860).
[46] E. Liais, letter to the editor, *Astr. Nach.,* no. 1246 (1860); and "Miscellaneous Intelligence," *Mon. Not. R.A.S.* 20:265 (1860).
[47] J. C. Adams, presidential address, *Mon. Not. R.A.S.* 35:232 (1876).
[48] G. B. Airy, address, *Mon. Not. R.A.S.* 37:246–47 (1877).
[49] *Ibid.*
[50] A. M. Clerke, *A Popular History of Astronomy During the Nineteenth Century* (4th ed.; London: Adam and Charles Black, 1902), p. 249.
[51] *Ibid.*, pp. 249–50.
[52] C. G. Abbot, *The Sun and the Welfare of Man* (Washington, D.C.: Smithsonian Institution, 1929), p. 267–69.

[53] W. W. Campbell, "The Closing of a Famous Astronomical Problem," *Publications of the Astronomical Society of the Pacific*, 21:103–115 (1908).
[54] Abbot, *op. cit.*, p. 268–69. C. W. Tombaugh, discoverer of Pluto, called my attention to H. C. Courten, D. W. Brown and D. B. Albert, "Ten Years of Solar Eclipse Comet Searches," a summary paper read at the annual meeting of the American Astronomical Society's Division for Planetary Science in Honolulu, Hawaii, Jan. 17–22, 1977, which reported that photographic observations of six solar eclipses "yielded data which indicate the possible existence of one or more relatively faint objects within twenty solar radii.... The object images indicate some alignment with the ecliptic and range from +9 to +7 equivalent visual magnitude."
[55] M. Grosser, "The Search for the Planet Beyond Neptune," in *Science in America Since 1820*, N. Reingold, ed. (New York: Science History Publications, 1976), p. 304.
[56] *Ibid.*, p. 305.
[57] D. P. Todd, "Preliminary Account of a Speculative and Practical Search for a Trans-Neptunian Planet," *American Journal of Science* 20:225–34 (1880).
[58] *Ibid.*
[59] G. Forbes, "On Comets," *Proceedings of the Royal Society of Edinburgh* 10:427 (1880).
[60] C. Flammarion, *Astronomie populaire* (Paris: 1879), p. 661. For his later statements see, for example, *Popular Astronomy*, trans. J. E. Gore (New York: D. Appleton, 1907), pp. 471–72.
[61] G. Forbes, "On an ultra-Neptunian planet," *Proc. Royal Soc. Edinburgh* 10:636–37 (1880).
[62] G. Forbes, "Additional Note on an ultra-Neptunian planet," *Proc. Royal Soc. Edinburgh*, 11:91-92 (1882).
[63] I. Roberts, "Photographic Search for a Planet Beyond Neptune," *Mon. Not. R.A.S.* 52:501–2 (1892).
[64] H.-E. Lau, "Planète inconnues," *Bull. Soc. Astron. France* 14:340–41 (1900).
[65] G. Dallet, "Contribution à la recherche des planètes située au dela l'orbite de Neptune," *Bull. Soc. Astron. France*, 15:266–71 (1901).
[66] T. Grigull, "Nouvelle contribution à la recherche d'une planète transneptunienne," *Bull. Soc. Astron. France*, 16:31–32, and 16:447–48 (1902).
[67] R. M. Ligondès, "Sur les planètes telescopiques," *Bull. Soc. Astron. France*, 15:358–61, 1901; and "Au sujet des planètes transneptuniennes," *Bull. Soc. Astron. France, 17:121*–22 (1903).
[68] T. J. J. See, "On the cause of the remarkable circularity of the orbits of the planets and satellites and on the origin of the planetary system," *Astr. Nach.*, no. 4308 (1909). *See also* See, *Researches on the Evolution of Stellar Systems* (Lynn, Mass: Thomas P. Nichols, 1910), pp. 375–76.
[69] A. Garnowsky, "Sur l'existence de quatres planètes transneptunienne," *Bull. Soc. Astron. France* 16:484 (1902).
[70] J. B. A. Gaillot, "Tables nouvelle des mouvements d'Uranus et de Neptune," *Analles de L'Observatoire de Paris*, vol. 28 (1909).
[71] W. H. Pickering, "A Search for a Planet Beyond Neptune," *Annals of the Astronomical Observatory of Harvard College* 61:113–162 (1909).
[72] J. B. A. Gaillot, "Contribution à la recherche des planètes ultra-neptuniennes," *Comptes Rendus* 148:754–58 (1909).
[73] The biographical material that follows here is taken primarily from A. Lawrence Lowell, *Biography of Percival Lowell* (New York: The Macmillan Company, 1935), and from W. G. Hoyt, *Lowell and Mars* (Tucson: University of Arizona Press, 1976).
[74] P. Lowell, *The Solar System* (Boston: Houghton, Mifflin and Company, 1903), p. 118.
[75] W. G. Hoyt, *op. cit.*, pp. 17–19, 97ff.
[76] *Ibid.*, pp. 27–38.
[77] *Ibid.*, pp. 105, 123–25.
[78] *Ibid.*, pp. 300–301.
[79] *Ibid.*, pp. 256–61, 262ff.
[80] *Ibid.*, pp. 127–50.
[81] *Ibid.*, 173–99.

The First Search for Planet X

[1] P. Lowell, *The Solar System* (Boston: Houghton, Mifflin and Company, 1903), p. 17.
[2] *Ibid.*, p. 114.
[3] P. Lowell, "Preliminary Investigation Into the Position of a Supposed Trans-Neptunian Planet," typewritten text with handwritten corrections and signature, dated Jan. 12, 1909, *Lowell Observatory Archives*.
[4] P. Lowell, "Other Worlds Than Ours," typewritten lecture text, *ca*. 1895, *Lowell Observ. Arch.*
[5] For example, Adams and Leverrier did not carry their calculations into the squares of the eccentricities of Uranus and the presumed trans-Neptunian planet. See P. Lowell, "Memoir on a Trans-Neptunian Planet," *Memoirs of the Lowell Observatory*, vol. 1, no. 1 (1915), p. 8.
[6] P. Lowell to C. O. Lampland, Feb. 9, 1906.
[7] C. O. Lampland, "Lowell Photographic Observations of Pluto in 1915, 1929 and 1930," text of paper read at December 1930 meeting of the American Astronomical Society in New Haven, Conn., *Lowell Observ. Arch.*
[8] A. L. Lowell, *Biography of Percival Lowell* (New York: The Macmillan Company, 1935), p. 192.
[9] G. R. Agassiz to V. M. Slipher, undated but *ca*. March 1930.
[10] For a detailed summary of Lowell's martian theories and observations, and the reaction to them in and out of astronomy, see W. G. Hoyt, *Lowell and Mars* (Tucson: University of Arizona Press, 1976).
[11] *Ibid.*, chaps. 9 and 11.
[12] P. Lowell to C. O. Lampland, Oct. 22, 1904.
[13] E. C. Pickering to P. Lowell, Feb. 8, 1905, and P. Lowell to C. Flammarion, Feb. 14, 1905.
[14] The Lawrence Fellowship was revived briefly in the 1930s to provide observers to continue the search for other possible trans-Saturnian planets after the discovery of Pluto.
[15] V. M. Slipher to J. A. Miller, April 3, 1905.
[16] V. M. Slipher to P. Lowell, Aug. 22, 1905. *See also* "Catalogue, Invariable Plane Plates, Lowell Observatory," and "Catalogue of Astronomical Negatives 1905–1906–1907–1908," in *Lowell Observ. Arch.*
[17] P. Lowell to V. M. Slipher, Oct. 17, 1905.
[18] P. Lowell to V. M. Slipher, Oct. 18, 1905.
[19] P. Lowell to V. M. Slipher, Nov. 2, 1905.
[20] P. Lowell to V. M. Slipher, Nov. 9, 1905.
[21] "Catalogue, Invariable Plane Plates...," and "Catalogue of Astronomical Negatives," *op. cit.*
[22] V. M. Slipher to P. Lowell, Dec. 8, 1905.
[23] R. G. Aitken, "Note on the Comets Discovered at the Lowell Observatory," *Publications of the Astronomical Society of the Pacific* 58:83–84 (1906).
[24] V. M. Slipher to R. G. Aitken, March 22, 1906.
[25] W. L. Leonard to Don Louie [sic] G. Leon, Jan. 13, 1906.
[26] P. Lowell to V. M. Slipher, telegram, Dec. 11, 1905.
[27] P. Lowell to J. C. Duncan, Dec. 14, 1905.
[28] P. Lowell to V. M. Slipher, Jan. 3, 1906.
[29] "Catalogue of Astronomical Negatives," *op. cit.*
[30] P. Lowell to V. M. Slipher, Jan. 31, 1906.
[31] P. Lowell to J. C. Duncan, March 16, 1906.
[32] V. M. Slipher to E. E. Barnard, April 16, 1906.
[33] V. M. Slipher to P. Lowell, April 16, 1906.
[34] P. Lowell to C. O. Lampland, April 24, 1906.
[35] P. Lowell to J. A. Brashear, April 24, 1906.
[36] P. Lowell to J. C. Duncan, May 9, 1906.
[37] P. Lowell to Max Wolf, June 26, 1906.

[38] V. M. Slipher to P. Lowell, March 28, 1906, and P. Lowell to V. M. Slipher and J. A. Miller, June 26, 1906.
[39] P. Lowell to V. M. Slipher, telegram, July 12, 1906.
[40] V. M. Slipher to P. Lowell, July 12 and 17, 1906.
[41] "Catalogue of Astronomical Negatives," op. cit.
[42] P. Lowell to V. M. Slipher, Sept. 11, 1906.
[43] P. Lowell to V. M. Slipher, Jan. 1, 1907.
[44] P. Lowell to V. M. Slipher, Jan. 3, 1907.
[45] V. M. Slipher to P. Lowell, Jan. 2, 1907.
[46] V. M. Slipher to P. Lowell, Jan. 8, 1907.
[47] P. Lowell to V. M. Slipher, Jan. 1, 1907.
[48] P. Lowell to V. M. Slipher, Jan. 12, 1907.
[49] V. M. Slipher to P. Lowell, Jan. 16, 1907.
[50] P. Lowell to V. M. Slipher, Jan. 24, 1907.
[51] P. Lowell to C. O. Lampland, Feb. 2, 1907.
[52] "Agreement" with David P. Todd, dated Feb. 28, 1907; Lowell Observ. Arch.
[53] P. Lowell to W. A. Cogshall, March 29, 1907. Actually the expedition was finally based at Alianza, Chile.
[54] "Catalogue of Astronomical Negatives," op. cit.
[55] P. Lowell to D. P. Todd, April 11, 1907.
[56] P. Lowell to Max Wolf, Jan. 3, 1907, and to A. Berberich, Feb. 7 and 20, 1907.
[57] "Catalogue of Astronomical Negatives," op. cit.
[58] P. Lowell to V. M. Slipher, Jan. 12, 1907.
[59] C. W. Tombaugh, "Reminiscences of the Discovery of Pluto," Sky and Telescope 19:264–70 (1960).
[60] P. Lowell to C. O. Lampland, March 16, 1906.
[61] P. Lowell to V. M. Slipher, Feb. 6, 1907.
[62] W. S. Harshman to P. Lowell, March 21, 1905.
[63] W. T. Carrigan to P. Lowell, March 22, 1905.
[64] A. L. Lowell, op. cit., p. 176.
[65] W. T. Carrigan, "A Method of Determining the Direction of the Sun's Motion in Space," Astronomical Journal 24:107–9 (1904); and (with E. D. Tillyer), "Elements of [Asteroid] (1903NF), Astronomical Journal 24:59 (1904).
[66] W. T. Carrigan to P. Lowell, April 11, 1905. Lalande recorded Neptune as a star on May 8 and 10, 1795.
[67] P. Lowell to W. T. Carrigan, March 7, 1906.
[68] P. Lowell to W. T. Carrigan, March 22, 1906.
[69] P. Lowell to W. T. Carrigan, Oct. 30, 1906.
[70] P. Lowell to W. T. Carrigan, Nov. 9, 1906.
[71] P. Lowell to W. T. Carrigan, Jan. 23, 1907.
[72] P. Lowell to W. T. Carrigan, Feb. 1, 1907.
[73] P. Lowell to W. T. Carrigan, Feb. 14, 1907.
[74] P. Lowell to W. T. Carrigan, March 7 and 13, 1908.
[75] P. Lowell to W. T. Carrigan, May 18, 1908.
[76] P. Lowell to C. O. Lampland, April 9, 1908.
[77] W. G. Hoyt, "William Henry Pickering's Planetary Predictions and the Discovery of Pluto," Isis 67:551–64 (1976).
[78] W. H. Pickering, "A Search for a Planet Beyond Neptune," Annals of the Astronomical Observatory of Harvard College 61:113 (1909).
[79] Ibid.
[80] W. H. Pickering, "The Next Planet Beyond Neptune," Popular Astronomy 36:143 (1928).
[81] W. H. Pickering, "A Photographic Search for Planet O," Annals Harvard Coll. Observ. 61:369 (1911).
[82] P. Lowell to W. H. Pickering, Nov. 16, 1908.
[83] P. Lowell to W. T. Carrigan, Nov. 19, 1908.
[84] P. Lowell to W. T. Carrigan, Nov. 13, 1908.
[85] P. Lowell to W. T. Carrigan, Nov. 17, 1908.
[86] P. Lowell to W. T. Carrigan, Nov. 18, 1908.

[87] P. Lowell to W. T. Carrigan, Nov. 19, 1908.
[88] *Ibid.*
[89] W. H. Pickering, "A Search for a Planet Beyond Neptune," *op. cit.*
[90] P. Lowell to W. T. Carrigan, Nov. 21, 24 and 28, 1908.
[91] P. Lowell to W. T. Carrigan, Dec. 1, 1908.
[92] P. Lowell to W. T. Carrigan, Dec. 11, 1908.
[93] G. Forbes, "The Comet of 1556; its possible breaking up by an unknown planet into three parts seen in 1843, 1880 and 1882," *Monthly Notices of the Royal Astronomical Society* 69:52 (1908).
[94] P. Lowell to W. T. Carrigan, Dec. 28, 1908.
[95] P. Lowell to W. T. Carrigan, Jan. 6, 1909.
[96] P. Lowell to W. T. Carrigan, Jan. 19, 1909.
[97] P. Lowell, "Preliminary Investigation...," *op. cit.*
[98] P. Lowell to E. C. Pickering, Feb. 9, 1909.
[99] P. Lowell to W. T. Carrigan, April 1, 1909.
[100] P. Lowell to W. T. Carrigan, May 3, 1909.
[101] W. T. Carrigan to P. Lowell, May 7, 1909.
[102] P. Lowell to W. T. Carrigan, May 11, 1909.
[103] P. Lowell, untitled handwritten drafts and 16-page typewritten text with handwritten corrections and signature, *ca.* June 1909, *Lowell Observ. Arch.*
[104] This book was based on a series of lectures Lowell delivered at the Massachusetts Institute of Technology in February and March 1909.
[105] The reflector was still not performing adequately a year later. See P. Lowell to Alvan Clark and Sons, Dec. 2, 1909, and to V. M. Slipher, July 18, 1910.
[106] P. Lowell, "The Canali Novae of Mars," *Lowell Observatory Bulletin No. 45* (1910).
[107] P. Lowell, "The Wisps of Saturn" *Lowell Observ. Bull. No. 44* (1910), and "Saturn's Rings," *Astronomische Nachrichten,* No. 4403 (1910). See also W. G. Hoyt, "Saturn, Seeing, and Percival Lowell," *Sky and Telescope* 50:28 (1975).
[108] W. G. Hoyt, *Lowell and Mars, op. cit.*, chap. 13.
[109] P. Lowell, "Motion of the Molecules in the Tail of Halley's Comet," *Lowell Observ. Bull. No. 48* (1910).
[110] P. Lowell, "The Hood of a Comet's Head," *Astron. J.*, 26:131–34 (1910); and "Velocities of Particles in the Tail of Halley's Comet," *Astron. J.*, 26:141–43 (1910).
[111] P. Lowell to A. C. D. Crommelin, Jan. 15, 1909.
[112] P. Lowell to A. C. D. Crommelin, July 28, 1909.
[113] P. Lowell to E. L. Williams, July 12, Aug. 2, 9, 15, and 28, 1910.
[114] W. H. Pickering, "A Search for a Planet Beyond Neptune," *op. cit.*; and D. P. Todd, "Preliminary Account of a Speculative and Practical Search for a Trans-Neptunian Planet," *American Journal of Science* 20:225–34 (1880). Lowell's annotated copy of Pickering's paper is in the Lowell Observatory Library.
[115] W. H. Pickering, "A Search for a Planet Beyond Neptune," *op. cit.*, and specifically the copy in the Lowell Library.
[116] *Ibid.*

The Second Search for Planet X

[1] C. O. Lampland to A. L. Lowell, Aug. 19, 1935, *Lowell Observatory Archives.*
[2] P. Lowell to C. O. Lampland, Sept. 14, 1914.
[3] P. Lowell, "Memoir on a Trans-Neptunian Planet," *Memoirs of the Lowell Observatory,* vol. 1, no. 1 (1915).
[4] W. L. Leonard to P. Lowell, April 21, 1915. Miss Leonard here advises Lowell that "Miss Williams wants *very* much to be left out of the X memoir—she is serious about it and has asked me to mention it to you.... She begs that she may be allowed to write *dele* after her name."
[5] P. Lowell, "Memoir on a Trans-Neptunian Planet," *op. cit.*, p. 3.
[6] *Ibid.*
[7] *Ibid.*, p. 7.
[8] *Ibid.*
[9] *Ibid.*, pp. 6, 8.

[10] *Ibid.* p. 4.
[11] *Ibid.*, pp. 7-8.
[12] *Ibid.*, pp. 102-4.
[13] *Ibid.*, pp. 5, 13, 23.
[14] *Ibid.*, pp. 6, 53.
[15] *Ibid.*, p. 7.
[16] *Ibid.*, pp. 24, 105.
[17] *Ibid.*, p. 9.
[18] *Ibid.*, pp. 8-9.
[19] *Ibid.*, p. 73.
[20] *Ibid.*, p. 98.
[21] E. L. Williams to P. Lowell, July 28, 1906, and P. Lowell to E. C. Marsh, June 29, 1908. *See also* general correspondence between Lowell and Miss Williams in the archives in 1905, 1907, and 1909.
[22] P. Lowell to E. L. Williams, Aug. 2, 1910.
[23] P. Lowell to E. L. Williams, July 12, Aug. 2, 9 and 15, 1910.
[24] P. Lowell to E. L. Williams, Aug. 28, 1910.
[25] P. Lowell to C. O. Lampland, Dec. 10, 1910.
[26] P. Lowell, "The Tores of Saturn," *Lowell Observatory Bulletin No. 32*, 1907; and "Memoir on Saturn's Rings," *Mem. of Lowell Observ.*, vol. 1, no. 2 (1915).
[27] P. Lowell, untitled handwritten drafts and 16-page typewritten text with handwritten corrections and signature, *ca.* June 1909, *Lowell Observ. Arch.*
[28] P. Lowell, "The Origin of the Planets," *Memoirs of the American Academy of Arts and Sciences*, vol. 14, no. 1 (1913).
[29] *Ibid.*
[30] P. Lowell to C. O. Lampland, telegram, March 13, 1911.
[31] C. O. Lampland to P. Lowell, March 15, 1911.
[32] W. L. Leonard to C. O. Lampland, March 22, 1911.
[33] C. O. Lampland to A. L. Lowell, Aug. 8, 1935.
[34] W. L. Leonard to C. O. Lampland, March 22, 1911.
[35] W. L. Leonard to V. M. Slipher, April 3, 1911.
[36] P. Lowell to C. O. Lampland, April 27, 1911.
[37] P. Lowell to E. L. Williams, June 12, 1911.
[38] P. Lowell to V. M. Slipher, telegram, July 3, 1911.
[39] P. Lowell to C. O. Lampland, telegram, July 8, 1911.
[40] P. Lowell to V. M. Slipher, telegram, July 13, 1911.
[41] P. Lowell to C. O. Lampland, telegram, July 27, 1911.
[42] E. C. Slipher, *The Photographic Story of Mars* (Cambridge: Sky Publishing Co., 1962), p. 61.
[43] P. Lowell, "Spectrographic Discovery of the Rotation Period of Uranus," *Lowell Observ. Bull. No. 53*, 1912. Slipher's value was subsequently confirmed both photometrically and spectrographically and stood for sixty-five years and until 1977 when observations with more spohisticated spectrographic equipment indicated a rotation period of 25 ± 4 hours.
[44] P. Lowell, "Precession of the Martian Equinoxes," handwritten text, dated Nov. 5, 1911, *Lowell Observ. Arch.*; and "Precession: And the Pyramids," *Popular Science Monthly* 20:449-60 (1912).
[45] P. Lowell to E. W. Brown, July 10, 1911; *see also* P. Lowell, "Libration and the Asteroids," *Astronomical Journal* 27:41-46 (1911).
[46] W. H. Pickering, "A Statistical Investigation of Cometary Orbits," *Annals of the Astronomical Observatory of Harvard College* 61:167 (1911).
[47] P. Lowell to C. O. Lampland, telegram, June 5, 1912.
[48] P. Lowell to C. O. Lampland, telegram, June 6, 1912.
[49] P. Lowell to C. O. Lampland, June 8, 1912.
[50] P. Lowell to V. M. Slipher and to C. O. Lampland, telegrams, June 12, 1912.
[51] P. Lowell to C. O. Lampland, June 12, 1912.
[52] P. Lowell to C. O. Lampland, telegram, June 14, 1912.
[53] P. Lowell to C. O. Lampland, telegram, June 20, 1912.
[54] P. Lowell to C. O. Lampland, June 29, 1912.

[55] P. Lowell to V. M. Slipher, telegram, July 10, 1912.
[56] P. Lowell to C. O. Lampland, July 10, 1912.
[57] P. Lowell to C. O. Lampland, Aug. 9, 1912.
[58] P. Lowell to C. O. Lampland, Aug. 24, 1912.
[59] W. L. Leonard to C. O. Lampland, Sept. 4, 1912.
[60] P. Lowell to C. O. Lampland, telegram, Sept. 12, 1912.
[61] P. Lowell to C. O. Lampland, Sept. 21, 1912.
[62] W. L. Leonard to V. M. Slipher, Nov. 8, 1912.
[63] W. L. Leonard to V. M. Slipher, April 22, 1914; P. Lowell to C. O. Lampland, Sept. 4, 1914; and P. Lowell to W. L. Leonard, telegram, Sept. 28, 1914.
[64] P. Lowell to Ruth B. Mark, Aug. 25, 1914; and W. L. Leonard to R. B. Jackson, April 23, 1915.
[65] W. L. Leonard to E. C. Slipher, Oct. 30, 1912.
[66] C. S. Lowell to V. M. Slipher, Nov. 8, 1912.
[67] W. L. Leonard to C. O. Lampland, Dec. 11, 1912. For details of Lowell's earlier illness, see W. G. Hoyt, *Lowell and Mars* (Tucson: University of Arizona Press, 1976), chap. 8.
[68] W. L. Leonard to V. M. Slipher, Dec. 12, 1912.
[69] P. Lowell to C. O. Lampland, telegram, Jan. 22, 1913.
[70] P. Lowell to C. O. Lampland, telegram, Jan. 22, 1913.
[71] P. Lowell to C. O. Lampland, telegram, Jan. 29, 1913.
[72] P. Lowell to C. O. Lampland, telegram, Feb. 3, 1913.
[73] P. Lowell to C. O. Lampland, telegram, Feb. 5, 1913.
[74] P. Lowell to C. O. Lampland, Feb. 8, 1913.
[75] P. Lowell to C. O. Lampland, Feb. 21, 1913.
[76] P. Lowell to E. M. Doe, telegram, Feb. 25, 1913.
[77] V. M. Slipher, "On the Spectrum of the Nebula in the Pleiades," *Lowell Observ. Bull. No. 55,* 1912; and V. M. Slipher to P. Lowell, Dec. 16 and 17, 1912, to J. C. Duncan, Dec. 29, 1912, and to W. W. Campbell, March 14, 1913. *See also* P. Lowell, "Epitome of Results at the Lowell Observatory—April, 1913–April, 1914," *Lowell Observ. Bull. No. 59,* 1914.
[78] V. M. Slipher to P. Lowell, Jan. 2, and Feb. 3, 1913; *see also* V. M. Slipher, "The Radial Velocity of the Andromeda Nebula," *Lowell Observ. Bull. No. 58,* undated.
[79] P. Lowell to V. M. Slipher, Feb. 8, 1913.
[80] V. M. Slipher to P. Lowell, April 12, May 14 and 16, 1913.
[81] W. W. Campbell to V. M. Slipher, Dec. 16, 1916, and V. M. Slipher to W. W. Campbell, undated reply. *See also* V. M. Slipher, "Structure and Motions of Spiral Nebulae," *Proceedings of the American Philosophical Society* 56:403–9 (1917). In recent years, of course, velocities representing substantial fractions of the speed of light have been recorded.
[82] W. L. Leonard to V. Emanuel, Oct. 8, 1913, quoting Lowell; and P. Lowell, "Nebular Motion," text of lecture to the Melrose Club, Boston, Nov. 23, 1915, *Lowell Observ. Arch.*
[83] E. L. Williams to P. Lowell, two telegrams, March 8, 1913.
[84] Lowell presented the memoir to the Academy April 23, 1913.
[85] P. Lowell to A. Gaillot, May 22, 1913.
[86] A. Gaillot to P. Lowell, June 20, 1913.
[87] A. Gaillot to P. Lowell, Aug. 7, 1913.
[88] P. Lowell to E. L. Williams, telegram, July 4, 1913.
[89] P. Lowell to C. O. Lampland, July 10, 1913.
[90] P. Lowell to C. O. Lampland, telegram, Aug. 21, 1913.
[91] P. Lowell to V. M. Slipher, telegram, Aug. 27, 1913.
[92] C. O. Lampland to P. Lowell, Sept. 25, 1912.
[93] P. Lowell to J. A. Miller, Nov. 24, 1913.
[94] E. L. Williams to P. Lowell, telegrams, Jan. 3 and 5, 1914.
[95] E. L. Williams to P. Lowell, telegram, undated but *ca.* mid-January 1914.
[96] P. Lowell to E. L. Williams, Jan. 29, 1914.
[97] E. L. Williams to P. Lowell, Feb. 19, and March 5, 1914.
[98] E. L. Williams to P. Lowell, Jan. 29, and Feb. 7, 1914.
[99] W. L. Leonard to V. M. Slipher, April 22, 1914.

[100] P. Lowell to C. O. Lampland, telegram, May 5, 1914.
[101] W. L. Leonard to C. O. Lampland, July 21, 1914; and P. Lowell to Mrs. C. O. Lampland, Aug. 8, 1914.
[102] P. Lowell to V. M. Slipher, Aug. 11, 1914.
[103] E. L. Williams to P. Lowell, July 10, 1914.
[104] P. Lowell to C. O. Lampland, Sept. 15, 1914.
[105] C. O. Lampland, untitled typewritten text with handwritten corrections, undated, but *ca.* 1930, *Lowell Observ. Arch.*
[106] P. Lowell to J. Trowbridge, Dec. 9, 1914.
[107] P. Lowell to C. O. Lampland, Dec. 21, 1914.

A Trans-Neptunian Planet

[1] P. Lowell to G. E. Pierce, Jan. 15, 1915.
[2] P. Lowell, "Memoir on a Trans-Neptunian Planet," *Memoirs of the Lowell Observatory,* vol. 1, no. 1 (1915).
[3] P. Lowell to J. Trowbridge, Dec. 9, 1914.
[4] In their correspondence over the years, Lowell addressed Storey as "Quadric" and Storey addressed Lowell as "Mars."
[5] P. Lowell to J. Trowbridge, Jan. 11, 1915.
[6] P. Lowell to E. B. Wilson, Feb. 4, 1915.
[7] P. Lowell to C. O. Lampland, undated telegram. In this, Lowell informs Lampland of his election to the Academy "last evening." There are a number of letters in the Lowell archives after 1910 relating to Lowell's efforts to secure Lampland's election.
[8] P. Lowell to E. L. Williams, Feb. 21, 1915.
[9] "Saturn," in "Planet Notes," *Popular Astronomy* 15:631 (1907); and "Saturn," in "Notes," *The Observatory* 30:427–28 (1907).
[10] "Saturn's Rings Are Falling In," *New York Times,* Nov. 9, 1907.
[11] "Astronomers Split on Saturn's Rings," *New York Times,* Nov. 11, 1907.
[12] *The New York Times,* Nov. 21, 1907.
[13] P. Lowell to C. O. Lampland, Nov. 20, 1907.
[14] P. Lowell, "The Tores of Saturn," *Lowell Observatory Bulletin No. 32,* 1907; "Saturn's Tores," *Scientific American,* December 1907; "The Tores of Saturn," *Philosophical Magazine* 15:468–77 (1908); and "The Tores of Saturn," *Pop. Astron.* 16:134–46 (1908).
[15] P. Lowell, "The Wisps of Saturn," *Lowell Observ. Bull. No. 44,* 1910; and "Saturn's Rings," *Astronomische Nachrichten,* no. 4403 (1910).
[16] P. Lowell, "Measures of Saturn—Ball, Rings and Satellites," *Lowell Observ. Bull. No. 66,* 1915.
[17] P. Lowell, "Saturn's Rings," *Lowell Observ. Bull. No. 68,* 1915.
[18] P. Lowell, "Memoir on Saturn's Rings," *Mem. Lowell Observ.,* vol. 1, no. 2 (1915).
[19] P. Lowell, "The Genesis of Planets," typewritten and handwritten drafts of text, dated April 27, 1916, *Lowell Observ. Arch.*
[20] P. Lowell to V. M. Slipher, Aug. 4, 1915.
[21] A. L. Lowell, *Biography of Percival Lowell* (New York: The Macmillan Company, 1935), pp. 193–94.
[22] E. C. Slipher, *The Photographic Story of Mars* (Cambridge: Sky Publishing Co., 1962), p. 61.
[23] "Mars at the Opposition of 1915–16," *Lowell Observatory Observation Circulars,* March 8, 9, 14, and 30, 1916; and "Aurora," *Lowell Observ. Obs. Circ.,* March 15, 1916.
[24] C. O. Lampland, "Lowell Photographic Observations of Pluto in 1915, 1929 and 1930," text of paper read at December 1930 meeting of the American Astronomical Society in New Haven, Conn.
[25] *Lowell Observatory Journal of Observations—Nine-Inch Swarthmore Photographic Telescope,* vol. 2, *Lowell Observ. Arch.*
[26] C. O. Lampland, *op. cit.*
[27] V. M. Slipher to J. Combs, July 7, 1930.
[28] A. C. D. Crommelin, quoted in the *Journal of the British Astronomical Association* 41:265 (1931).

[29] C. O. Lampland to J. A. Miller, June 28, 1915.
[30] P. Lowell to C. O. Lampland, telegram, July 21, 1915.
[31] C. O. Lampland to P. Lowell, Aug. 15, 1915.
[32] C. O. Lampland to J. A. Miller, Sept. 15, 1915.
[33] *Lowell Observatory Journal of Observations*, vol. 3, *Lowell Observ. Arch.*
[34] C. O. Lampland, untitled, undated typewritten text with handwritten corrections, *Lowell Observ. Arch.* From internal evidence, this was clearly written during the controversy following the discovery of Pluto in 1930.
[35] *Lowell Observatory Journal of Observations*, 4 vols, *Lowell Observ. Arch.*
[36] R. L. Putnam and V. M. Slipher, "Searching out Pluto," *Scientific Monthly*, 34:5 (1932).
[37] P. Lowell to V. M. Slipher, telegram, June 30, 1916.
[38] P. Lowell to V. M. Slipher, June 2, 1916.
[39] P. Lowell to C. O. Lampland, May 23, and July 16, 1916.
[40] P. Lowell to V. M. Slipher, telegram, July 28, 1916.
[41] W. G. Hoyt, *Lowell and Mars* (Tucson: University of Arizona Press, 1976), chap. 15.
[42] *Ibid.*; see also A. L. Lowell, *op. cit.*, p. 193–94.
[43] P. Lowell, "Memoir on a Trans-Neptunian Planet," *op. cit.*
[44] *Ibid.*
[45] *Ibid.*, p. 6.
[46] P. Lowell, *The Evolution of Worlds* (New York: The Macmillan Company, 1909), pp. 125–26.
[47] P. Lowell, "Memoir on a Trans-Neptunian Planet," *op. cit.*.
[48] *Ibid.*, p. 20.
[49] *Ibid.*, p. 69.
[50] *Ibid.*, pp. 69–70.
[51] *Ibid.*, p. 32.
[52] *Ibid.*
[53] *Ibid.*, p. 33.
[54] *Ibid.*, p. 33–34.
[55] *Ibid.*, pp. 34–35.
[56] *Ibid.*, pp. 98–99.
[57] *Ibid.*, pp. 99–100.
[58] *Ibid.*, pp. 61, 97.
[59] *Ibid.*, pp. 98–99.
[60] *Ibid.*, pp. 99–100.
[61] *Ibid.*, p. 70.
[62] *Ibid.*, pp. 100–101.
[63] *Ibid.*, p. 101.
[64] *Ibid.*
[65] *Ibid.*
[66] *Ibid.*
[67] *Ibid.*, pp. 101–3.
[68] *Ibid.*, p. 103.
[69] *Ibid.*
[70] *Ibid.*, pp. 103–4.
[71] *Ibid.*, pp. 104–5.

Lowell's Legacies

[1] C. O. Lampland to J. A. Miller, Nov. 1, 1917.
[2] C. O. Lampland to E. Strömgren, Oct. 8, 1918.
[3] C. O. Lampland to T. Cooke & Sons, Ltd., and to Chance Brothers, June 16, 1919.
[4] G. Lowell to J. Metcalf, July 1, 1919.
[5] V. M. Slipher to R. L. Putnam, March 9, 1927.
[6] P. Lowell, last will and testament, dated Feb. 21, 1913, on file in Coconino County (Arizona) Superior Court, with codicils dated March 1, 1913; May 11, 1914; and Nov. 22, 1914.
[7] F. Greenslet, *The Lowells and Their Seven Worlds* (Boston: Houghton, Mifflin and Company, 1945), pp. 368–73.

[8] R. L. Putnam to V. M. Slipher, Feb. 26, 1927.
[9] R. L. Putnam to V. M. Slipher, March 15, 1927.
[10] R. L. Putnam to S. L. Boothroyd, Aug. 31, 1927.
[11] J. A. Miller to V. M. Slipher, March 19, 1927.
[12] C. O. Lampland, "Observed Changes in the Structure of the 'Crab' Nebula (N.G.C. 1952), *Publications of the Astronomical Society of the Pacific* 33:79-84 (1921); and "Crab in Taurus," *Popular Astronomy* 29:247 (1921).
[13] V. M. Slipher to S. I. Bailey, March 19, 1921.
[14] J. C. Duncan, "Changes Observed in the Crab Nebula in Taurus," *Proceedings of the National Academy of Sciences* 7:179-80 (1921); see also N. U. Mayall and J. H. Oort, "Further Data Bearing on the Identification of the Crab Nebula with the Supernova of 1054 A.D.," *Publ. Astr. Soc. Pac.* 54:95 (1942).
[15] V. M. Slipher to E. A. Fath, March 19, 1921.
[16] Obituary of Earl C. Slipher, in "General Notes," *Publ. Astr. Soc. Pac.* 76:367 (1964).
[17] *Ibid.; see also* J. S. Hall, "V. M. Slipher's Trailblazing Career," *Sky and Telescope* 39:84 (1970); and J. C. Duncan, "Carl Otto Lampland," *Publ. Astr. Soc. Pac.* 64:293-96 (1952).
[18] Obituary of Earl C. Slipher, *op. cit.;* J. S. Hall, *op. cit.;* and J. C. Duncan, *op. cit.*
[19] V. M. Slipher to C. S. Lowell, June 4 and 22, 1922. The Hamiltons were married June 2, 1922, and eventually moved to Jamaica, B.W.I., where Hamilton was associated for some years with planet-hunter W. H. Pickering.
[20] G. H. Hamilton, "Mars During the Opposition of March 1918," *Lowell Observatory Bulletin No. 82* (1919); and "Martian Observations, 1920, Emphasizing Meteorological Conditions," *Lowell Observ. Bull. No. 83* (1920). But *see also* G. P. Serviss, "George Hall Hamilton's Latest Observations of Mars Prove That It Is Fitted to Support a Form of Life Equal to Ours," New York *Evening Journal*, Dec. 29, 1920.
[21] W. W. Coblentz to C. O. Lampland, Feb. 16, March 8 and 20, 1920; C. O. Lampland to W. W. Coblentz, Nov. 18, 1920 and April 8, 1922; W. W. Coblentz to V. M. Slipher, Feb. 15, June 2, July 23, and Dec. 6, 1921.
[22] V. M. Slipher to W. W. Coblentz, May 12, 1922; and Coblentz' telegraphic reply, May 26, 1922.
[23] V. M. Sliper to C. S. Lowell, June 19 and 21, 1922.
[24] Among the extensive correspondence involving Lampland, Coblentz and Menzel in the Lowell Observatory archives, *see* especially C. O. Lampland to D. H. Menzel, Dec. 12, 1925; to G. R. Agassiz, June 6, 1925; to V. M. Slipher, Aug. 27, 1925; and to W. W. Coblentz, Dec. 5, 1926. *See also* D. H. Menzel, "Water Cell Transmissions and Planetary Temperatures," *Astrophysical Journal* 58:65-74 (1923); and, with W. W. Coblentz and C. O. Lampland, "Planetary Temperatures Derived From Water Cell Transmissions," *Ap. J.* 63:177-87 (1926).
[25] V. M. Slipher to W. W. Coblentz, Nov. 25, 1926; to G. K. Burgess, Dec. 31, 1926; G. K. Burgess to V. M. Slipher, Dec. 14, 1926, and Jan. 19, 1926; and C. O. Lampland to W. W. Coblentz, Dec. 5, 1926. The latter letter by Lampland was in reply to one sent by Coblentz, Nov. 20, 1926, which does not appear to have been preserved in the Lowell Observatory Archives.
[26] W. W. Coblentz and C. O. Lampland, "Radiometric Measurements of Mars," *Publ. Astr. Soc. Pac.* 36:272-74 (1924).
[27] E. Pettit and S. B. Nicholson, "Measurements of the Radiation From the Planet Mars," *Pop. Astron.* 32:601-8 (1924).
[28] H. N. Russell to C. O. Lampland, March 3, 1926.
[29] R. J. Trumpler, "Visual and Photographic Observations of Mars," *Publ. Astr. Soc. Pac.* 36:263-69 (1924). *See also* V. M. Slipher to E. C. Payne, Aug. 28, 1926; and to J. H. Jeans, Aug. 21, 1930.
[30] V. M. Slipher, "Spectrum Observations of Mars," *Publ. Astr. Soc. Pac.* 36:261-62 (1924).
[31] V. M. Slipher to G. Lowell, Oct. 4, 1924.
[32] W. S. Adams and C. St. John, "An attempt to detect water vapor and oxygen lines in the spectrum of Mars with the registering microphotometer," *Publ. Astr. Soc. Pac.* 37:158-59 (1925).

[33] F. W. Very, "Measurement of the Intensification of the Aqueous Bands in Mars," *Lowell Observ. Bull. No. 36*, 1909; and "Quantitative Measurement of the Intensity of Great B in the Spectrum of Mars," *Lowell Observ. Bull. No. 41*, 1909. Astronomers, however, do not consider the Slipher-Very or Adams-St. Johns work definitive as to water vapor or oxygen on Mars. The discovery of water vapor on Mars usually is credited to H. Spinrad, G. Munch and L. Kaplan ["Spectrographic Determination of Water Vapor on Mars," *Ap. J.* 137:1319 (1963)], who used, incidentally, Lowell's "velocity shift" method.

[34] V. M. Slipher to H. Wetherald, March 9, 1926.

[35] V. M. Slipher to E. C. Payne, Aug. 28, 1926.

[36] V. M. Slipher to H. Hankin, Dec. 31, 1926; *see also* V. M. Slipher to H. McEwen, Dec. 28, 1926, for comments in a similar vein.

[37] E. C. Slipher, *The Photographic Story of Mars* (Cambridge: Sky Publishing Co., 1962), p. 70.

[38] V. M. Slipher to F. O. Grover, Jan. 23, 1923.

[39] V. M. Slipher, "The Spectrum of Venus," *Lowell Observ. Bull. No. 84*, 1921. The rotation of Venus, as determined by radar studies in the 1960s, has turned out to be very long indeed—243 earth days and in a retrograde direction.

[40] V. M. Slipher to G. Lowell, March 22, and May 5, 1921. Neptune's rotation is now believed to be about 15.5 hours, but with a considerable error factor, and is direct.

[41] V. M. Slipher to E. W. Scott, April 28, 1924, and to C. S. Lowell, May 20, 1924.

[42] V. M. Slipher to P. W. Glover, undated, but *ca.* late 1927.

[43] V. M. Slipher, "The Lowell Observatory Eclipse Expedition," *Pop. Astron.* 30:346–47 (1918); and "The spectrum of the Corona as observed by the Expedition From the Lowell Observatory at the Total Eclipse of June 8, 1918," *Ap. J.* 55:73–84 (1922). *See also* correspondence between V. M. Slipher, C. S. Lowell and J. A. Miller, March–June 1918; and between Slipher and D. Roberts and W. W. Campbell, Feb.–Aug. 1923, L.O.A.

[44] H. Shapley and H. D. Curtis, "The scale of the universe," *Bulletin of the National Research Council*, 2:217 (1921). For descriptions of the overall "great debate," *see* O. Struve and V. Zebergs, *Astronomy of the Twentieth Century* (New York: The Macmillan Company, 1962), pp. 416ff and 441ff; and R. Berendzen, R. Hart and D. Seeley, *Man Discovers the Galaxies* (New York: Science History Publications, 1976), *passim*.

[45] V. M. Slipher, "Spectrographic observations of nebulae," abstract, *Pop. Astron.* 23:21–24 (1915); *see also* text of this paper in the *Lowell Observ. Arch.* as read to the Seventeenth (Evanston, Ill.) meeting of the American Astronomical Society, Aug. 14, 1914. It is interesting to note that Edwin P. Hubble, then a young astronomer at Yerkes Observatory, attended this American Astronomical Society meeting at Evanston.

[46] V. M. Slipher, "The structure and motions of spiral nebulae," *Proceedings of the American Philosophical Society* 56:403–10 (1917).

[47] A. S. Eddington to V. M. Slipher, Nov. 11, 1921. Eddington here requests Slipher's full list of radial velocities, noting: "I do not trouble about measures which merely duplicate yours as I know the agreement is in general quite satisfactory."

[48] V. M. Slipher to J. A. Miller, March 19, 1918.

[49] V. M. Slipher, "Two Nebulae With Unparalleled Velocities," *Lowell Observ. Obs. Circ.*, Jan. 17, 1921. Of course, velocities representing significant fractions of the speed of light are now known.

[50] *See*, for examples, correspondence between V. M. Slipher and A. S. Eddington, Nov. 11, 1921, and Feb. 5 and March 8, 1922; K. Lundmark, March 11, 1924; G. Strömberg, Sept. 30, Oct. 7 and 11, Nov. 29, and Dec. 8, 1924; W. S. Adams, May 29 and 31, 1930; and E. P. Hubble, April 11, 1930, and July 22 and Aug. 24, 1932.

[51] E. P. Hubble to V. M. Slipher, March 6, 1953.

[52] E. P. Hubble, *The Realm of the Nebulae* (New Haven: Yale University Press, 1936), pp. 102–5, 115–16.

[53] V. M. Slipher to P. Lowell, cablegram, May 25, 1914; *see also* V. M. Slipher, "The Detection of Nebular Rotation," *Lowell Observ. Bull. No. 62*, 1914.

[54] V. M. Slipher, "The spectrum and velocity of the nebula N.G.C. 1068 (M77)," *Lowell Observ. Bull. No. 80*, 1917.

[55] An excellent discussion of van Maanen's work and its role in the "great debate" can be found in Berendzen et al., *Man Discovers the Galaxies, op. cit.*, pp. 108–31. See also N. S. Hetherington, "The Simultaneous 'Discovery' of Internal Motions in Spiral Nebulae," *Journal of the History of Astronomy* 6:115–25 (1975).
[56] J. C. Duncan to V. M. Slipher, July 14, 1916.
[57] V. M. Slipher to A. C. Gifford, Oct. 10, 1924.
[58] E. P. Hubble, "Angular rotations of spiral nebulae," *Ap. J.* 81:334–35 (1935); and A. van Maanen, "Internal motions in spiral nebulae," *Ap. J.* 81:336–37 (1935).
[59] E. P. Hubble, "The direction of rotation in spiral nebulae," *Ap. J.* 97:112–18 (1943); and V. M. Slipher, "The direction of rotation in spiral nebulae," *Science* 99:144–45 (1944).
[60] J. H. Moore to C. O. Lampland, June 27, 1921.
[61] C. O. Lampland to A. van Maanen, June 1, 1921.
[62] V. M. Slipher to J. C. Duncan, Feb. 7, 1913.
[63] P. Lowell, "Aurora," *Lowell Observ. Obs. Circ.*, March 15, 1916.
[64] V. M. Slipher, "Spectral Evidence of a Persistent Aurora," *Lowell Observ. Bull. No. 76*, 1916.
[65] V. M. Slipher, "On the General Auroral Illumination of the Sky and the Wave-Length of the Chief Auroral Line," *Ap. J.* 49:266–75 (1919).
[66] V. M. Slipher, "Emission of the Spectrum of the Night Sky," abstract, *Publ. Amer. Astr. Soc.* 6:241–42 (1931); "On the Interpretation of the Auroral Spectrum," *Pop. Astron.* 38:94 (1931); and "Preliminary note on the spectrum of the zodiacal light," *Lowell Observ. Obs. Circ.*, Feb. 20, 1931.
[67] W. H. Pickering, "The Trans-Neptunian Planet," *Annals of the Harvard College Observatory* 82:49–59 (1919); and "Perturbations of Neptune," *Harvard College Observatory Observation Circular No. 215*, May 15, 1919.
[68] O. H. Truman and G. H. Hamilton to Los Angeles *Times*, telegram, Dec. 30, 1919; and "New Planet May Be Added to Sun Staff," *Coconino Sun* (Flagstaff), Jan. 2, 1920.
[69] P. W. Gifford, letter to the editor, *Pop. Astron.* 28:251 (1920).
[70] S. I. Bailey, "Joel Hastings Metcalf," *Pop. Astron.* 33:493–95 (1925).
[71] R. L. Putnam and V. M. Slipher, "Searching Out Pluto—Lowell's Trans-Neptunian Planet," *Scientific Monthly* 34:5 (1932).

The Third Search for Planet X

[1] W. G. Hoyt, "W. H. Pickering's Planetary Predictions and the Discovery of Pluto," *Isis* 67:551–64 (1976).
[2] W. H. Pickering, "A Search for a Planet Beyond Neptune," *Annals of the Astronomical Observatory of Harvard College* 61:113 (1909); and "The Next Planet Beyond Neptune," *Popular Astronomy* 36:143 (1928).
[3] W. H. Pickering, "Planet U and the Orbits of Saturn and Jupiter," *Pop. Astron.* 40:69 (1932).
[4] W. H. Pickering, "The Three Planets Beyond Neptune," *Pop. Astron.* 36:417 (1928).
[5] W. H. Pickering, "Planet P, Its Orbit, Position and Magnitude. Planets S and T," *Pop. Astron.* 39:385 (1931).
[6] *Ibid.*
[7] S. T. Cargill to Lowell Observatory, Nov. 17, 1930; *Lowell Observatory Archives*.
[8] W. H. Pickering, "Planet U...," *op. cit.*
[9] W. H. Pickering, "Planet P...," *op. cit.*
[10] W. H. Pickering, "A Search for a Planet...," *op. cit.*
[11] W. H. Pickering, "Perturbations of Neptune," *Harvard College Observatory Circular 215*, May 15, 1919; and "The Trans-Neptunian Planet," *Annals Harvard Coll. Observ.* 82:51 (1919).
[12] W. H. Pickering, "The Next Planet Beyond...," *op. cit.*
[13] *Ibid.*
[14] V. M. Slipher to H. H. Turner, April 20, 1930.
[15] R. L. Putnam to V. M. Slipher, Feb. 26, 1927.
[16] R. L. Putnam to V. M. Slipher, March 9, 1927.
[17] R. L. Putnam to V. M. Slipher, March 15, 1927.

[18] R. L. Putnam to V. M. Slipher, April 26, 1927.
[19] R. L. Putnam to V. M. Slipher, April 29, 1927.
[20] R. L. Putnam to V. M. Slipher, June 21, 1927.
[21] R. L. Putnam to C. O. Lampland, June 27, 1927.
[22] R. L. Putnam to V. M. Slipher, June 30, 1927.
[23] V. M. Slipher to R. L. Putnam, July 14, 1927.
[24] V. M. Slipher to Philadelphia Gear Works, June 20 and 25, and July 6, 1927.
[25] V. M. Slipher to R. L. Putnam, July 14, 1927.
[26] R. L. Putnam to V. M. Slipher, Aug. 9, 1927.
[27] V. M. Slipher to R. L. Putnam, Aug. 16, 1927.
[28] R. L. Putnam to V. M. Slipher, Aug. 31, 1927.
[29] C. O. Lampland to C.A.R. Lundin, June 22, 1925.
[30] R. L. Putnam to Alvan Clark and Sons (attn: C.A.R. Lundin), Nov. 15, 1927.
[31] V. M. Slipher to R. L. Putnam, Dec. 3, 1927.
[32] V. M. Slipher to W. A. Cogshall, Dec. 3, 1927.
[33] C. A. R. Lundin to V. M. Slipher, Feb. 9, 1928.
[34] V. M. Slipher to R. L. Putnam, Feb. 18, 1928.
[35] V. M. Slipher to R. L. Putnam, March 23, 1928.
[36] R. L. Putnam to V. M. Slipher, March 28, 1928.
[37] V. M. Slipher to H. N. Russell, May 12, 1928; and to J. A. Miller, May 29, 1928.
[38] V. M. Slipher to R. L. Putnam, Aug. 20, 1928.
[39] V. M. Slipher to R. L. Putnam, Sept. 30, 1928.
[40] R. L. Putnam to V. M. Slipher, Jan. 22, 1929.
[41] V. M. Slipher to H. N. Russell, May 6, 1928.
[42] V. M. Slipher to H. N. Russell, May 12, 1928.
[43] H. N. Russell to R. L. Putnam, May 16, 1928.
[44] W. G. Hoyt, "Historical Note: Astronomy on the San Francisco Peaks," *Plateau* 47:113–17 (1975). Slipher does not record exactly where these observations were made, noting only that it was at "nearly 13,000 feet," which would put them on or near the summit of Humphreys Peak, their highest point.
[45] V. M. Slipher to G. Lowell, Oct. 2, 1926.
[46] G. Lowell to V. M. Slipher, telegram, Oct. 6, and letter, Oct. 20, 1926.
[47] W. G. Hoyt, "Historical Note...," *op. cit.*
[48] V. M. Slipher to G. Lowell, Oct. 29, and Dec. 13, 1926.
[49] C. O. Lampland to Supervisor of Radio, San Francisco, Calif., April 19, 1927; and to Bernard H. Linden, Feb. 2, 1928.
[50] W. G. Hoyt, "Historical Note...," *op. cit.*
[51] R. L. Putnam to V. M. Slipher, March 6, 1928.
[52] H. S. Colton to V. M. Slipher, May 17, 1928; and G. Gammage to V. M. Slipher, Aug. 1, 1927. The first letter announces a meeting to discuss the possibility of establishing a museum in Flagstaff, and to hear Henry Norris Russell talk on "Life on Other Planets."
[53] V. M. Slipher to C. O. Lampland, June 22, 1928; and C. O. Lampland to V. M. Slipher, July 8, 1928.
[54] V. M. Slipher to W. de Sitter, and to A. S. Eddington, June 23, 1928.
[55] V. M. Slipher to R. L. Putnam, March 9, 1928.
[56] Authors during the decade, some of them writing two or more books on astronomy, included J. C. Duncan, E. A. Fath, H. N. Russell, R. S. Dugan and J. Q. Stewart, H. H. Turner, Hector McPherson Jr., Mary Proctor, W. M. Smart, J. H. Jeans, A. S. Eddington, C. Payne, Harlow Shapley and G. E. Hale.
[57] *See* V. M. Slipher to R. L. Putnam, June 3, 1928 and Nov. 11, 1929, for examples. Some drafts of several chapters are in the Lowell Observatory Archives.
[58] R. L. Putnam to V. M. Slipher, night letter, March 17, 1930.
[59] E. C. Slipher, *The Photographic Story of Mars* (Cambridge: Sky Publishing Co., 1962).
[60] E. P. Hubble to V. M. Slipher, Oct. 5, 1928.
[61] R. L. Putnam to V. M. Slipher, Sept. 12, 1928.
[62] E. P. Hubble to V. M. Slipher, Nov. 2, 1928.
[63] V. M. Slipher to W. A. Cogshall, Dec. 11, 1928.
[64] V. M. Slipher to R. L. Putnam, Jan. 12, 1929.
[65] R. L. Putnam to V. M. Slipher, Jan. 22, 1929.
[66] V. M. Slipher to J. H. Worthington, Jan. 16, 1929.

[67] V. M. Slipher to J. Cochrane, Jan. 5, 1929.
[68] V. M. Slipher to R. L. Putnam, Jan. 12, 1929.
[69] V. M. Slipher to J. A. Miller, Jan. 21, 1929.
[70] R. L. Putnam to V. M. Slipher, Jan. 22, 1929.
[71] C. A. R. Lundin to R. L. Putnam, Jan. 25, 1929.
[72] R. L. Putnam to C. A. R. Lundin, Jan. 31, 1929.
[73] C. A. R. Lundin to R. L. Putnam, Feb. 1, 1929.
[74] R. L. Putnam to C. A. R. Lundin, Feb. 4, 1929.
[75] Lowell Observatory to R. L. Putnam, telegram, Feb. 6, 1929.
[76] V. M. Slipher to R. L. Putnam, Feb. 7, 1929.
[77] R. L. Putnam to V. M. Slipher, Feb. 14, 1929.
[78] V. M. Slipher to E. C. Slipher, Feb. 11, 1929.
[79] Lowell Observatory to R. L. Putnam, telegram, undated, but probably Feb. 19, 1929.
[80] R. L. Putnam to V. M. Slipher, telegram, Feb. 20, 1929.
[81] C. O. Lampland to V. M. Slipher, Feb. 28, 1929. Lampland was at Princeton University under an exchange program by which Princeton astronomer R. S. Dugan worked at Lowell Observatory that spring. Lampland, incidentally, apparently found eastern living expensive, for by mid-April he was running out of money and by early May he declared himself "flat broke." *See* C. O. Lampland to V. M. Slipher, April 16, and May 9, 1929.
[82] V. M. Slipher to C. O. Lampland, March 3, 1929.
[83] V. M. Slipher to R. L. Putnam, March 9, 1929.
[84] R. L. Putnam to V. M. Slipher, March 18, 1929.
[85] C. O. Lampland to V. M. Slipher, April 10, 1929.
[86] R. L. Putnam to C. A. R. Lundin, May 24, 1929.
[87] V. M. Slipher to F. E. Ross, March 23, 1929.
[88] V. M. Slipher to C. S. Lowell, March 14, 1929.
[89] V. M. Slipher to C. O. Lampland, April 5, 1929.
[90] "13-inch Photographic Record Book" (Lowell Observatory Journal of Observations), vol. 1 (1929), 1, *Lowell Observ. Arch.*

A Young Man From Kansas

[1] V. M. Slipher to H. N. Russell, May 12, 1928, *Lowell Observatory Archives*.
[2] V. M. Slipher to H. E. Knight, May 9, 1924.
[3] R. S. Richardson to V. M. Slipher, July 19, 1924; and V. M. Slipher to R. S. Richardson, July 21, 1924.
[4] V. M. Slipher to W. A. Cogshall, Dec. 11, 1928.
[5] V. M. Slipher to R. L. Putnam, Dec. 31, 1928.
[6] V. M. Slipher to R. L. Putnam, Jan. 12, 1929.
[7] R. L. Putnam to V. M. Slipher, Jan. 14, 1929.
[8] C. W. Tombaugh, personal communication, Nov. 6, 1974.
[9] C. W. Tombaugh, "The Sun's New Trans-Neptunian Planet," *Science News Letter,* 17:179 (1930).
[10] *American Men of Science,* (11th edition; New York: R. R. Bowker Company, 1967), p. 5434.
[11] V. M. Slipher to R. L. Putnam, Feb. 7, 1929.
[12] V, M. Slipher to R. L. Putnam, March 9, 1929.
[13] C. W. Tombaugh, "Reminiscences of the Discovery of Pluto," *Sky and Telescope* 19:264–70 (1960).
[14] R. L. Putnam and V. M. Slipher, "Searching Out Pluto—Lowell's Trans-Neptunian Planet X," *Scientific Monthly* 34:5 (1932); and C. W. Tombaugh, *op. cit.*, and "The Discovery of Pluto," *Astronomical Society of the Pacific Leaflet 209,* 1946 [reprinted in Harlow Shapley, ed., *Source Book in Astronomy 1900–1950* (Cambridge: Harvard University Press, 1960), 69–74] and "The Trans-Neptunian Planet Search," in G. P. Kuiper and B. Middlehurst, eds., *The Solar System* (Chicago: University of Chicago Press, 1961), pp. 12–30.
[15] C. W. Tombaugh, "Reminiscences...," *op. cit.*
[16] *Ibid.*
[17] V. M. Slipher to Cramer Dry Plate Company, Feb. 20, 1929.

[18] V. M. Slipher to C. O. Lampland, March 21, 1929.
[19] "13-Inch Photographic Record Book" (Lowell Observatory Journal of Observations), vol. 1 (1929), hereinafter cited as "13-Inch Log." *See* entries of April 11 and May 3, 1929. *See also* Tombaugh, personal communication, Nov. 6, 1974.
[20] R. L. Putnam and V. M. Slipher, *op. cit.*
[21] *Ibid.*
[22] *Ibid. See also* "13-Inch Log," *op. cit.*
[23] R. L. Putnam and V. M. Slipher, *op. cit.*
[24] "13-Inch Log," *op. cit.*
[25] C. O. Lampland, "Lowell Photographic Observations of Pluto in 1915, 1929, and 1930," text of paper read to the American Astronomical Society annual meeting, New Haven, Conn., December 1930, *Lowell Observ. Arch.*
[26] "13-Inch Log," *op. cit.*
[27] C. W. Tombaugh, personal communication, Nov. 6, 1974.
[28] V. M. Slipher to C. O. Lampland, April 15, 1929.
[29] *Ibid.*
[30] V. M. Slipher to C. O. Lampland, June 20, 1929.
[31] V. M. Slipher to K. Lundmark, Nov. 13, 1929.
[32] C. O. Lampland to V. M. Slipher, May 9, 1929.
[33] C. W. Tombaugh, personal communication, Nov. 6, 1974.
[34] *Ibid.*
[35] C. W. Tombaugh, "Reminiscences...," *op. cit.*
[36] *Ibid.*
[37] C. W. Tombaugh, personal communication, Nov. 6, 1974.
[38] *Ibid.*
[39] C. W. Tombaugh, "The Discovery of Pluto," *op. cit.*
[40] C. W. Tombaugh, personal communication, Nov. 6, 1974.
[41] *Ibid.*
[42] C. W. Tombaugh, personal communication, Nov. 7, 1974.
[43] C. W. Tombaugh, personal communication, Nov. 6, 1974.
[44] C. W. Tombaugh, "Reminiscences...," *op. cit.*
[45] C. W. Tombaugh, personal communication, Nov. 7, 1974.
[46] V. M. Slipher to R. L. Putnam, Oct. 15, 1929.
[47] R. L. Putnam to V. M. Slipher, Nov. 7, 1929.
[48] V. M. Slipher to R. L. Putnam, Nov. 8, 1929.
[49] C. W. Tombaugh, "Reminiscences...," *op. cit.*
[50] "13-Inch Log," *op. cit.*
[51] *Ibid.*
[52] C. W. Tombaugh, handwritten note on dust jacket of Pluto discovery plate No. 171; *Lowell Observatory 13-Inch Plate Vault*, hereinafter cited as "Tombaugh note."
[53] *Ibid.*
[54] C. W. Tombaugh, "Reminiscences...," *op. cit.*
[55] *Ibid., see also* "13-Inch Log," *op. cit.*
[56] C. W. Tombaugh, "Reminiscences...," *op. cit.*
[57] R. L. Putnam and V. M. Slipher, *op. cit.*
[58] *Ibid.*
[59] C. W. Tombaugh, "The Sun's New...," *op. cit.*
[60] C. W. Tombaugh, "Reminiscences...," *op. cit.*
[61] V. M. Slipher to R. L. Putnam, Feb. 23, 1930.
[62] R. L. Putnam to V. M. Slipher, Feb. 28, 1930; and A. L. Lowell to R. L. Putnam, March 1, 1930.
[63] R. L. Putnam to V. M. Slipher, telegram, letter, March 3, 1930.
[64] V. M. Slipher to R. L. Putnam, telegram, March 4, 1930.
[65] R. L. Putnam to V. M. Slipher, March 6, 1930.
[66] R. L. Putnam to V. M. Slipher, March 11, 1930.
[67] "13-Inch Log," *op. cit.*
[68] Tombaugh "note," *op. cit.*
[69] J. A. Miller to V. M. Slipher, March 14, 1930.
[70] J. C. Duncan to V. M. Slipher, telegram, March 9, 1930.

[71] V. M. Slipher to R. L. Putnam, telegram, March 16, 1930.
[72] V. M. Slipher to W. A. Cogshall, March 8, 1930.
[73] W. A. Cogshall to V. M. Slipher, April 20, 1930.
[74] C. S. Lowell to V. M. Slipher, March 9, 1930. C. W. Tombaugh recalled in a personal communication received by the author on Jan. 16, 1978, that "I met Mrs. Lowell once when she made a visit to the Observatory in the early 1930s (I think). She came to the room where I was blinking plates and chatted a few minutes. I remember being very uncomfortable about it. She remarked that years earlier, she spotted Planet X on one of the plates and that V. M. Slipher would not acknowledge it. It was obvious to me that she had not the least understanding of what was involved. It was then that I realized what a terrible shadow the older men had to live under. I had tremendous respect for Roger Lowell Putnam, and enjoyed his annual visits to Flagstaff."
[75] R. L. Putnam and V. M. Slipher, *op. cit.*
[76] *Harvard Announcement Card 108*, March 13, 1930.
[77] V. M. Slipher to Harvard Observatory, telegram, March 12, 1930 (10:00 p.m. Flagstaff time).
[78] V. M. Slipher, "The Discovery of a Solar System Body Apparently Trans-Neptunian," *Lowell Observatory Observation Circular,* March 13, 1930.
[79] *Ibid.*

Pluto: The Early Controversies

[1] W. G. Hoyt, "W. H. Pickering's Planetary Predictions and the Discovery of Pluto," *Isis* 61:551–64 (1976).
[2] *See*, for example, D. Rawlins, "The Mysterious Case of the Planet Pluto," *Sky and Telescope* 36:160 (1968).
[3] "The Discovery of New Planets," *Science* 71:x (1930).
[4] V. M. Slipher to C. S. Lowell, March 16, 1930.
[5] K. P. Williams and T. B. Gill to V. M. Slipher, telegrams, March 14, 1930.
[6] R. L. Putnam and V. M. Slipher, "Searching Out Pluto—Lowell's Trans-Neptunian Planet X," *Scientific Monthly* 34:5 (1932).
[7] R. L. Putnam to V. M. Slipher, telegram, March 13, 1930.
[8] H. Shapley to V. M. Slipher, March 15, 1930.
[9] *Ibid.*
[10] V. M. Slipher to C. S. Lowell, March 16, 1930.
[11] A. O. Leuschner to Lowell Observatory, telegram, March 13, 1930.
[12] A. O. Leuschner to E. C. Slipher, March 21, 1930.
[13] G. van Biesbroeck to V. M. Slipher, March 18, 1930. *See also* U.S. Naval Observatory to Lowell Observatory, and S. B. Nicholson to Lowell Observatory, March 15, 1930.
[14] R. L. Putnam to V. M. Slipher, March 15, 1930.
[15] R. L. Putnam to V.M. Slipher, telegram, March 15, 1930.
[16] V. M. Slipher to R. L. Putnam, telegram, March 16, 1930 and reply, March 17, 1930.
[17] V. M. Slipher to R. L. Putnam, undated draft apparently mailed March 21, 1930, along with a similarly worded letter to Putnam's mother (P. Lowell's sister Elizabeth), Mrs. William L. Putnam.
[18] R. L. Putnam to V. M. Slipher, March 28, 1930.
[19] *The New York Times* to Lowell Observatory, telegram, March 23, 1930.
[20] H. Shapley to R. L. Putnam, April 5, 1930.
[21] V. M. Slipher to J. A. Miller, undated.
[22] J. A. Miller to V. M. Slipher, telegram, undated.
[23] C. W. Tombaugh, personal communication, Nov. 6, 1974.
[24] *Harvard Observatory Announcement Card 112*, April 7, 1930; and E. C. Bower and F. L. Whipple, "Preliminary Elements and Ephemeris of the Lowell Observatory Object," *Lick Observatory Bulletin 421*, April 15, 1930.
[25] H. Shapley, "Trans-Neptunian Planet," in "Planet Notes," *Popular Astronomy* 38:295 (1930).
[26] R. L. Putnam to V. M. Slipher, telegram and letter, April 7, 1930.
[27] *Harvard Observatory Announcement Card 121*, April 8, 1930.

[28] *Ibid.*, April 14, 1930.
[29] *Ibid. See also* V. M. Slipher, *Lowell Observatory Observation Circular,* May 1, 1930.
[30] R. L. Putnam to V. M. Slipher, telegram, April 12, 1930.
[31] R. L. Putnam to V. M. Slipher, April 14, 1930.
[32] *The New York Times,* April 14, 1930.
[33] *Ibid.*
[34] R. L. Putnam to V. M. Slipher, telegram, April 15, 1930.
[35] R. L. Putnam to V. M. Slipher, April 15, 1930.
[36] V. M. Slipher to J. A. Miller, telegram, April 14, 1930.
[37] J. A. Miller to V. M. Slipher, telegram, April 16, 1930.
[38] V. M. Slipher to J. A. Miller, April 19, 1930.
[39] V. M. Slipher to R. L. Putnam, April 19, 1930.
[40] V. M. Slipher to A. O. Leuschner, April 18, 1930.
[41] A. O. Leuschner to V. M. Slipher, April 28, 1930.
[42] A. O. Leuschner, "The Astronomical Romance of Pluto," *Publications of the Astronomical Society of the Pacific* 44:197 (1932); C. W. Tombaugh, personal communication, Jan. 16, 1978.
[43] V. M. Slipher to R. L. Putnam, April 19, 1930.
[44] R. L. Putnam to V. M. Slipher, April 25, 1930.
[45] W. A. Cogshall to V. M. Slipher, April 29, 1930.
[46] *Lowell Observ. Obs. Circ.,* May 1, 1930.
[47] *Ibid.*
[48] *Ibid.*
[49] A. C. D. Crommelin to V. M. Slipher, April 16, 1930.
[50] V. M. Slipher to A. C. D. Crommelin, April 23, 1930.
[51] A. C. D. Crommelin, report to meeting of April 30, 1930, *Journal of the British Astronomical Association* 40:248 (1930).
[52] *Harvard Observatory Announcement Card 131,* June 5, 1930.
[53] W. S. Adams to V. M. Slipher, May 29, 1930.
[54] V. M. Slipher to W. S. Adams, May 31, 1930.
[55] V. M. Slipher to J. A. Miller, May 31, 1930.
[56] J. A. Miller to V. M. Slipher, June 12, 1930.
[57] E. C. Slipher to V. M. Slipher, telegram, June 12, 1930.
[58] V. M. Slipher to R. L. Putnam, August 23, 1930; and R. L. Putnam to V. M. Slipher, August 26, 1930.
[59] C. O. Lampland, typewritten text of paper read before the annual meeting of the American Astronomical Society in New Haven, Connecticut, December 1930; Lowell Observatory Archives.
[60] *Harvard Observatory Announcement Card 148,* Feb. 24, 1931.
[61] *Ibid.*, No. 133, June 16, 1930. *See also* S. B. Nicholson and N. U. Mayall, "Position, Orbit and Mass of Pluto," *Contributions From the Mount Wilson Observatory 417,* 1931; and in *Astrophysical Journal* 73:1 (1931). Crommelin, "Images of Pluto on Plates Exposed in 1919," *Journ. B.A.A.* 49:298 (1930), compares his orbit with Nicholson's and Mayall's.
[62] *Harvard Observatory Announcement Card 137,* June 20, 1930.
[63] E. C. Bower, "On the Orbit and Mass of Pluto With an Ephemeris for 1931–32," *Lick Observ. Bull. 437,* 1931.
[64] E. C. Bower and F. L. Whipple, "Elements and Ephemeris of the Lowell Observatory Object (Pluto)—Second Paper," *Lick Observ. Bull. 427,* 1930; F. Zagar, "Die Bohn Plutos mit Berücksichtigung der Störungen," *Astronomische Nachrichten* 240:335 (1930); and E. C. Bower, "On the Orbit and Mass of Pluto With an Ephemeris for 1931–32," *op. cit.*
[65] C. S. Lowell to V. M. Slipher, March 15, 1930.
[66] W. E. D. Stokes, Jr., to Lowell Observatory, March 15, 1930. What dictionary, however, he did not say. The word is not listed in the 1976 edition of *Webster's New Twentieth Century Dictionary (unabridged).*
[67] Louise Brown to V. M. Slipher, March 15, 1930.
[68] Vivian Shirley to R. L. Putnam, March 15, 1930.

[69] E. P. Hollis to V. M. Slipher, March 16, 1930. Hollis was an amateur astronomer from Swampscott, Massachusetts, who maintained a 12-inch reflecting telescope at this time near Cameron, Arizona, an Indian trading post fifty-four miles north of Flagstaff. *See* E. P. Hollis to V. M. Slipher, July 19, 1930; Slipher's reply, August 11, 1930; and V. M. Slipher to J. A. Anderson, August 20, 1930.
[70] E. G. Murtagh to Lowell Observatory, June 24, 1930.
[71] E. G. Jackson to ''P. Lowell,'' undated.
[72] Editorial, *The New York Times*, March 15, 1930.
[73] R. L. Putnam to V. M. Slipher, March 15, 1930.
[74] R. L. Putnam to V. M. Slipher, March 17, 1930.
[75] V. M. Slipher to Mrs. W. L. Putnam, undated. See's name for one of his planets was actually ''Chronos,'' but Slipher had misread See's letter; *see* V. M. Slipher to R. L. Putnam, May 10, 1930.
[76] H. Shapley to V. M. Slipher, March 20, 1930.
[77] C. S. Lowell to V. M. Slipher, March 9, 1930.
[78] C. S. Lowell to V. M. Slipher, telegram, March 14, 1930.
[79] C. S. Lowell to V. M. Slipher and C. O. Lampland, March 15, 1930.
[80] H. Shapley to R. L. Putnam, March 15, 1930.
[81] H. F. Ashurst to Lowell Observatory, March 31, 1930.
[82] R. L. Putnam to V. M. Slipher, March 27, 1930.
[83] R. L. Putnam to V. M. Slipher, March 28, 1930.
[84] J. Stokely to V. M. Slipher, telegram, April 2, 1930.
[85] *Lowell Observ. Obs. Circ.*, May 1, 1930.
[86] V. M. Slipher, typewritten text of statement, undated, *Lowell Observ. Arch.*
[87] R. L. Putnam, typewritten statement, undated, *Lowell Observ. Arch.*
[88] R. L. Putnam and V. M. Slipher, ''Searching Out Pluto,'' *op. cit.*
[89] *Lowell Observ. Obs. Circ.*, May 1, 1930.
[90] R. L. Putnam, statement, *op. cit.*
[91] R. L. Putnam and V. M. Slipher, ''Searching Out Pluto,'' *op. cit.*
[92] *Associated Press* dispatch, datelined Milan, Italy, March 25, 1930, published in *The New York Times*, March 26, 1930.
[93] *Lowell Observ. Obs. Circ.*, May 1, 1930.
[94] J. A. Miller to V. M. Slipher, July 29, 1930.
[95] C. S. Lowell to V. M. Slipher, Jan. 1, 1931.
[96] W. G. Hoyt, *op. cit.*
[97] W. H. Pickering to Lowell Observatory, postcard, April 12, 1930.
[98] W. H. Pickering, ''The Trans-Neptunian Planet,'' *Pop. Astron.* 38:292 (1930).
[99] W. H. Pickering, ''The Mass and Density of Pluto—Are the claims that it was predicted by Lowell justified?'' *Pop. Astron.* 39:2 (1931).
[100] W. H. Pickering, ''Planet *P*, Its Orbit, Position and Magnitude. Planets *S* and *T*,'' *Pop. Astron.* 39:385 (1931).
[101] A. J. Cannon, ''William Henry Pickering 1858–1938,'' *Science* 87:179 (1938).

The Longer Controversy

[1] A. C. D. Crommelin, ''The Discovery of Pluto,'' *Monthly Notices of the Royal Astronomical Society* 91:380 (1931).
[2] L. Abbott to V. M. Slipher, telegram, March 13, 1930.
[3] J. Stokely to V. M. Slipher, telegram, March 13, 1930.
[4] H. Shapley to R. L. Putnam, March 15, 1930.
[5] R. L. Putnam to V. M. Slipher, March 28, 1930.
[6] W. H. Pickering, ''The Trans-Neptunian Planet,'' *Popular Astronomy* 38:285 (1930).
[7] *Ibid.*
[8] W. H. Pickering, ''Supplementary Note,'' *Pop. Astron.* 38:293 (1930).
[9] *Ibid.*
[10] W. H. Pickering to V. M. Slipher, postcard, April 12, 1930.
[11] V. M. Slipher to R. L. Putnam, May 10, 1930.
[12] V. M. Slipher to R. L. Putnam, May 10, 1930.
[13] W. H. Pickering, ''The Trans-Neptunian Comet,'' *Pop. Astron.* 38:341 (1930).

[14] P. Fox to V. M. Slipher, July 11, 1930.
[15] A. C. D. Crommelin to V. M. Slipher, May 8, 1930.
[16] A. C. D. Crommelin, "Pluto," *Journal of the British Astronomical Association* 41:116 (1930).
[17] A. C. D. Crommelin to V. M. Slipher, April 16, 1930.
[18] A. C. D. Crommelin, "Pluto, the Lowell Planet," *Journ. B.A.A.* 40:265 (1930).
[19] *Ibid.*
[20] A. C. D. Crommelin to V. M. Slipher, May 20, 1930.
[21] E. W. Brown, "On the Prediction of Trans-Neptunian Planets From the Perturbations of Uranus," *Proceedings of the National Academy of Sciences* 16:364 (1930).
[22] E. W. Brown, "On a Criterion for the Prediction of an Unknown Planet," *Mon. Not. R.A.S.* 92:80 (1931); and W. H. Pickering, "A Reply to Professor Brown's Criticism of My Views on Pluto," *Pop. Astron.* 40:519 (1932).
[23] E. W. Brown, "On the Prediction of Trans-Neptunian Planets," *op. cit.*
[24] *Ibid.*
[25] *Ibid.*
[26] *Ibid.*
[27] *Ibid.*
[28] *Ibid.*
[29] A. C. D. Crommelin, "Pluto, the Lowell Planet," *op. cit.*
[30] V. M. Slipher to R. L. Putnam, May 1, 1930.
[31] V. M. Slipher to R. L. Putnam, June 24, 1930.
[32] V. M. Slipher to R. L. Putnam, July 18, 1930.
[33] R. L. Putnam to V. M. Slipher, July 24, 1930.
[34] R. L. Putnam to V. M. Slipher, Sept. 19, 1930.
[35] A. C. D. Crommelin, "Pluto, the Lowell Planet," *op. cit.*
[36] Report of meeting of June 25, 1930, *Journ. B.A.A.* 40:292 (1930).
[37] R. L. Putnam and V. M. Slipher, "Searching Out Pluto—Lowell's Trans-Neptunian Planet," *Scientific Monthly* 34:5 (1932).
[38] H. N. Russell, "Planet X," *Scientific American* 143:20 (1930).
[39] H. N. Russell, "How Pluto's Orbit was Figured Out," *Sci. Amer.* 143:364 (1930).
[40] H. N. Russell, "More About Pluto," *Sci. Amer.* 143:446 (1930).
[41] *Ibid.*
[42] *Ibid.*
[43] H. N. Russell, appendix, in A. L. Lowell, *Biography of Percival Lowell* (New York: The Macmillan Company, 1935), pp. 203–5.
[44] *Ibid.*
[45] W. H. Pickering, "The Mass and Density of Pluto—Are the claims that it was predicted by Lowell justified?" *Pop. Astron.* 39:2 (1931).
[46] *Ibid.*
[47] *Ibid.*
[48] *Ibid.*
[49] *Ibid.*
[51] A. C. D. Crommelin, "The Discovery of Pluto," *op. cit.*
[52] W. H. Pickering, "The Discovery of Pluto," *Mon. Not. R.A.S.* 91:812 (1931).
[53] *Ibid.*
[54] E. C. Bower, "On the Orbit and Mass of Pluto, With an Ephemeris for 1931–32," *Lick Observatory Bulletin* 437, 1931.
[55] *Ibid.*
[56] E. W. Brown, "On the Criterion for the Prediction of an Unknown Planet," *op. cit.*
[57] E. W. Brown, "Observation and Gravitational Theory in the Solar System," *Publications of the Astronomical Society of the Pacific* 44:21 (1932).
[58] *Ibid.*
[59] *Ibid.*
[60] R. L. Putnam and V. M. Slipher, *op. cit.*
[61] A. O. Leuschner, "The Astronomical Romance of Pluto," *Publ. Astr. Soc. Pac.* 44:197 (1932).
[62] *Ibid.*
[63] *Ibid.*

[64] V. Kourganoff, "La Part de la Mécanique Céleste dans la découverte de Pluton (These du doctorat-es-Mathematique Dediée a la Mémoire de Percival Lowell)," hereafter cited as Thesis. This was first published in the *Bulletin Astronomique* 12:147–258; 271–301; and 303–41 (1940); and later by Gauthier-Villars, Paris, pp. 1–187. Citations here are from the Gauthier-Villars edition.
[65] V. Kourganoff, "Nouvelles données sur l'histoire de la découverte de Pluton," *Ciel et Terre* 60:180–95 (1944).
[66] G. Reaves, "Kourganoff's Contributions to the History of the Discovery of Pluto," *Publ. Astr. Soc. Pac.* 63:49–60 (1951).
[67] M. Grosser, "The Search for a Planet Beyond Neptune," in N. Reingold, ed., *Science in America Since 1820* (New York: Science History Publications, 1976) pp. 322.
[68] Kourganoff, *Thesis, op. cit.*, p. 2.
[69] *Ibid.*, p. 170.
[70] *Ibid.*, pp. 2–3. See also Reaves, *op. cit.*, p. 52.
[71] *Ibid.*, pp. 3, 41, 58–59.
[72] *Ibid.*, p. 75.
[73] *Ibid.*, pp. 4–5, 113ff.
[74] W. H. Pickering, "The Assumed Planet Beyond Neptune," *Pop. Astron.*, 77:545 (1909).
[75] Kourganoff, *Thesis, op. cit.*, p. 131.
[76] *Ibid.*, pp. 139–40.
[77] *Ibid.*, p. 140.
[78] *Ibid.*, p. 3.
[79] Reaves, *op. cit.*, p. 57.
[80] Kourganoff, *Thesis, op. cit.*, pp. 148–50.
[81] *Ibid.*, p. 146.
[82] *Ibid.*, pp. 5, 150–67.
[83] *Ibid.*, p. 6. Leverrier, replying to the "happy accident" argument in the discovery of Neptune, declared: "Such a mistake, if it takes for a moment the place of truth, would bring about a deep discouragement among those who devote themselves to the progress of science." See *Comptes Rendus*, 27:274 (1848).
[84] *Ibid.*, pp. 169–70.
[85] P. K. Seidelmann, W. J. Klepczinski, R. L. Duncombe and E. S. Jackson, "The Mass of Pluto," *Astronomical Journal* 76:488 (1971).
[86] D. Brouwer, "Current Problems of Pluto," *Sky and Telescope*, 9:103–5 (1950).
[87] L. R. Wylie, "A Comparison of Newcomb's Tables of Neptune With Observation 1795–1938," *Publications of the U.S. Naval Observatory*, 2nd series, vol. 15, part 1, 1942.
[88] D. Brouwer, *op. cit.*
[89] W. J. Eckert, D. Brouwer and G. M. Clemence, "Coordinates of the Five Outer Planets 1653–2060," *Astronomical Papers of the American Ephemeris and Nautical Almanac*, vol. 12, 1951.
[90] G. M. Clemence and D. Brouwer, "The Motions of the Five Outer Planets," *Sky and Telescope* 10:83–86 (1951).
[91] G. P. Kuiper, "The Diameter of Pluto," *Publ. Astr. Soc. Pac.* 62:133–37 (1950).
[92] D. Rawlins, "The Mysterious Case of the Planet Pluto," *Sky and Telescope* 35:160–62 (1968).
[93] I. Halliday, R. Hardie, O. G. Franz and J. B. Priser, "An Upper Limit for the Diameter of Pluto," *Astron. J.* 70:676–77 (1965); and W. L. Sanders, "A near occultation of a star by Pluto," *Publ. Astr. Soc. Pac.* 77:298–99 (1965).
[94] R. L. Duncombe, W. J. Klepczinski and P. K. Seidelmann, "Orbit of Neptune and Mass of Pluto," *Astron. J.* 73:830–35 (1968).
[95] D. Rawlins, "The Great Unexplained Residual in the Orbit of Neptune," *Astron. J.* 75:856–57, 1970.
[96] P. K. Seidelmann *et al, op. cit.*
[97] D. P. Cruikshank, C. B. Pilcher and D. Morrison, "Pluto: Evidence for Methane Frost," *Science* 194:835–37 (1976). Dr. Cruikshank has informed me (1979) that further photometric observations indicate a diameter for Pluto of between 1200 and 1800 kilometers, only slightly larger than the largest asteroid, Ceres.

[98] Harold Ables, personal communications, July 6–7, 1978. Dr. Ables, director of the U.S. Naval Observatory's Flagstaff station, kindly supplied me with details of Charon's discovery prior to the public announcement on July 7, 1978, days before the final revision of this manuscript was due to be returned to the publisher.
[99] J. W. Christy and R. Harrington, "The Satellite of Pluto," *Astron. J.* 83:1005–1008 (1978).
[100] *Ibid.*

More Planets X?

[1] V. M. Slipher to R. L. Putnam, April 19, 1930.
[2] C. W. Tombaugh, holographic note, dated June 9, 1930, on evelope of Pluto discovery plate No. 171, Lowell Observatory Archives (13-inch plate vault).
[3] C. W. Tombaugh, "The Trans-Neptunian Planet Search," in *The Solar System*, G. P. Kuiper and B. Middlehurst, eds. (Chicago: The University of Chicago Press, 1961), vol. 3, pp. 12–30.
[4] V. M. Slipher, "The Trans-Neptunian Planet Search," *Proceedings of the American Philosophical Society* 79:435 (1938).
[5] R. L. Putnam to H. N. Russell, Sept. 18, 1930.
[6] H. N. Russell to R. L. Putnam, Oct. 9, 1930.
[7] C. W. Tombaugh, "The Trans-Neptunian Planet Search," *op. cit.*
[8] C. W. Tombaugh, holographic note, *op. cit.*
[9] "Orbit of Planet X; the Ottawa Object," *Journal of the Royal Astronomical Society of Canada* 24:241 (1930).
[10] A. C. D. Crommelin, "Detection of Another Distant Object," *British Astronomical Association Circular No. 91*, May 4, 1930.
[11] A. C. D. Crommelin to V. M. Slipher, May 8, 1930.
[12] Tombaugh subsequently received an M.A. degree from Kansas in 1939, and an honorary D. Sc. from what is now Northern Arizona University in 1960.
[13] C. W. Tombaugh, "The Trans-Neptunian Planet Search," *op. cit.*
[14] *Ibid.* See also C. W. Tombaugh "Reminiscences of the Discovery of Pluto," *Sky and Telescope* 19:264–70 (1960).
[15] *Ibid.*
[16] *Ibid.*
[17] *Ibid.*
[18] H. L. Giclas, "The Comet and Minor Planet Program at Lowell Observatory," paper read at the 22nd Colloquium on Asteroids, Comets and Meteoric Matter of the International Astronomical Union, Nice, France, April 4–6, 1972. *See also* Giclas, *Lowell Proper Motion Survey, Northern Hemisphere* (Flagstaff: The Lowell Observatory, 1971).
[19] W. G. Hoyt, "W. H. Pickering's Planetary Predictions and the Discovery of Pluto," *Isis* 67:551–64 (1976).
[20] P. Lowell, "The Origin of the Planets," *Memoirs of the American Academy of Arts and Sciences*, vol. 14, no. 1, 1913; and "Memoir on a Trans-Neptunian Planet," *Memoirs of the Lowell Observatory*, vol. 1, no. 1, 1915.
[21] "Une planète transplutoniene," *Bulletin de la Société astronomique de France* 60:188–89 (1946); and W. Strubell, "Existenmöglichkeit eines Transplutonische Planeten," *Die Sterne* 3:70–72 (1952).
[22] A. C. D. Crommelin, "Pluto, the Lowell Planet," *Journal of the British Astronomical Association* 40:265 (1930).
[23] D. Brouwer, "The Orbit of Pluto over a long period of time," *Theory of Orbits in the Solar System and Stellar Systems*, International Astronomical Union Symposium 25, Thessalonica, Greece, 1964 (New York: Academic Press, 1966), pp. 227–29; and C. J. Cohen and E. C. Hubbard, "The Orbit of Pluto," The *Observatory* 85:43–44 (1965).
[24] W. H. Pickering, "Planet *P*, Its Orbit, Position and Magnitude. Planets *S* and *T*," *Popular Astronomy* 39:385 (1931).
[25] *Associated Press* dispatch, datelined Moscow, U.S.S.R., published in the Arizona [Phoenix]*Republic*, June 8, 1975.

[26] H. H. Kritzinger, "Transpluto, hypothetische Elemente," *Nachrichtenblatt der Astronomische Zentralstelle* 8:4 (1954).
[27] *Ibid.*; see also R. A. Naef, "Hypothetische Elemente eines Transpluto," *Orion: Mitteilungen der Schweizerischen Astronomische* 4:484 (1955).
[28] H. H. Kritzinger, "Transpluto, hypothetische Elemente," *Nach. Astron. Zent.* 11:4 (1957).
[29] H. H. Kritzinger, "Transpluto, hypothetische elliptische Elemente," *Nach. Astron. Zent.* 13:3 (1959).
[30] R. S. Richardson, "An attempt to determine the mass of Pluto from its disturbing effect on Halley's Comet," *Publ. Astr. Soc. Pac.* 54:19–23 (1942).
[31] J. L. Brady, "The Effect of a Transplutonian planet on Halley's Comet," *Publ. Astr. Soc. Pac.* 84:314–22 (1972).
[32] *Ibid.*
[33] *Ibid.*
[34] *Ibid.*
[35] P. K. Seidelmann, B. G. Marsden and H. L. Giclas, "Note on Brady's Hypothetical Trans-Plutonian Planet," *Publ. Astr. Soc. Pac.* 84:858–64 (1972).
[36] P. Goldreich and W. R. Ward, "The Case Against Planet X," *Publ. Astr. Soc. Pac.* 84:737–42 (1972).
[37] A. R. Klemola and E. A. Harlan, "Search for Brady's Hypothetical Trans-Plutonian Planet," *Publ. Astr. Soc. Pac.* 84:736 (1972).
[38] R. L. Putnam and V. M. Slipher, "Searching Out Pluto—Lowell's Trans-Neptunian Planet," *Scientific Monthly* 34:5 (1932).
[39] S. H. J. Wanrooy, "Pluto ein Planetoid?" *Die Sterne*, 18:85–86 (1937).
[40] J. Ashbrook, "Kowal's Strange Slow-Moving Object," *Sky and Telescope* 55:4–5 (1978).
[41] *Ibid.*
[42] T. Gehrels, personal communication, Feb. 28, 1978.
[43] "Prof. Yamamoto's Suggestion on the Origin of Pluto," *Kwasan* (Kyoto) *Observatory Bulletin* 288, 1934.
[44] R. A. Lyttleton, "On the Possible Results of an Encounter of Pluto With the Neptunian System," *Monthly Notices of the Royal Astronomical Society* 97:108–115 (1936).
[45] G. P. Kuiper, "The Formation of the Planets," *Journ. R.A.S. Canada* 50:171–73 (1956); and "Further Studies on the Origin of Pluto," *Astrophysical Journal* 125:287–89 (1957).
[46] M. F. Walker and R. Hardie, "A photometric determination of the rotational period of Pluto," *Publ. Astr. Soc. Pac.* 67:224–31 (1955).
[47] Harold Ables, personal communications, July 6–7, 1978.
[48] W. J. Luyten, "Pluto Not a Planet?" *Science* 123:896–97 (1956).
[49] G. P. Kuiper, "The Formation of Planets," *op. cit.*
[50] *Ibid.*
[51] H. Ables, personal communications, July 6–7, 1978. See also J. W. Christy and R. Harrington, "The Satellite of Pluto," *Astron. J.* 83:1005–1008 (1978).

BIBLIOGRAPHY

The literature of the history of planetary discovery, and particularly of the discovery of Pluto, is predominately a periodical literature contained in astronomical journals, institutional annals and memoirs, and similar publications printed in various languages and in various countries over the past two hundred years. While there are a few excellent books available that describe the discoveries of Uranus, the asteroids, and/or Neptune, there has been no definitive history of the discovery of Pluto. Nor has the whole history of planetary discovery been explored at length in a single volume, although most textbooks and general histories of astronomy include brief summaries of its major events.

The selected bibliography that follows, therefore, is necessarily short, and is intended primarily as a guide for further reading. Two periodicals should be especially noted as they represent important summaries of the analytical work of Leverrier and Lowell regarding their respective hypothetical trans-Uranian and trans-Neptunian planets. These are:

Urbain J. J. Leverrier. "Recherches sur le mouvement de la planète Herschel (dites Uranus)." *Connaissance des Temps pour 1849, Additions.* Paris: 1846.

Percival Lowell. "Memoir on a Trans-Neptunian Planet." *Memoirs of the Lowell Observatory,* vol. 1, no. 1, 1915.

Adams, William G., ed. *The Scientific Papers of John Couch Adams*. 2 vols. London: 1896.

Alexander, Arthur F. O'D. *The Planet Uranus*. New York: American Elsevier Publishing Company, Inc., 1965.

Bell, Eric. T. *Men of Mathematics*. New York: Simon and Schuster, 1937.

Clerke, Agnes M. *A Popular History of Astronomy During the Nineteenth Century*. 4th edition. London: Adam and Charles Black, 1902.

Dreyer, John L. E., ed. *The Scientific Papers of Sir William Herschel*. 2 vols. London: 1912.

Gould, Benjamin A., Jr. *Report on the History of the Discovery of Neptune*. Washington, D.C.: The Smithsonian Institution, 1850.

Grosser, Morton. *The Discovery of Neptune*. Cambridge: Harvard University Press, 1962.

Herschel, John F. W. *Outlines of Astronomy*. 5th edition, London: Longmans, Green (1865).

Hoyt, William G. *Lowell and Mars*. Tucson: University of Arizona Press, 1976.

Lowell, A. Lawrence. *Biography of Percival Lowell*. New York: The Macmillan Company, 1935.

Lowell, Percival. *Mars*. London: Longmans, Green and Company, 1895.

_____. *Mars and Its Canals*. New York: The Macmillan Company, 1906.

_____. *Mars as the Abode of Life*. New York: The Macmillan Company, 1908.

_____. *The Evolution of Worlds*. New York: The Macmillan Company, 1909.

_____. *The Solar System*. Boston: Houghton, Mifflin and Company, 1903.

Lubbock, Constance A. *The Herschel Chronicle*. Cambridge: The University Press, 1933.

Roth, Günter D. *The System of Minor Planets*. Alex Helm, trans. New York: D. Van Nostrand Company, Inc., 1962.

Shapley, Harlow, and Helen E. Howarth, eds. *Source Book in Astronomy 1900–1950*. Cambridge: Harvard University Press, 1960.

Sidgwick, J. B. *William Herschel: Explorer of the Heavens*. London: Faber and Faber, Ltd., 1953.

Watson, Fletcher G. *Between the Planets*. Philadelphia: The Blakiston Company, 1941.

INDEX

Page numbers for illustrations are in italics.

Abbot, Charles Greeley, 74, 172
Abbott, Lyle, 222
Académie des Sciences de France. *See* French Academy of Sciences
Adams, John Couch, 7, 18, 37, 39–40, *41*, 42–43, 45–48, 50–51, 53–58, 60–61, 65–67, 69–70, 73, 85, 95, 106, 133–35, 138–40, 194, 221–23, 231, 240
   analytical methods of, 40, 45–48, 85, 106, 133–35
   role in Neptune controversies, 7, 39, 54–58, 66–67, 69
   trans-Uranian planet prediction, 7, 42, 46–47, 50, 53, 63
Adams, Walter S., 153, 172, 213
Agassiz, George Russell, 81, 86
Airy, Sir George Biddell, 7, 35–40, *36*, 42–43, 47–48, 50–58, 61, 73
   observations of Uranus, 37, 56
   role in Neptune controversies, 7, 53–58
Aitken, Robert G., 91
Alexander, A. F. O'D., 2, 19; *The Planet Uranus* (1965), 2, 19
Allegheny Observatory, 91, 127
Altona Observatory, 61–62
Alvan Clark and Sons, 82, 92, 103, 168, 176
American Academy of Arts and Sciences, 66, 86, 98, 113, 122, 125–27
American Association for the Advancement of Science, 236
American Astronomical Society, 214, 217, 235
American Philosophical Society, 249
Amherst College Observatory, 18-inch refractor, 94
Amory, Harcourt, 145
Andromeda, Great Nebula in (M31). *See* Nebulae
Annales de L'Observatoire (Impériale) de Paris, 71, 79, 101
Appollonius of Perga, 4
Arago, François, 39, 43, 45, 53–54
Ariel, 22. *See also* Uranus, satellites of
Aristarchus of Samos, 3
Arizona State College. *See* Northern Arizona University
Arizona State Teachers College. *See* Northern Arizona University
Arizona, University of, 173, 258
Ashurst, Henry Fountain, 216
Assurbanipal, 3
Asteroids, 2–3, 7–8, 27–30, 71, 92, 113, 119, 132–33, 188, 195, 201, 207–10, 212, 217, 251, 253, 258, 260. *See also* specific name
   discovery of, 27–28
   early observations of, 27–30
   intra-mercurial, 71–72
   Kirkwood gaps in, 113
   relation to Bode's law of, 28
Astraea, 30
Astronomers Royal. *See* specific name

Astronomical Journal 61, 76, 96
Astronomical "seeing." *See* Seeing, astronomical
Astronomical Society of the Pacific, 237
Astronomie populaire (Flammarion), 77
Astronomische Nachrichten, 39, 62, 73, 128
Astronomy, general, 17–18, 60–61, 172, 178–79
Athenaeum, 53–55, 58
Atlantic Monthly, 81
Atmosphere, 3, 81, 159
Aubert, Alexander, 12, 14–15
Aurorae, 19, 130, 159, 170–71, 185. *See also* Slipher, V. M., discoveries of

Babinet, Jacques, 76
Bailey, Solon I., 147, 160
Banks, Sir Joseph, 12, 20–24, 28
Barnard, Edward Emerson, 92, 214
Bath Literary and Philosophical Society, 10, 17
Berberich, Anton, 92, 94
Berlin Academy, 52
Berlin Observatory, 20, 39, 52; 9-inch Frauenhofer refractor, 52
Bessel, Friedrich W., 34, 38, 46, 62
Between the Planets (Watson), 2
Biography of Percival Lowell (A. L. Lowell), 96, 232
"Black Friday," 189
"Blink" comparator. *See* Zeiss "blink" comparator
Bode, Johann Elert, 20, 24–29, 32–33, 61
Bode's law, 24–29, 37–38, 40, 46–47, 49, 56, 75, 232, 253; in discovery of Neptune, 61
Bond, George P., 63
Bond, William Cranch, 60
Bonnet, Charles, 25
Boothroyd, Samuel, 146
Boscovich, Abbé Ruggiero Giuseppe, 18–19, 32
Bouvard, Alexis, 33–38, 46, 48, 69, 71, 99–100, 139
Bouvard, Eugene, 38, 56
Bower, Ernest Clare, 205–6, 214–15, 224–25, 233–36, 241, 243
Bowlker, Mrs. R. W. (P. Lowell's sister, Katherine), 122
Boyden Station. *See* Harvard College Observatory
Bradley, Dr. James, 34
Brady, Joseph L., 255–57
Brashear, John A., 91–92, 127. *See also* Lowell Observatory, telescopes of
Bremiker, Carl, 52
Brera Observatory, 20
British Association for the Advancement of Science 35, 40, 51

British Astronomical Association, 212–13, 230, 249
Brouwer, Dirk, 241–*42*, 244
Brown, Constance, 190
Brown, Ernest William, 226–31, *227*, 232–40, 256
Brucia, 30
Burckhardt, Johann Karl, 28, 34, 45
Bureau des Longitudes, 53
Burgess, George K., 151
Burland, M. S., 249
Burney, Venetia, 218
Burnham, S. W., 127
Butler, Howard Russell, 193

Cacciatore, Niccolo, 37
California Institute of Technology, 173
Cambridge Observatory, 6, 35–37, 46, 50, 55, 58–59; 11.75-inch Northumberland refractor, 50, 59
Cambridge University, 39–40, 42, 46, 70
Campbell, William Wallace, 74, 127
"Canals" of Mars. *See* Mars
Cannon, Annie J., 219
"Capture Theory." *See* See, Thomas Jefferson Jackson
Carbon dioxide, in planetary atmospheres, 133
Carrigan, William T., 95–103, 106, 112
Carte photographique du ciel, 72
Cartesian "cosmic fluid," 36
Ceres, 7, 28–30, 212
Cerro Tololo Inter-American Observatory, 247
Cetus, nebulae in. *See* Nebulae
Challis, Rev. James, 6–7, 39–40, 50–59, 61
Charon, 8, 193, 198, *245*–46, 254, 260–61
Chebotarev, Gleb, 255
Chiron, 258, 261
Chlorophyll. *See* Mars
Christy, James L., *246*–47
Clairaut, Alexis Claude, 36
Clark, Alvan and Sons. *See* Alvan Clark and Sons
Clemence, G. M., 242, 244
Clerk Maxwell, James, 128
Clerke, Agnes M., 45, 74
Coblentz, W. W., 151–52, 159, 172
Cogshall, Wilbur A., *79*, 87, 94, 150, 169, 173, 179, 183, 210–*11*
Coma Berenices, nebula in. *See* Nebulae
Comets, 3, 9–11, 18–21, 23, 35–37, 43, 45, 49, 76–78, 83–84, 91, 101, 103–4, 112, 201, 204, 207–10, 224–25, 232, 251, 254–58. *See also* specific name
Pluto as a, 204, 207–9, 212, 224–25, 232

Uranus as a, 9–11, 18–21, 23
  use in planetary prediction, 36–37,
    77–78, 83–84, 104, 112, 254–57
Commensurability of periods, 65–66,
    113–14, 118, 122, 127–29, 134–36,
    141. *See also* Lowell, P., concepts of
Comptes Rendus, 45, 66, 72
Connaissance des Temps, 45
Copernicus, Nicholas, 3–4
Copley Medal, 20–21, 57
Cornell University, 22
Crab nebula (M1). *See* Nebulae
Cracow Observatory, 206
Crommelin, A. C. D., 103, 130, 212–14,
    221, 225–26, 228–30, 232–35, 241,
    243–44, 249, 254
Cruikshank, Dale P., 245
Curtis, Heber D., 156

Dallet, Gabriel, 78
Darquier, A., 20
D'Arrest, Heinrich Louis, 52
De La Hire, Philippe, 45
De la Hire's Comet, 45
Delambre, J. B. J., 33
De Medici, Cosmo II, 7
De Saron, J. B. G. Bouchart, 18
De Sitter, Willem, 156, 172
De Vico, Francesco, 45
De Vico's Comet, 45
Discovery, accidental nature of, 5–6
Discovery of Neptune, The (Grosser), 2,
    33, 71
Doe, Edward M., 121
Dominion Observatory, report of "Ottawa
    object," 249
Douglass, Andrew Ellicott, 81, 173
Duncan, John Charles, 87–92, *88*, 119,
    147, 157, 195, 201
Duncombe, R. L., 244

Eckert, W. J., 242, 244
Eclipse, 16, 155
Ecliptic, 4, 85, 94, 121, 125, 182, 189
Eddington, Sir Arthur S., 172
Edmondson, Frank K., 250
Edwards, Earl A., 120, 124, 132
Elliot, James L., 22
Enceladus, 128. *See also* Saturn,
    satellites of
Encke, Johann Franz, 52–53, 59, 61
Encke's Division, 59. *See also* Saturn,
    rings of
Ernestine Observatory, 26
Euler, Leonard, 20
Evolution of Worlds, The (Lowell, P.),
    103, 134
Expanding universe, theory of, 122, 156
Extraterrestrial life, 154–55. *See also*
    Mars, life on

Fath, Edward A., 147
Faye, Hervé, 45, 72
Faye's comet, 45
Fecker, J. W., 164, 166–68
Ferdinand IV, king of Sicily and Naples,
    7, 28
Ferguson, James, 76
Fixlmillner, Rev. Placidus, 33
Flammarion, Camille, 76–77, 83–84, 87,
    95, *109*, 112; *Astronomie populaire*
    (1879), 77
Flamsteed, Sir John, 6, 20, 33–34, 42, 46,
    49, 111, 136, 236
Flemming, Friedrich Wilhelm, 38
Forbes, George, 77, 83–84, 95, 101,
    234, 253
Fox, Philip, 225
French Academy of Sciences, 39, 43, 48,
    51, 53–54, 62, 68, 72
French Astronomical Society, 78, 229

Gaillot, Jean Baptiste Aimable, 66, 78–80,
    106, 109, 111–12, 117, 119, 120, 122,
    124, 136–39, 226, 234, 239
Galaxies, 122, 159. *See also* Nebulae,
    Milky Way
Galileo, 3, 7
Galle, Johann Gottfried, 5–6, 39, 52–54,
    57, 61–63, 84
Garnowsky, Alexander, 78
Gauss, Carl Friedrich, 4, 18, 28, 31, 53, 95
Gay-Lussac, Joseph Louis, 43
Gehrels, Tom, 258
George III of England, 7, 11, 15, 23–24
Giclas, Henry L., 250, 252, 256–57
Gifford, A. C., 158
Gifford, P. W., 160
Gill, Thomas B., 120, 124, 130–32, *131*,
    160, 200
Goldreich, Peter, 257
Gould, Benjamin Apthorp, 2, 34, 36,
    38–39, 45, 48–49, 51, 59, 61,
    66–71, *67*
Gravitation, law of, 3, 17, 19, 28, 35, 49
"Great debate," 156–58
Great Depression, 189
Greenwich Observatory. *See* Royal
    Observatory (Greenwich)
Grigull, Theodore, 78
Grosser, Morton, 2, 33–34, 35, 40, 50, 53,
    71, 75, 78; *The Discovery of Neptune*
    (1962), 2, 33, 71
Grubb, Sir Howard, 167–68

Hale, George Ellery, 78, 172–74
Hale Observatories. *See* Mount Wilson
    Observatory, Palomar Mountain
    Observatory
Hall, John S., *259*
Halley's Comet, 35–37, 103, 208, 255–57

Index 295

Hamilton College Observatory, 76
Hamilton, George Hall, 150, 160
Hansen, Peter Andreas, 36
Hardie, R., 260
Harding, Karl, 28, 30
Harlan, E. A., 257
Harrington, Robert, 246–47
Harshman, Walter S., 95
Harvard College Observatory, 61, 87, 91, 99, 101, 104, 113, 163–64, 196, 198, 200–201, 203–4, 206, 210, 214, 216, 224; Boyden Station (Arequipa, Peru), 24-inch Bruce doublet, 99; 15-inch refractor, 61
Harvard University, 39, 60–61, 63, 80–81, 84, 95, 119, 164, 229
Hawaii, University of, 245
Hegel, Georg Friedrich Wilhelm, 27
Henroteau, C. F., 249
Heraclides, 3
Herschel, Caroline (W. Herschel's sister), 10, 12, 24
Herschel Chronicle, The (Lubbock), 2
Herschel, Sir John F. W. (W. Herschel's son), 39, 50, 53–55, 66–67, 69–71, 76, 98–99, 134, 136, 222; *Outlines of Astronomy* (1849), 76
Herschel, Sir William, 1–2, 5–7, 9–32, *13*, 39, 45, 53–54, 115, 196
 biographical, 12, 15–17
 discoveries of satellites, 22–23
 discovery of Uranus, 1, 5–6, 9–26, 115, 196
 early interest in astronomy, 16–17
 honors, 15, 20–21, 23–25
 impact on astronomy, 15
 as a musician, 15–16
 observing techniques of, 10–15, 21
 telescopes of, 9, 12, *14*, 16–17, 23–24
Hewitt, Anthony, 247
Hidalgo, 253
Hind, John Russell, 6, 76
Histoire Céleste Français (Lalande), 62–63
Hornsby, Rev. Thomas, 10–11, 17
Hubbard, Joseph, 62
Hubble, Edwin P., 172–74, 251; velocity-distance relationship, 157
Humason, Milton, 157, 213–14
Hussey, Rev. Thomas J., 35–37, 56
Hygea, 76

Iapetus, 259. *See also* Saturn, satellites of
Imagination, scientific, 6–7
Indiana University, 87, 94, 150, 169, 195, 210, 214
Infrared astronomy, 22
Institute of Theoretical Astronomy, 255
International Astronomical Union, 171–72

Intra-mercurial planet. *See* Leverrier; "Vulcan"
Invariable plane, 4, 87, 92–94, 106, 115
Ishtar, 3
Isis, 75
"Island universes," 156–57. *See also* "Great debate"

Jackson, E. S., 244
Jackson, John, 204
Janssen, Pierre Jule César, 73
Jayne Foundation, 200
Juno, 30
Jupiter, 3, 21, 33–34, 45–46, 65, 82, 100, 104, 108, 129, 133, 162–63, 209–10, 212–13, 218, 250, 258; ring of, 22; satellites of, 3, 65, 108, 129, 133, 209–10, 212–13

Kansas, University of, 181, 250
Kant, Immanuel, 25
Keeler, James E., 128
Kelvin, Lord (William Thompson), 72
Kepler, Johannes, 3–4, 25, 110, 247
Kirkwood, Daniel, 113
Kirkwood Observatory, 169
Kitt Peak National Observatory, 244
Klemola, A. R., 257
Klepczinski, W. J., 244
Königstuhl Observatory, 214
Kourganoff, Victor, 237–41, 247, 256, 261
Kowal, Charles T., 258, 261
Kritzinger, H. H., 255
Kuiper, Gerard P., 22, 241, 243–44, 259–60

Lagrange, Joseph Louis, 43, 48, 95, 140
Lalande, Joseph J. F., 6, 18–19, 27, 33, 62–63, 236, 241–42, 244, 261
Lamont, Dr., 6
Lampland, Carl Otto, 86–89, *88*, 92, 95, 105–6, 112, 115, 117–28, 130, 132–33, 142–43, *145*, 147, 149–52, 154, 158–59, 161, 167–68, 171–72, 176–77, 182, 184–85, 192–96, 205–6, 209, 214, 216, 241
 biographical, 149–52, 158–59, 161
 comparative photography of nebulae, 133, 147, 158–59
 as a Lowellian, 150, 154
 observations of Pluto, 192–93
 planetary photography of, 86, 114
 radiometric observations of, 151–52, 159
 role in planet X searches, 89, 105–6, 112, 115, 117–28, 130, 132–33, 142–43, 147, 167–68, 171–72, 176–77, 182, 184–85, 192–96, 205–6, 209, 214, 216, 241
Langley, Samuel Pierpont, 74

Laplace, Pierre Simon, 4, 18, 29, 43, 65, 81, 128, 140; *Méchanique Céleste*, 31
"Laplacian librations." *See* Commensurability
Lassell, William, 22, 58–59, 66
Lau, Hans-Emil, 77, 106, 226, 239
Lawrence Fellowships. *See* Lowell Observatory
Least squares, method of, 18–19, 110–12, 119, 123, 136–37, 140, 228
Legendre, Adrien-Marie, 18
Lemonnier, Pierre Charles, 6, 33–34, 136
Leonard, Wrexie Louise (P. Lowell's secretary), *88*, 91, 115, 119
Lescarbault, Dr., 72–74
Leuschner, Armin O., 201, *203*, 205, 207–9, 229, 236–37, 258
Leverrier, Urbain Jean Joseph, 5, 7, 18, 37, 39, 43, *44*, 45–46, 48–50, 60–74, 76, 84–85, 95, 98–100, 106, 108–12, 133–40, 194, 215, 221–23, 228, 231, 237–40
  analytical methods of, 43, 46, 48–50, 85, 106, 108–12, 228
  discovery of Neptune, 52–53
  honors, 57–58
  intra-mercurial planet, 43, 71–74. *See also* "Vulcan"
  role in Neptune controversies, 7, 39, 54–58, 63–71
  trans-Uranian planet prediction, 7, 51–53, 63
Lexell, Anders Johann, 18–20, 32
Lexell's "lost" comet, 11, 45
Liais, E., 73
Lick Observatory, 74, 91, 127, 151–52, 154, 156, 205, 244, 257; 20-inch double astrograph, 257
Light, of night sky. *See* Night sky
Ligondès, Vicomte du, 78
"Lilienthal detectives," 27
Lowell, Abbott Lawrence (P. Lowell's brother), 80, 86, 96, 167–68, 175–76, 183, 193–94, 216, 232; *Biography of Percival Lowell* (1935), 96, 232
Lowell, Amy (P. Lowell's sister), 80
Lowell and Mars (Hoyt), 2
Lowell, Mrs. Constance Savage Keith (P. Lowell's wife), 98, *143*–44, 151, 155, 176–77, 196, 200–201, 216, 218
Lowell, Elizabeth (P. Lowell's sister, Mrs. William L. Putnam), 80, 146, 215
Lowell, Guy (P. Lowell's cousin), *144*–46, 153, 160, 163–64, 170–72
Lowell Institute, 93, 98
Lowell, Katherine (P. Lowell's sister, Mrs. R. W. Bowlker), 80, 122
Lowell Observatory, 1–2, 6–8, 80–82, 85, 96, 98, 105, 118, 123–24, 129, 133, 142, 144–47, 149–52, 154–55, 160, 163–64, 167–71, 173–74, 176–81, 189–90, 194–96, 200–205, 207–8, 212–18, 220, 222–24, 229–32, 241, 248–50, 252, 256–57, 260
  astronomical staff of, 149–50
  discovery of Pluto, 1–2, 130, 190, 192–215, 241, 257
  eclipse expeditions of, 155
  financial status of, 144–46, 189–90
  founding of, 80–81, 98
  Lawrence Fellowships of, 87, 92, 94, 147, 250
  mountain station of, 170–71
  observational work of, 86–87, 94, 98, 103, 118, 121–22, 127–30, 133, 147, 149–53, 155, 159, 252
  planet X searches of, 84–125, 130–33, 142–44, 159–61, 164–70, 174–89, 248–52
  radiometry at, 150–52, 159
  South American expedition of, 82, 94
  telescopes of: 5-inch Brashear camera, *88*, 91–92; 5-inch Cogshall Camera, 183, 188, 190, 192, 195; 5-inch Voigtlander camera, 89–92; 6.75-inch Roettger camera, 91; 9-inch Brashear (Sproul) doublet, 123, 130–32, *131*, 175; 12-inch (Harvard) refractor, 81; 13-inch Lawrence Lowell refractor, 160, 164, *166*–70, 174–77, 179–88, *186*, 190, 192–94, 209, 250; 15-inch Petitdidier reflector, 159, 171; 18-inch Brashear refractor, 81; 24-inch Clark refractor, 82, 89, 92, 123, 160, 168, 192–94, 200, 212; 40-inch (42-inch after 1925) Clark reflector, 98, 103, *114*–15, *117*–18, 123, 147, 151, 168, 185, 192–94, 196, 212
Lowell, Percival, 1–2, 5–6, 18, 60, 65–66, 71, 78, 80–81, 83–85, *88*, 93, 98, 101–3, 106–14, *107*, *109*, 122, 126–42, 144–47, 150–54, 160–64, 170–72, 174–77, 179, 183–84, 193–98, 200–201, 204, 207–10, 212–13, 215–17, 220–41, 244, 247, 250, 253–54, 256, 261. *See also* Planet X, Lowell's predictions of
  biographical: childhood, 80; death, 129, 142; early interest in astronomy, 80–81; education, 60, 80–81; family, 80; financial status, 81; illness of, 85, 120; marriage, 98; as Orientalist, 81; provisions of will of, 144–46; as a traveler, 81, 124
  concepts of: on celestial mechanics, 106, 108–9, 114; on commensurability of periods, 65–66, 113–14; on discoveries in science, 103, 133; on imagination in science, 6, 136; on nebulae, 122; on planetary evolution, 103, 113; on solar system, 83, 209–10

Lowell, Percival (*continued*)
  Mars theory, 80–82, 103, 152–53
  planetary law of, 113–14
  planet X: analytical searches for, 106, 108–14, 134–37, 232–33; belief in, 1, 83–84, 160, 170; predictions of, 83–84, 101–3, 126–40, 174, 184, 193–98, 207, 253, 256; memoir on, 106, 110–12, 124–26, 132–41, 184, 220–21, 233
  works by: *The Evolution of Worlds* (1909), 103, 134; *Mars* (1895), 81; *Mars and Its Canals* (1906), 93; *Mars as the Abode of Life* (1908), 93, 98; *The Solar System* (1903), 83
Lowell Prize, 176, 196
Lubbock, Constance A., 2, 10, 18; *The Herschel Chronicle* (1933), 2
Lundin, C. A. R., *114*, *116*, 168–70, 174–76
Lundmark, Knut, 185
Luyten, W. J., 260
Lyttleton, Raymond, A., 258, 260

McDonald Observatory, 82-inch reflector, 242
Magnification, telescopic, 10, 12, 14–15
Mariner Project, 172
Mars, 3, 80–82, 86–87, 89, 91–92, 94, 97–99, 103, 118, 122, 129–30, 133, 150–55, 172, 210, 212, 258. *See also* Lowell, P.
  atmosphere of, 98, 133, 153–54, 172
  "canals" of, 81–82, 86, 89, 92, 98, 103
  chlorophyll on, 86, 153
  life on, 81–82, 86, 133, 153–54
  Lowell's theory of, 80–82, 86, 89, 133, 150, 152–54, 172
  markings on, 81, 130, 154
  observations of: photographic, 86, 92, 94, 118, 129, 150, 152; radiometric, 150–52; spectrographic, 86, 98, 150, 153; visual, 86, 97, 103, 118, 150, 152
  physical conditions on, 150–54, 172
  public interest in, 81, 86
Mars (P. Lowell), 81
Mars and Its Canals (P. Lowell), 93
Mars as the Abode of Life (P. Lowell), 93, 98
Marsden, Brian G., 256, 258
Maskelyne, Rev. Dr. Nevil, 10–12, 17, 19
Massachusetts Institute of Technology, 83, 86, 112
Mathematics, 1; role in planetary discovery, 3–5
Maury, Matthew Fontaine, 51, 62
Mauvais, Félix Victor, 63
Mayall, Nicholas U., 214, 230
Mayer, Christian, 32
Mayer, Tobias, 32–34

Méchain, Pierre François André, 18–19
Méchanique Céleste (Laplace), 31
"Mediccean stars," 7
Mendeliev's periodic law, 113
Menzel, Donald H., 151
Mercury, 3, 8, 43, 58, 71–73, 82, 155, 210, 233, 253
Merope. *See* Nebulae, in Pleiades
Messier, Charles, 12, 18
Metcalf, Rev. Joel, 99, 144, 160, 164
Meteors, 3, 30–31, 72, 76–77, 83–84
Meudon Observatory, 225
Milky Way, 156–57, 185, 190, 250
Miller, Alice F., 196
Miller, John A., 87, 123, 130–31, 142, 147, 150, 174, 195, 201–2, 204–8, 210, 213, 218, 225, 230
Millis, Robert L., 22
Mimas, 128. *See also* Saturn, satellites of
Minor planets. *See* Asteroids
Miranda, 22. *See also* Uranus, satellites of
Misterium Cosmographicum (Kepler), 25
Monatliche Correspondenz, 27
Moore, J. H., 158
Morrison, David D., 245
Mount Wilson Observatory, 147, 150, 152–57, 160, 172–74, 194, 202, 213–14, 229–30, 232, 241, 260; 100-inch Hooker reflector, 155; 60-inch reflector, 147, 153; 10-inch Cooke triplet, 214
Museum of Northern Arizona, 171

National Academy of Sciences, 156, 226
National Research Council, 178, 249
National Research Fund, 170
Nautical Almanac Office. *See* United States Naval Observatory
Nebulae, 12, 121–22, 133, 147, 155–59. *See also* Hubble; Slipher, V. M., discoveries of; Van Maanen
  in Andromeda (M31), 122, 157
  in Cetus: N. G. C. 584, 156; N. G. C. 1068, 157
  in Coma Berenices (M99), 158
  comparative photography of, 133, 157–58
  controversy over nature of, 122, 156–59
  Crab Nebula (M1), 147–49, 158
  in Pleiades, 121
  radial velocities of, 122, 156–58
  rotation of spiral, 133, 157–58
  in Ursa Major: M81, 157; M101, 157–58
  velocity-distance relationship, 157
  in Virgo (N. G. C. 4594), 122, 157
Nebular hypothesis, 81
Neptune, 1, 5–8, 25, 37–39, 49, 51–71, 76, 78–79, 81, 84, 91, 96, 98, 104, 106, 111, 162–63, 174–75, 193, 196–97, 209, 217–19, 221–23,

228–31, 233, 235–36, 238–41,
244–45, 251, 254–61. *See also*
Adams, J. C.; Airy; Leverrier
analytical searches for, 39–52
commensurability of period, 65–66, 114
discovery of, 49, 51–54
distance of, 61, 63–65, 67. *See also*
Bode's law
early orbits for, 60–63
mass of, 63, 66, 68–69
name of, 53
Pluto as satellite of, 258–60
post-discovery controversies, 53–58,
60–71, 81, 133–36, 229
predictions of, 42, 46–47, 50–53
pre-discovery observations of, 6, 61–63,
96, 241, 261
"ring" of, 58–59
satellites of, 58–59, 66, 258–60
spectrographic observation of, 155
in trans-Neptunian planet problem, 155
Newcomb, Simon, 99–101, 127–28, 244
Newton, Hubert, 77
Newton, Sir Isaac, 3, 36, 49
New York Times, The, 127–28, 204,
207, 215
Nicholson, Seth B., 152, 172, 202, 214,
229–30, 241
Nicolai, F. B. G., 37
Night sky, light of, 159, 171, 185. *See also*
Slipher, V. M.
Northern Arizona Society of Science and
Art, 171
Northern Arizona State Teachers College.
See Northern Arizona University
Northern Arizona University, 176, 181, 196
Novae, 195

Oberon, 22. *See also* Uranus, satellites of
Observatories. *See* under specific name
O'Conner, Johnson, 120
Olbers' comet, 256
Olbers, Heinrich Wilhelm Matthias, 28–30
Oriani, Barnabo, 20, 27
"Ottawa object," 249–50
Outlines of Astronomy (J. Herschel), 69,
76, 98
Oxford University Observatory, 10

Pallas, 28, 29
Palomar Mountain Observatory, 48-inch
Schmidt reflector, 258; 200-inch Hale
reflector, 173–74, 213, 243
Parallax, 10–11
Paris Observatory, 39, 63, 71, 96, 101, 237
Parsons and Company, 167
Peirce, Benjamin, 39, 60, 63–67, *64*,
69–71, 80, 84, 113, 134–35, 229

Perturbations, planetary, in celestial
mechanics, 4–5, 34–40, 43, 45–46,
48–50, 85, 96–103, 106, 108–12, 117,
119–20, 122, 124, 134–41, 160,
162–63, 207, 220, 222–23, 226–28,
231, 233, 235–36, 238–41, 244, 247,
249, 261
Peters, C. H. F., 76
Peterson, Adolf Cornelius, 61–62
Pettit, Edison, 152, 172
Philolaus, the Pythagorean, 3
Philosophical Magazine, 128
Philosophical Transactions, 11–12, 22
Piazzi, Giuseppe, 7, 27–29
Pickering, Edward Charles
(W. Pickering's brother), 87, 101
Pickering, William Henry, *79*–81, 98–102,
104, 118, 159–63, 189, 198, 214,
218–19, 222–26, 231, 232–34,
237–39, 241, 247, 253–55
analytical methods of, 98–99, 102, 104,
161, 226, 238–39
criticism of Lowell's work, 218–19, 223,
232–33
planetary predictions of: planet O,
79–80, 98–99, 104, 118, 159–63, 198,
214, 221–26, 230, 232–34, 238–39,
241, 253; planet P, 118, 161–62,
218–19, 254–55; planet Q, 118,
161–62; planet R, 118, 161–62; planet
S, 161–62; planet T, 161–62, 253;
planet U, 161–62, 253
reaction to Pluto discovery, 218–19,
222–23
relationship with P. Lowell, 99, 223, 234
role in Pluto controversies, 198, 218–19,
221–26, 230, 232–34, 237–39, 241
Pilcher, Carl B., 245
Pillans, James, 53
Planetary motion, laws of, 3–4, 17,
110, 247
Planets, 1, 8, 84, 96, 100, 209–10, 258; in
antiquity, 1; outer, 84, 96, 100,
209–10; problem of definition of, 8,
210, 258. *See also* specific name
predictions of new, 1, 3–5, 7, 25, 35–38,
71–72, 76–80, 83–84, 98–99,
101–3, 118, 121, 189, 256; by
Adams, 7, 42, 47; by Babinet, 76; by
Bode, 26; by Brady, 256; by Dallet,
78; by Flammarion, 77; by Forbes, 77,
101; by Gaillot, 80; by Garnowsky,
78; by P. W. Gifford, 160; by Grigull,
78; by Kritzinger, 255; by Lau, 77; by
Leverrier, 5, 7, 43, 51–52, 71–72; by
P. Lowell, 1, 5, 35, 83–85, 101–3,
121; by W. H. Pickering, 79–80,
98–99, 118, 168, 254–55; by Schütte,
255; by See, 78; by Sevin, 253; by
Todd, 76–77

Index 299

Planet Uranus, The (Alexander), 2, 19
Planet X, 1–2, 6, 60, 83–103, 105–27, 130, 132–33, 137–40, 142, 146–47, 150, 160, 161–64, 169–70, 176–79, 182–85, 188–90, 194–98, 200–201, 204, 207–10, 212, 216, 218, 220–22, 224–36, 238–39, 241, 247, 250, 253, 258, 261. *See also* Lowell, P.; Lowell Observatory, planet X searches of; Pluto
   comparisons with Pluto, 140, 194–98, 201, 207–9, 220–39, 241, 247, 256
   early searches for, 83–103, 105–25, 130, 132–33
   final search for, 6, 177, 178–79, 182–85, 188–90, 192–97, 250
   Lowell's predictions of, 83–84, 101–3, 140–41, 184, 194–98, 253
   prediction of, by Brady, 256–57
   role in Pluto controversies, 140, 193–98, 200–201, 204, 207–10, 212, 220–39, 247
Plato, 1
Pleiades. *See* Nebulae
Pluto, 1–3, 5–8, 35, 60, 86–87, 94–95, 105, 115, 130, 140, 161–62, 172, 181, 184, 189, 190–98, *191*, 200–210, 212–46, *245*, 247–49, 252–61. *See also* Tombaugh
   as an asteroid, 207–9, 258
   as a comet, 204, 207–9, 224–25, 230, 232, 258
   diameter of, 193, 225, 233, 242–45
   discovery of, 130, 172, 190–200, 248
   distance of, 193, 196, 231, 254
   early observations of, 190, 192–93, 196, 201, 204–5
   early orbits for, 194–95, 200–208, 210, 212
   later orbits for, 212–15, 234
   as Lowell's planet X, 94, 194–98, 200–201, 207–10, 212, 220–21, 225–38, 247
   magnitude of, 94, 192–94, 196, 210, 225, 235
   mass of, 209–10, 221, 225–26, 232–33, 235, 238, 240–47, 254–55, 261
   name of, 198, 215–19, 231
   as Neptunian satellite, 258–60
   as Pickering's planet O, 161, 221, 222–25, 230, 232–34
   photometry of, 245, 260
   post-discovery positions of, 130, 192–93, 196, 202–5
   pre-discovery observations of, 86, 184, 214, 232
   public reaction to, 198, 200–201, 215–19
   rotation of, 259–60
   satellite of (Charon), 8, 193, 198, *245*–47, 254, 260–61

   telescopic appearance of, 94, 192–93, 196–97, 208–9, 212
   theories of origin of, 258–60
Pons-Brooks Comet, 256
Populaire Astronomie (von Maedler), 38
Popular Astronomy, 128, 219, 223–24, 232
Precession, 118
Princeton University, 151, 252
Proper motion, of stars, 252
"Proto-planets," 260
Pulkowa Observatory, 53
Putnam, George, 145
Putnam, Michael C. J., *259*
Putnam, Roger Lowell, 80, 146, 163–64, 166–77, 179, 181, 183, 189–90, 193–95, 200, 203–10, 214–18, 222, 224, 229–31, 235, *259*
   biographical, 164, 166–70
   role in planet X search, 164–70, 174–77, 193
   role in Pluto controversies, 193–95, 200, 203–10, 214–18, 222, 224, 229–31, 235
Putnam, Mrs. William L. *See* Lowell, Elizabeth
Putnam, William L. (P. Lowell's brother-in-law), 145–46

Radial velocity, of nebulae. *See* Nebulae
Radiometry, astronomical, 150–53, 172
Rawlins, Dennis, 33, 244, 256, 261
Reaves, Gibson, 237, 239
Report on the History of the Discovery of Neptune (Gould), 2, 36
"Resisting medium," theory of, 78
Richardson, Robert S., 178, 255
Ring systems. *See* Jupiter; Saturn; Uranus
Ritchey, G. W., 147
Roberts, Sir Isaac, 77
Roche limit, 162
Ross, Frank E., 176, 214
Roth, Gunter D., 2
Royal Academy of Sciences (Gottingen), 38, 40
Royal Astronomical Society, 39, 55, 57–58, 72, 217, 233
Royal Observatory (Greenwich), 10, 12, 23, 37, 40–41, 43, 46, 50–51, 73, 96, 99, 101, 109, 136, 204
Royal Observatory (Uccle), 213–14, 225–26, 230
Royal Society, 10–12, 17, 20–23, 29, 57
Royal Society of Edinburgh, 77
Russell, Henry Norris, 152, 170, 173, 178, 221, 230–32

Saint John, Charles, 153
Satellites, 3–4, 7, 21–22, 30, 58–59, 127–29, 163, 209. *See also* specific name

Saturn, 3, 18, 21–22, 26, 33–34, 46,
    48–49, 58–59, 65, 82, 98, 100, 103,
    108, 113, 122–23, 127–29, 224, 251,
    256, 259
  Lowell's observations of, 82, 98, 103,
    122–23, 127–29
  rings of, 21–22, 58–59, 103, 113,
    127–29
  rotation of, 129
  satellites of, 65, 108, 113, 128–29, 259
Schiaparelli, Giovanni Virginio, 81, 152
Schlesinger, Frank, 173
Schroeter, Johann Heironymous, 24,
    27–28, 30
Schütte, K., 255
Schwartzenbrenner, Boniface, 35
Schwassman-Wachmann Comet, 201
Science, 174, 200
Science Service, 206–7, 217, 222
Scientific American, 128, 230, 235
Scientific Monthly, 235
See, Thomas Jefferson Jackson, 79, 215,
    234; "Capture Theory," 78
Seeing, astronomical, 12–15, 190, 192
Seidelmann, P. K., 244, 256
Sevin, M. E., 253, 255
Shapley, Harlow, 156, 164, 168, 173,
    200–201, 203–4, 206, 208, 216, 222
Sidgwick, J. B., 2
Sigma Xi, 103, 133
Skjellerup's Comet, 155
Sky and Telescope, 242
Slipher, Earl C., 87, *89*, 92–94, 97,
    121–22, 128–29, 149, 150, 154, 172,
    192, 196, 208, 214
Slipher, Vesto Melvin, 86–89, *88*, 91–93,
    98, 115, 117–18, 121–30, 133, 144–51,
    148, *149*, 153–60, 163–64, 166–79,
    181–85, 189, 192–98, 200–201,
    203–10, 212–18, 220, 222, 224–26,
    229–31, 235, 248–50, 252, 257
  biographical, 149–50
  discoveries of: nebular radial velocities,
    122, 156–57; nebular rotation,
    157–58; permanent aurora, 130, 159;
    reflection nebulae, 121
  as a Lowellian, 150, 153–55
  role in Pluto controversies, 192–96,
    200–201, 204–10, 212–15, 220, 222,
    224–26, 229–31, 235
  role in Pluto discovery, 89, 130, 163–64,
    166–79, 181–85, 189, 192–96,
    200–201
Smart, W. M., 2, 57, 63, 70–71
Smithsonian Astrophysical Observatory,
    256
Smithsonian Institution, 74, 172
Société astronomique de France. *See*
    French Astronomical Society
Solar System, The (P. Lowell), 83

Speculum metal, 9, 16–17
Spiral nebulae. *See* Nebulae
Sproul Observatory, 87, 123, 130–31, 175,
    195, 206; 9-inch Brashear doublet,
    123, 130–33, *131*, 175
Steward Observatory, 173
Stewart, R. M., 249
Stokely, James, 206–7, 217, 222
Stokes, W. E. D. Jr., 215
Storey, W. E., 127
Strömgren, Elis, 143
Struve, F. G. W., 53
Struve, Otto, 128
Students' Observatory, University of
    California, 201
Sun, 221–23
Supernova, 147
Swarthmore College, 87, 123, 195
Swift, Lewis, 74
Sykes, Stanley, 164–65, 169
System of Minor Planets (Roth), 2

Tables astronomiques (Bouvard), 34, 69,
    71, 99–100, 139
Telescopes. *See* specific observatory
*Theoria Motus Corporum
    Coelestium . . .* (Gauss), 31
Theories, of planets, 4–5, 43, 48, 98–99,
    109, 117, 136–39, 231, 244
Thompson, William (Lord Kelvin), 72
Three-body problem, 4–5
Timaeus (Plato), 1
Tisserand, François Félix, 106
Titania, 22. *See also* Uranus, satellites of
Titius-Bode Law. *See* Bode's law
Titius, Johann Daniel, 25–26
Todd, David Peck, 75–77, 82, 84, 95, 98,
    104, 234, 253
Tombaugh, Clyde William, 1–2, 5–6, 8,
    94, 177, 179–90, *180*, *187*, 192–96,
    200, 205, 209, 212, 216, 245, 248–*51*,
    252, 257
  biographical, 179, 181
  discovery of Pluto, 190–96, 245
  early interest in astronomy, 179
  honors, 212
  observing techniques of, 182–90, 192
  post-Pluto planet search, 248–51
Tores. *See* Saturn, rings of
Trans-Neptunian planet, P. Lowell memoir
    on, 106, 110–12, 124–27, 133–41
Triton, 58–59. *See also* Neptune,
    satellites of
Trowbridge, John, 125–27
Truman, Harry S., 164
Truman, Orley Hosmer, 160
Trumpler, Robert J., 152
Tucker, Herbert H., 120
Turner, Herbert H., 163, 218

Index 301

Umbriel, 22. *See also* Uranus, satellites of
United States Coast Survey, 39, 51, 60
United States Forest Service, 169, 171
United States Naval Observatory, 51, 62, 76–77, 85, 95, 198, 202, 241, 244–45, 261; discovery of Pluto satellite, 8, 198, 245–47; Nautical Almanac Office, 85, 95; telescopes: 61-inch astrometric reflector (Flagstaff), 245; 26-inch Clark refractor (Washington), 77
Uranus, 1–2, 5–7, 9–40, 45–51, 53–56, 58, 61, 64–66, 68–71, 77–80, 82, 84–85, 95–103, 106, 108–15, 117, 122, 134–39, 162, 193, 196–97, 209, 217, 226–28, 231–33, 235–36, 238–41, 245–47, 251, 254, 261. *See also* Herschel, Sir William
  in Bode's law, 25–26
  Bouvard's tables of, 33–35, 46, 48, 70
  as a comet, 9–11, 18–21, 23
  diameter of, 23
  discovery of, 1, 5–6, 9–25, 115, 196
  early orbits for, 18–20, 32–33, 76
  name of, 7, 20, 24–25, 53
  observations of, 11, 20, 23–24, 37, 47, 118
  pre-discovery observations of, 6, 20, 32–34, 96, 111, 232, 239
  radius vector of, 37, 40, 42, 46–48, 50
  rings of, 22, 58
  rotation of, 118
  satellites of, 22, 58
  theories of, 34, 98–100, 109, 117

Valz, J. E. B., 37
Van Biesbroeck, George, 202, 213–14, 230
Van Maanen, Adriaan, 157–58
Variable stars, 119, 132–33, 189, 195, 250
Venus, 3, 16, 21, 82, 155
Very, Frank W., 153
Vesta, 30
Virgo, nebula in. *See* Nebulae

Von Maedler, Johann Heinrich, 38
Von Wolff, Christian Freiherr, 25
Von Zach, Baron Franz Xavier, 26–28
Voyager Project, 22
"Vulcan," 43, 72–74, 215

Walker, Merle F., 260
Walker, Sears Cook, 39, 51, 60–63, 65–66, 68–69
Walsh, John, 23
Wanrooy, S. N. J., 258
Ward, W. R., 257
Wartmann, Louis Francis, 37
Watson, Fletcher G., 2
Watson, James C., 74
Watson, Sir William, 10–12, 17, 20, 29
Weatherford, J. W., 170–71
Wellesley College, 87, 195
Whipple, Fred L., 205–6, 214, 224–25, 233
William Herschel—Explorer of the Heavens (Sidgwick), 2
Williams, Elizabeth Langdon, 106, 112–15, 117, 119–20, 122–25, 127, 150
Williams, Kenneth P., 87, *89*, 94, 195, 200
Wolf, Max, 30, 92, 94, 132, 142, 185
World War I, 132, 142–43
World War II, 237, 250
Worthington, James R., 174
Wylie, Lloyd R., 241, 244

Yale University, 77, 226, 229, 241
Yamamoto, Issei, 258
Yerkes Observatory, 92, 127, 176, 202, 204, 214, 216

Zagar, F., 214
Zeiss "blink" comparator, 95, 115, 117, 182, 185–86, 188–90, 192, 195, 248–49, 250, *251*, *252*
Zodiacal light, 157, 185